AQA Geography A2

SERIES EDITORS: ROSS • DIGBY

BAYLISS • COLLINS • CHAPMAN

D0320192

OXFORD
UNIVERSITY PRESS

Great Clarendon Street, Oxford OX2 6DP

Oxford University Press is a department of the University of Oxford. It furthers the University's objective of excellence in research, scholarship, and education by publishing worldwide in

Oxford New York

Auckland Cape Town Dar es Salaam Hong Kong Karachi Kuala Lumpur Madrid Melbourne Mexico City Nairobi New Delhi Shanghai Taipei Toronto

With offices in

Argentina Austria Brazil Chile Czech Republic France Greece Guatemala Hungary Italy Japan Poland Portugal Singapore South Korea Switzerland Thailand Turkey Ukraine Vietnam

Oxford is a registered trade mark of Oxford University Press in the UK and in certain other countries

© Oxford University Press 2011

Authors: Simon Ross, Bob Digby, Tim Bayliss, Lawrence Collins, Russell Chapman

Database right Oxford University Press (maker)

First published 2011

All rights reserved. No part of this publication may be reproduced, stored in a retrieval system, or transmitted, in any form or by any means, without the prior permission in writing of Oxford University Press, or as expressly permitted by law, or under terms agreed with the appropriate reprographics rights organization. Enquiries concerning reproduction outside the scope of the above should be sent to the Rights Department, Oxford University Press, at the address above

You must not circulate this book in any other binding or cover and you must impose this same condition on any acquirer

Third party website addresses referred to in this publication are provided by Oxford University Press in good faith and for information only and Oxford University Press disclaims any responsibility for the material contained therein

British Library Cataloguing in Publication Data

Data available

ISBN 978-0-19-913545-5

10 9 8 7 6 5 4 3

Printed by Vivar Printing Sdn Bhd., Malaysia

Paper used in the production of this book is a natural, recyclable product made from wood grown in sustainable forests. The manufacturing process conforms to the environmental regulations of the country of origin.

Bob Digby would like to thank Graham Yates, Phillippa Fine, and Geography students at Kingston Grammar School for help with research.

Tim Bayliss and Lawrence Collins would like to thank Hellen Hornby for her invaluable expertise on Sheffield's Blue Loop.

Lawrence Collins would like to thank Sally, Oliver, and Megan Collins for their patience and support.

Russell Chapman would like to thank Maria Dunn for passing on news items at every opportunity, Hannah Chapman for her help with research, and Linda Chapman for her support and constructive criticism.

Books to read, music to listen to, and films to see compiled by Dan Cowling.

Acknowledgements

The publisher and authors would like the thank the following for permission to use photographs and other copyright material:

Cover: Llin Sergey/Shutterstock; **p.6:** Kyodo/XinHua/Xinhua Press/Corbis; **p.8t:** Science Photo Library; **p.8b:** Marie Tharp 1977/2003; **p.11:** Scott Camazine/Photolibrary; **p.13:** Worldsat International/Science Photo Library; **p.15:** Howell Williams/NOAA; **p.19t:** Paul Harris/Getty Images; **p.19b:** David Parker/Science Photo Library; **p.20:** T Alipalo/UNEP/Topham; **p.23t:** J.D. Griggs/USGS; **p.23b:** Ed Freeman/Getty Images; **p.25:** Skyscan/Science Photo Library; **p.26:** pdtnc/Shutterstock; **p.28:** Andrew Woodley/Alamy; **p.29t:** Swisseduc.ch; **p.29b:** NASA; **p.30:** Anguilar Price/Alamy; **p.31t:** Pichugin Dmitry/Shutterstock; **p.31b:** Federico Meneghetti/Photolibrary; **p.35:** EastVillage Images/Shutterstock; **p.37:** www.buildchange.org; **p.41:** Rex Features; **p.42:** STR New/Reuters; **p.44:** Stefano Carofei/Rex Features; **p.45:** Alessandra Tarantino/AP Photo; **p.48t:** STR/AFP/Getty Images; **p.48b:** Master Sgt. Val Gempis/US Airforce photo; **p.49:** 3777190317/Shutterstock; **p.50:** REUTERS/KYODO Kyodo; **p.65:** Oyvind Martinsen/Alamy; **p.66:** Oli Scarff/Staff/Getty Images; **p.70:** University of Dundee/Science Photo Library; **p.76:** Rex Features; **p.81:** Florin Iorganda/Reuters; **p.82:** Jeff Schmaltz, MODIS Rapid Response Team, NASA/GSFC; **p.85:** Jeremy Woodhouse/Masterfile; **p.86t:** Courtesy of architect/Aga Khan Award for Architecture; **p.86b:** Ho New/Reuters; **p.87t:** Mario Tama/Staff/Getty Images; **p.87b:** STR New/Reuters; **p.88:** Reinhard Krause/Reuters; **p.89:** Simon Ross; **p.90:** Paul Grover/Rex Features; **p.93;** Simon Ross; **p.94:** Simon Ross; **p.96:** R.J. Benson; **p.99:** Pete Ryan/National Geographic Stock; **p.101:** Phillip Carr/Photolibrary; **p.102:** Reinhard Krause/Reuters; **p.103:** Jacques Jangoux/Photolibrary; **p.106:** Anup Shah/Nature Picture Library; **p.108:** David Hoffman/Photographers Direct; **p.111:** Amy Tseng/Shutterstock; **p.114:** Paul B. Moore/Shutterstock; **p.118:** Colin Harris/LightTouch Images/Alamy; **p.119:** Forestry Commission; **p.120:** Michael Busselle/Corbis; **p.124t:** Blickwinkel/Alamy; **p.124b:** Yukihiro Fukuda/Alamy; **p.125:** Garry DeLong/Alamy; **p.127:** NASA; **p.128:** Jacques Jangoux/Science Photo Library; **p.129:** Dani-Jeske Dani-Jeske/Photolibrary; **p.130:** Ricardo Funari/BrazilPhotos; **p.131:** Alexandre Meneghini/AP Photo; **p.132:** Nigel Pavitt/Photolibrary; **p.133:** Russell and Rachael Collins; **p.135:** MKimages/Alamy; **p.138:** Paul Ridsdale/Alamy; **p.139:** Tim Bayliss; **p.141:** Tim Bayliss; **p.142:** Daniel Berehulak/Getty Images; **p.144:** Ho New/Reuters; **p.146:** Tim Bayliss; **p.147:** Tim Bayliss; **p.162b:** Adrian Fisk; **p.164:** GoSeeFoto/Alamy; **p.165:** David R. Frazier Photolibrary, Inc./Alamy; **p.166:** Robert Wallis/Corbis; **p.167:** Manor Photography/Alamy; **p.169:** Cameron Davidson/Corbis; **p.171:** Robert Landau/Alamy; **p.172:** Shepard Sherbell/Corbis SABA; **p.175:** Bob Digby; **p.176l:** English Heritage.NMR Aerofilms Collection; **p.176r:** Christopher Pillitz/In Pictures/Corbis; **p.177:** Supershoot/Alamy; **p.178:** Bob Digby; **p.179:** Bob Digby; **p.180:** Dan Atkin/Alamy; **p.181l:** Bob Digby; **p.181r:** Bob Digby; **p.183:** Warren King/Alamy; **p.184:** Trevor Llewelyn; **p.185:** Frank May/dpa/Corbis; **p.186l:** Bob Digby; **p.186r:** Bob Digby; **p.187:** Jonathan CK Webb/webbaviation.co.uk; **p.193:** Rungroj Yongrit/epa/Corbis; **p.195:** Bob Digby; **p.198:** Randy Faris/Corbis; **p.201:** Yasuyoshi Chiba/AFP/Getty Images; **p.202t:** Ariadne Van Zandbergen/The Africa Image Library; **p.202b:** Russell Chapman; **p.205:** Victor Englebert/Photographers Direct; **p.207:** ildogesto/Shutterstock; **p.208:** Michael Christopher Brown/Corbis; **p.210:** Philippe Hays/Still Pictures; **p.211:** SCPhotos/Alamy; **p.212:** Photographers Direct; **p.214:** Dan Vincent/Alamy; **p.219:** Russell Chapman; **p.221:** Russell Chapman; **p.227t:** Jason Hawkes, Aerial Photographer; **p.227b:** Eric Tourneret; **p.228:** Stringer/AFP/Getty Images; **p.229:** oorka/Shutterstock; **p.232:** Kelly-Mooney Photography/Corbis; **p.235:** Russell Chapman; **p.236:** Telegraph Media Group Limited 2006/Mike Pflanz; **p.237:** Chinatopix/Associated Press; **p.238:** CCPHO/AP; **p.241:** Juan Manuel Borrero/Nature Picture Library; **p.246:** Corporal Adrian Harlen RLC/epa/Corbis; **p.248t:** Lynsey Addario/VII Network/Corbis; **p.248b:** Commercial Estates Group; **p.249:** Immo Klink; **p.250:** AP/Lefteris Pitarakis; **p.254:** Jonathan Hordle/Rex Features; **p.255:** Stefan Rousseau/PA Archive/Press Association Images; **p.257:** Scott Hortop/Alamy; **p.258:** Reuters/Dylan Martinez; **p.259:** Trackair/BAA Airports Limited; **p.260:** Greg Balfour Evans/Alamy; **p.262:** PhotoXpress/ZUMA Press/Corbis; **p.263:** H. Mark Weidman Photography/Alamy; **p.264:** Kevin Britland/Alamy; **p.266:** AFP/Getty Images; **p.267t:** Collection International Institute of Social History, Amsterdam; **p.268:** EPA/Mohammed Jalil; **p.269:** Burhan Ozbilici/AP Photo; **p.270:** Paul Hackett/In Pictures/Corbis; **p.272:** DBimages/Alamy; **p.274:** Getty Images; **p.276:** S. Sabawoon/epa/Corbis; **p.278:** Reuters/STR New; **p.279:** Ray Tang/Rex Features; **p.281:** Ashley Cooper/Corbis; **p.283:** The British National Party; **p.286:** Borderlands/Alamy; **p.288:** Karen Kasmauski/Corbis; **p.292:** Simon Ross; **p.293:** Simon Ross; **p.295:** Simon Ross; **p.296:** Simon Ross; **p.301(all):** Tim Bayliss; **p.302:** Tim Bayliss; **p.307t:** Stock Connection Blue/Alamy; **p.307b:** Robert Pratta/Reuters.

Illustrations are by Barking Dog Art
Design by John Dickinson Design

Every effort has been made to contact copyright holders of material reproduced in this book. Any omissions will be rectified in subsequent printings if notice is given to the publisher.

How to succeed at A2

- Use this textbook. Read what's in the news as well. Some issues provoke different viewpoints – so get as many angles on a topic as you can. Keep a record of sources from which you get information – you might want to check the book, article, or web page later.
- In class, take part in questions, answers, and discussions. Get involved in whatever goes on. Question when you don't understand something.
- Sometimes your classwork will be in rough notes and won't need to be copied up neatly. However, do keep legible work. Don't leave it more than a week to tidy up your files!
- Keep up with work – essays, exercises, past questions, or prescribed notes – as it is set; don't leave it so that work builds up. Always meet deadlines.

Researching and reading further

Keep up to date! Read journals and newspapers – look at them online, or use newspaper Apps on your mobile phone. Wikipedia is useful for basics, but go further! Get to know websites of different research organisations – you'll find good sources by searching 'research in tectonic hazards' rather than just 'tectonic hazards'. Be specific – for example, searching 'data on tectonic hazards' will bring up the CRED database for tectonic hazards. Visit libraries, too – at school or in your local city or university.

Geography and careers

Geography can lead to careers in finance, management, the media, and advertising. Plenty of solicitors, accountants, and IT specialists have a Geography qualification. And some careers use Geography directly, like teaching, environmental work, land management, and planning.

Contents

This book uses:
LIC – low-income country
MIC – middle-income country
HIC – high-income country

1 Plate tectonics and associated hazards

Natori, north-east coast of Japan, 11 March 2011

What's happening in the photo?
What will happen next?
What caused this to happen?
What will the people be doing?
Will people re-build the town in the same place?

Introduction

The driving force responsible for the Earth's spectacular mountain ranges, deep ocean trenches, and broad undersea abyssal plains is tectonic activity. Developed over recent decades, the theory of plate tectonics helps us understand the formation of the Earth's major physical features and the processes responsible for natural hazards such as earthquakes and volcanoes.

Earthquakes and volcanoes, and associated hazards such as tsunamis, have claimed the lives of millions of people. In 2010, over 230 000 perished as a result of a single earthquake in Haiti. In 2011, a massive earthquake and tsunami devastated Japan's east coast.

In this chapter you will learn about the scientific evidence behind the theory of plate tectonics. You will find about the causes and impacts of earthquakes and volcanic eruptions, and the responses of people to the hazard threat they pose.

Books, music, and films

Books to read

The Planet in a Pebble by Jan Zalasiewicz
Earth on Fire by Bernhard Edmaier
The Day the Island Exploded by Alexandra Pratt
Richter 10 – Taming the Earthquakes by Arthur Clarke & Mike McQuay
Volcano – Nature's Inferno National Geographic, 1997

Music to listen to

'Tsunami' by Manic Street Preachers
'Ring of Fire' by Johnny Cash
'Sonification' New Earthquake Warning Systems on YouTube
'Volcano' by Jimmy Buffet

Films to see

Dante's Peak
Volcano
Earthquake

About the specification

'Plate tectonics and associated hazards' is one of three Physical Geography options in Unit 3 Contemporary Geographical Issues – you have to study at least one.

This is what you have to study:

Plate movement

- The structure of the Earth.
- The theory of plate tectonics: convection currents and sea-floor spreading.
- Evidence for the theory of plate tectonics: continental drift and palaeomagnetism.
- Plate margins: destructive, constructive, and conservative margins.
- Tectonic processes: seismicity and vulcanicity.
- Landforms associated with tectonic activity: young fold mountains, rift valleys, ocean ridges, deep sea trenches, and island arcs.
- Hot spots associated with plumes of magma, and their relationship to plate movement.

Vulcanicity

- Variations in the type and frequency of volcanic activity in relation to types of plate margin and types of lava.
- Forms of intrusive activity: dykes, sills, and batholiths.
- Minor forms of extrusive activity: geysers, hot springs, and boiling mud.
- Major forms of extrusive activity: types of volcanoes.
- Two case studies of recent volcanic events (from within the last 30 years) from contrasting areas of the world.

Seismicity

- Earthquakes: causes and characteristics, including focus and epicentre; seismic waves, and earthquake measurement.
- Tsunamis: causes and characteristics.
- Two case studies of recent seismic events (from within the last 30 years) from contrasting areas of the world.

- For all case studies, you should study:
  - the nature of the volcanic or seismic hazard
  - the impact of the event
  - the management of the hazard and responses to the event.

In this section you will learn about:

- the different layers of the Earth

Our amazing Earth

> *"Suddenly, from behind the rim of the moon ... there emerges a sparkling blue and white jewel, a light, delicate sky-blue sphere laced with slowly swirling veils of white, rising gradually like a small pearl in a thick sea of black mystery. It takes more than a moment to fully realize this is Earth . . . home."*
>
> **Edgar Mitchell, United States astronaut describing Earth as seen from space**

Figure 1.1 *Our amazing Earth – seen from Apollo 17 in 1972. What lies within?* ▲

Earth is quite simply amazing! Every detail of our planet challenged peoples' imaginations long before astronauts took photographs from space that changed the way we thought about our planet (Figure 1.1). Now, some 350 million *Google Earth* users on home computers zoom over mountains, cities and even into our own streets just as generations before us wondered at great canyons, volcanoes and the mysteries of the ocean depths (Figure 1.2). It has always been human nature to think about and try to understand our planet.

The structure of the Earth

Over 2000 years ago, the Greek philosopher Plato was considering the structure of the Earth. But it wasn't until 1692 when Edmond Halley (of comet fame) first proposed a theory to describe the Earth's structure. He suggested that it was made up of hollow spheres – rather like Russian nesting dolls. Halley considered that each sphere was actually habitable – a comical idea today.

While the Earth appears to be a perfect sphere when seen from space, it is in fact a **geoid**. This means that it bulges around the Equator and is flatter at the poles. The cause of this shape is **centrifugal forces**, generated by the Earth's rotation, which fling the semi-molten interior outwards, just like children on a roundabout!

- **Centrifugal forces** are forces that cause something to move away from its centre of rotation.

The crust

The Earth's outer shell is called the **crust** – this is the layer we live on. The crust varies in thickness, from around 5 to 10 km beneath the oceans to nearly 70 km under the continents. But just how thick is this? Some say that if the Earth was an apple the crust would be as thin as its skin. Others say that if the Earth was an egg its crust would be thinner than the eggshell! In other words, its average thickness, relative to the Earth in total, is thin – very, very thin.

Figure 1.2 *The Earth without oceans – a map showing the world's ocean floor* ▼

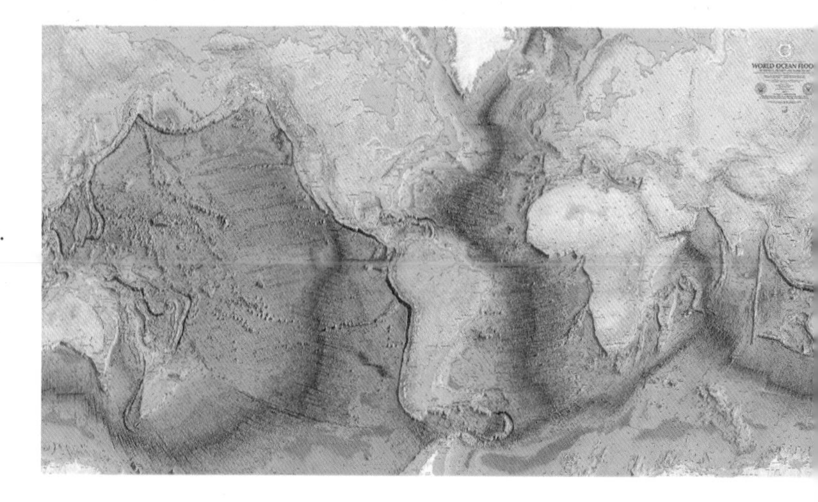

There are two types of crust:

- Oceanic – an occasionally broken layer of basalt rocks known as **sima** (because they are made up of **si**lica and **ma**gnesium).
- Continental – bodies of mainly granite rocks known as **sial** (because they are made up of **si**lica and **al**uminium).

Sima is found under the oceans, while sial forms the continental land masses. The lower boundary of sial grades into the upper part of the sima. While sial is much thicker than the oceanic sima, it is also less dense.

The lithosphere

Together, the crust and **upper mantle** are known as the **lithosphere**. It is in this zone that the tectonic plates are formed.

The mantle

As Figure 1.3 shows, the **mantle** is the widest section of the Earth – it is 2900 km thick. Due to the great heat and pressure within this zone, the mainly silicate rocks are in a thick, liquid state which becomes denser the deeper you go.

The rocks in the upper part of the mantle are solid and sit on top of the **asthenosphere**, a layer of softer, almost plastic-like rock. The asthenosphere can move very slowly carrying the lithosphere on top. Densities within the mantle increase as you go down into the **lower mantle**.

> Learning about the inner structure of the Earth is tricky – the depth and hot temperatures mean that we cannot drill into or physically see most of the Earth. Instead, scientists map the interior by watching how seismic waves, either from earthquakes or man-made blasts, travel through the various layers.

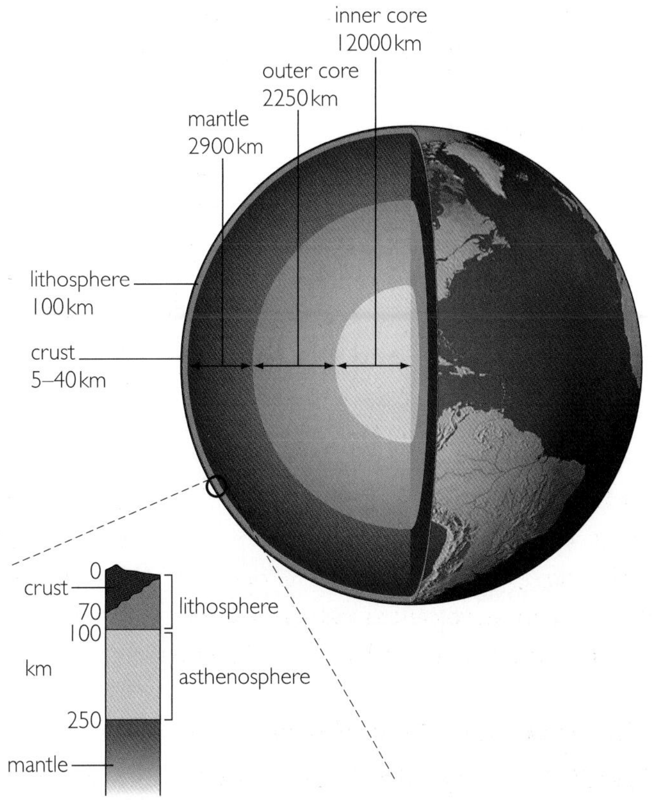

Figure 1.3 *Cross-section showing the internal structure of the Earth* ▲

The core

The **core** is the centre and the hottest part of the Earth – temperatures can reach above 5000 °C. It is mostly made of iron and nickel and is 4 times denser than the crust!

The core is actually made up of two parts. The **outer core** is semi-liquid and consists mostly of iron. The **inner core** is solid and is made up of an iron-nickel alloy. It is thought that as the Earth rotates, the liquid outer core spins, creating the Earth's magnetic field.

The phenomenal heat at the core generates **convection currents** within the mantle above. These currents spread very slowly within the asthenosphere – they are critical in explaining the movement of the tectonic plates.

> **Did you know?**
>
> *The Earth is 4600 million years old. But just how old is this? Compress it into 24 hours and humans have only occupied the last few seconds!*

> - **Convection** is the transfer of heat in a liquid or gas in a circular motion.

ACTIVITIES

1. Study Figure 1.2.
   - **a** Find the Atlantic Ocean sea bed. Describe the relief (the physical landscape) of the sea bed between Africa and South America.
   - **b** What types of rock would be found here?
2. In no more than 100 words, describe the structure of the Earth.
3. Do you think a compass will work anywhere on the surface of the Earth? Explain your answer.

1.2 The theory of plate tectonics

In this section you will learn about:

- continental drift and the theory of plate tectonics
- evidence from different sciences for continental drift
- how tectonic theory developed from new knowledge of the oceans
- palaeomagnetism – the science of fossil compasses
- old and young tectonic plates
- convection currents and tectonic plate movement

Continental drift and the theory of plate tectonics

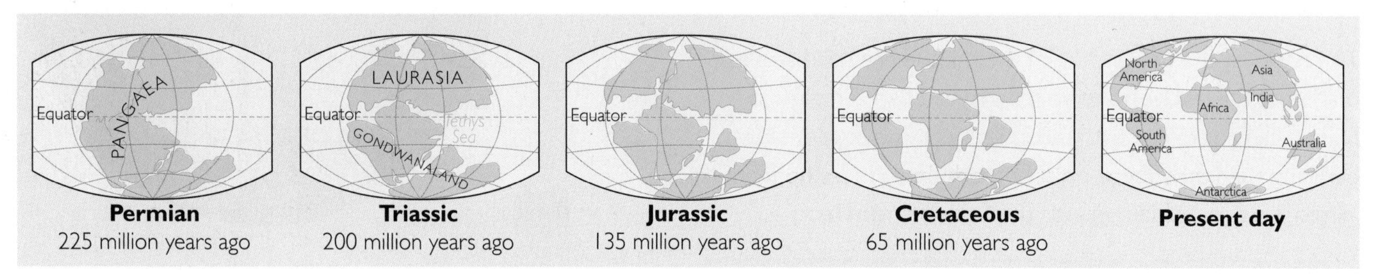

Figure 1.4 *Wegener's theory of continental drift* ▲

Plate tectonic theory proved to be one of the great scientific U-turns of the twentieth century. In fact, possibly the most important scientist associated with what we now know as tectonics – an intellectual rebel called Alfred Wegener – was widely ridiculed for his theories. Yet Wegener's 'continental drift' ideas have since evolved and been refined into a theory now accepted as providing the fundamental explanation for the development of the Earth's surface.

As far back as 1620, people noted the close fit between the east coast of South America and the west coast of Africa – but no theory existed to explain why. Then in 1912, the German scientist Alfred Wegener proposed his explanation – the controversial **theory of continental drift**. Wegener suggested that all the present continents were originally joined together to form a single supercontinent called Pangaea before drifting apart into Laurasia in the north and Gondwanaland in the south. Laurasia and Gondwanaland then broke up to form the continental arrangement we know today (Figure 1.4). But this arrangement of the continents, of course, continues to slowly – very, very slowly – change.

Wegener's theory was based on evidence from several sciences. The evidence included:

- the jigsaw fit of South America and Africa (Figure 1.5), with less obvious fits elsewhere
- matching rock sequences (of age and type) linking northwest Scotland and eastern Canada
- coal (formed only under warm, wet conditions) found beneath the Antarctic ice cap. Only if Antarctica was once positioned in warmer latitudes can this be explained
- Permian fossil brachiopods in India match those in Australia
- a unique Permian fossil reptile called mesosaurus is found only in southwest Africa and Brazil
- glacial deposits and scratches on rocks (called striations) in Brazil match those in West Africa.

Figure 1.5 *Wegener's jigsaw fit – notice how the two land masses fit together almost perfectly* ▼

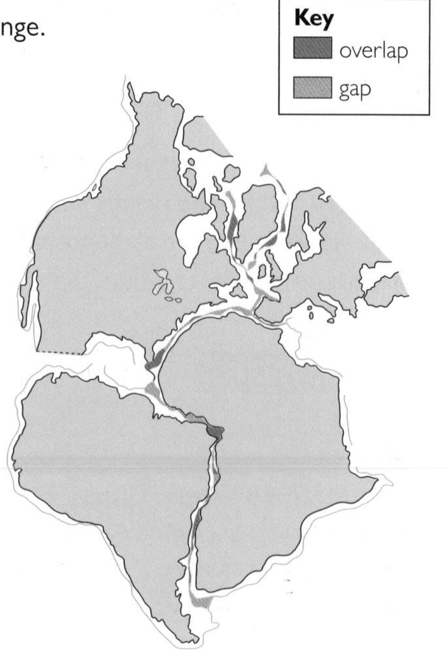

How tectonic theory developed

Wegener's theories did not explain how continental movement occurs – hence the widespread controversy about his ideas. But improvements in rock and fossil dating, and the discoveries of marine fossils in some of the world's great mountain chains, supported his idea of uplift and folding caused by colliding continental masses (Figure 1.6).

In the 1950s, nuclear submarines began monitoring and mapping the ocean floors with great accuracy. The Mid-Atlantic Ridge, an underwater mountain chain, was discovered, as were similar features in the Pacific. These ridges together with **deep ocean trenches** were found to be seismically active – in other words, prone to earthquakes.

Figure 1.6 *Marine fossil brachiopods (a clam-like organism) found in mountain rocks* ▲

In 1962 an American geologist named Harry Hess updated Wegener's ideas. He studied the age of the rocks on the Atlantic Ocean floor, finding the youngest in the middle and the oldest nearest the USA and the Caribbean. With the newest rocks still being formed in Iceland, this was compelling evidence that the Atlantic sea floor was spreading outwards from the centre, a concept known as **sea floor spreading**. The rate of spreading was estimated at about 5cm a year.

Palaeomagnetism – fossil compasses

Studies of **palaeomagnetism** effectively confirm sea-floor spreading. Palaeomagnetism involves the study of the history of changes in the Earth's magnetic field (polarity). Every 400 000 years or so, the Earth's magnetic field switches polarity, causing the magnetic north and south poles to swap. Once cooled and solidified, particles of iron oxide called magnetite, which were formed in lava and then erupted onto an ocean floor, record the Earth's magnetic orientation of that time. These rocks then become, in effect, fossil compasses.

> **Did you know?**
>
> Alfred Wegener was not a geologist, but an astronomer working as a meteorologist. He named Pangaea from the Greek for 'All Lands'.

The switching of magnetic fields was measured by magnetometers towed behind ships in the Atlantic Ocean. The results showed a mirror-imaged pattern of 'switches', or reversals, in the Earth's magnetic field as it moves away from the Mid-Atlantic Ridge. (Notice on Figure 1.7 that the pattern is exactly symmetrical on either side of the ridge.) The oldest lavas were furthest from the ridge and showed a repeated reversal of north and south poles. The resulting striped pattern is rather like a mirrored magnetic supermarket bar code. It confirms that the Atlantic floor is very slowly spreading from the mid-ocean ridge – with progressively older rocks the further you go from it.

Of course, this would suggest that the Earth would be getting bigger were it not for the discovery of deep ocean trenches where the ocean floor was being pulled downwards (**subducted**) and destroyed. With all this mounting evidence, plate tectonic theory was refined and has become universally accepted.

Figure 1.7 *The Atlantic Ocean's magnetic bar code* ▼

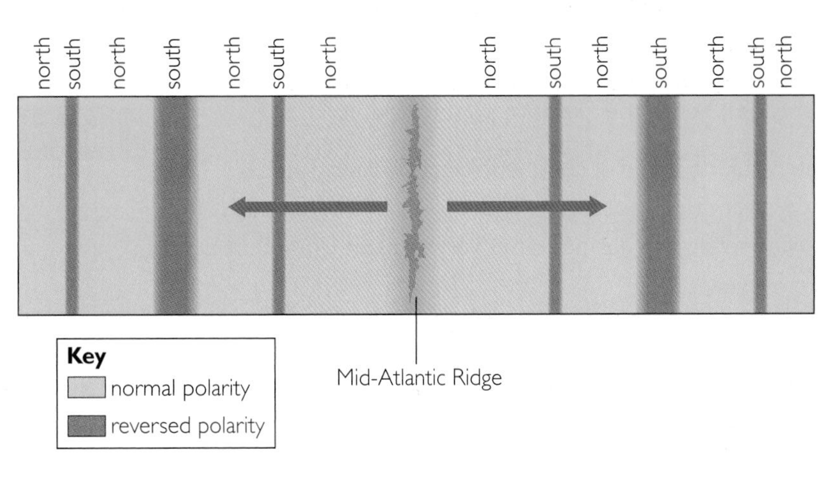

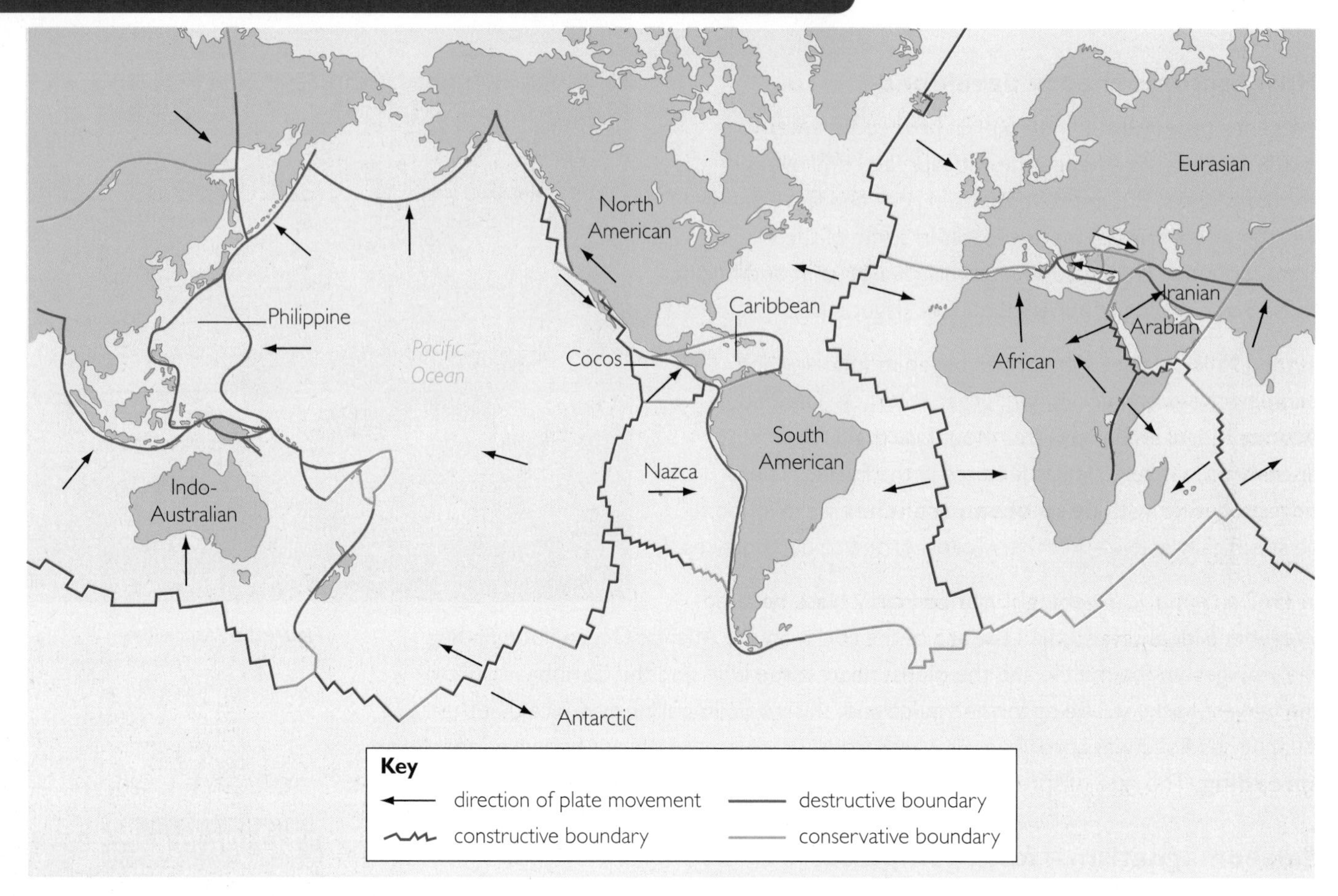

Figure 1.8 *Tectonic plate margins and plate movement* ▲

Tectonic plates

The Earth's surface is made up of seven major and several minor **tectonic plates** (Figure 1.8). Each plate is an irregularly shaped 'raft' of lithosphere effectively 'floating' on the plastic asthenosphere underneath. There are two types of plate material:

- Continental (mainly sial) over 1500 million years old.
- Oceanic (mainly sima) less than 200 million years old.

Continental plates are permanent and may extend far beyond the margins of current land masses. They will not sink into the asthenosphere because of their relatively low density. Denser oceanic plates, in contrast, are continually being formed at ocean ridges and destroyed at ocean trenches – hence their relatively young age.

The resulting jigsaw of continental and oceanic plates is always very slowly changing, with the plates continuously moving relative to each other at varying rates from 2 cm to 16 cm a year.

Did you know?

Tectonic plates move at about the same rate as your fingernails grow – an average of a little more than 3 – 4 centimetres per year.

Did you know?

Alfred Wegener never lived to see his theory of continental drift believed. During his lifetime it was criticised and ridiculed with some critics going as far as to call it 'utter rot'.

The plates are driven by vast, slow moving flows of heat called **convection currents** within the mantle. Radioactive decay within the core of the Earth generates exceptional temperatures. Hot spots around the core heat the lower mantle, creating thermal currents which rise towards the surface before spreading in the asthenosphere, then cooling and sinking again (Figure 1.9).

Tectonic plates can move sideways, towards each other or away from each other. They cannot overlap at their boundaries, so must either push past each other, be pushed upwards on impact, or be forced downwards into the asthenosphere and destroyed by melting.

No gaps can occur between plates, so if they are moving apart, new oceanic plates must be formed. In fact, because the Earth is neither expanding nor shrinking, any new oceanic plate that is formed must be compensated for by the subduction and destruction of an older plate elsewhere. As a result, it is at plate boundaries where most of the world's major landforms occur, relating to the earthquake, volcanic and mountain zones located along these great faults.

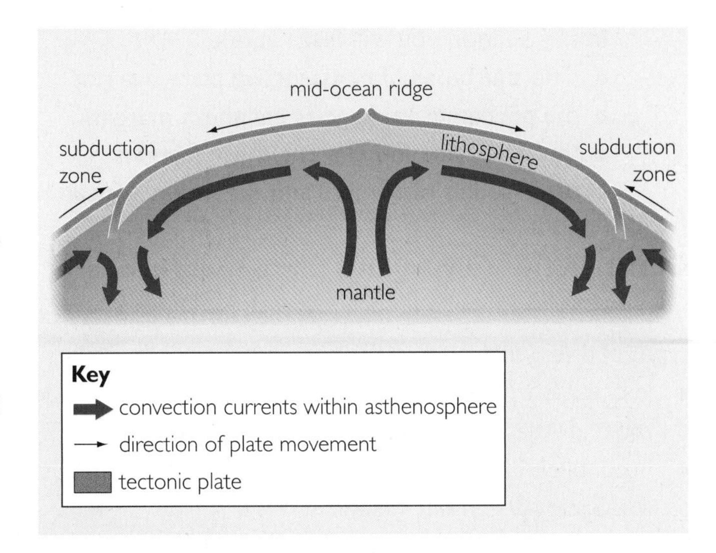

Figure 1.9 *Convection currents and tectonic plate movement* ▲

ACTIVITIES

1. For this activity you will need a blank outline map of the world.
   - **a** Use an atlas to help you locate and write boxed annotations to identify the evidence suggesting continents have 'drifted' over time.
   - **b** Now colour-code your boxes to separate Wegener's earlier evidence from the more recent proof.
   - **c** In your opinion, how strong is this evidence?
2. Draw a series of simple sketches to illustrate the concept of palaeomagnetism. In what way does this form compelling evidence of sea floor spreading, and therefore support the idea of continental drift?

Internet research

Use the Internet to source an animation of continental drift. Include this in an illustrated presentation on 'The Evidence for the Theory of Plate Tectonics'.

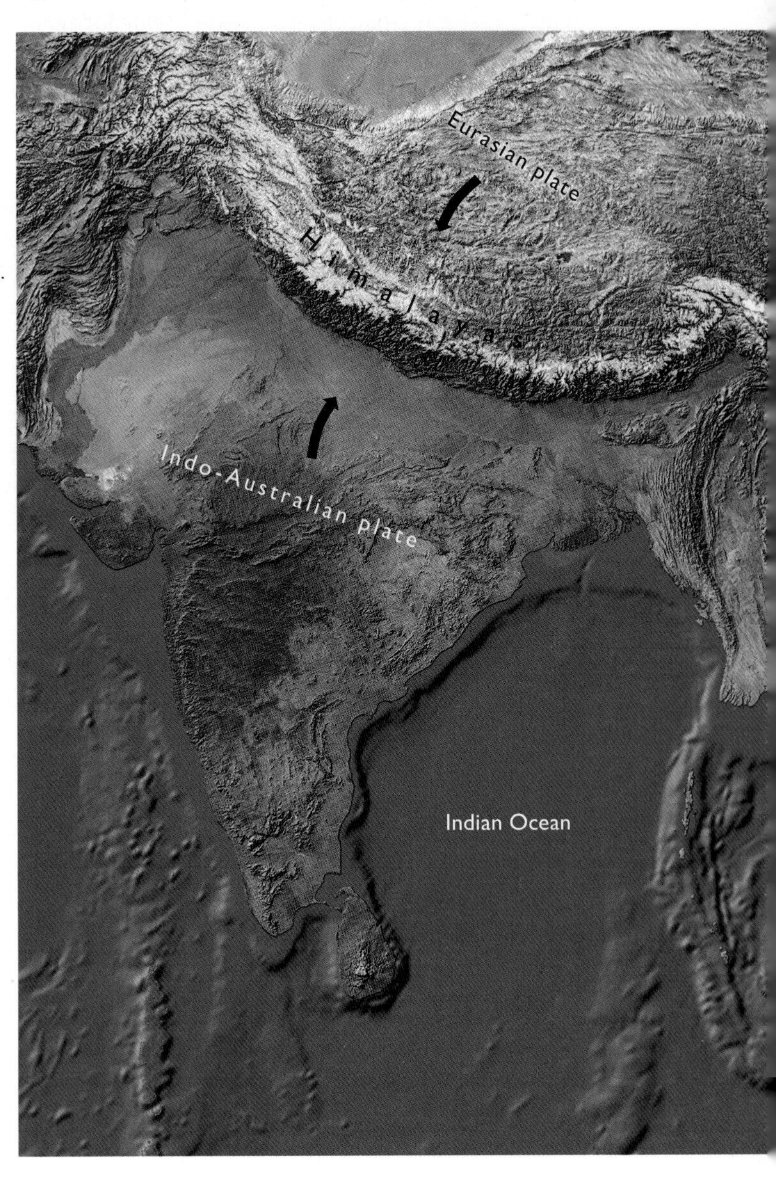

Figure 1.10 *The Himalayas – a collision between the Indo-Australian and Eurasian plates* ▲

Constructive plate margins

In this section you will learn about:

- different types of constructive plate margins
- the processes taking place at these margins
- the landforms found along these margins
- case studies of Surtsey and the Great African Rift Valley

Constructive (divergent) plate margins

When two plates separate (diverge) they form a **constructive margin**. There are two types of divergence:

- in oceanic areas, seafloor spreading occurs on either side of mid-ocean ridges – such as the Mid-Atlantic Ridge.
- in continental areas, stretching and collapsing of the crust creates **rift valleys** – such as the Great African Rift Valley.

It is at constructive plate margins that some of the youngest rocks on the Earth's surface are to be found. This is because new crust is being formed as the gap created by the spreading plates is filled by magma rising from the asthenosphere. As the magma cools, it solidifies to form dense new basaltic rock.

Mid-ocean ridges

Oceanic divergence forms chains of submarine mountain ridges which extend for thousands of kilometres across the ocean floor. If we could drain the oceans of water they would look like giant, bending spinal cords snaking along the constructive plate margins! Regular breaks called **transform faults** cut across the ridges – similar to how our discs break up our spines. (See Figure 1.2 on page 8.)

These faults occur at right angles to the plate boundary separating sections of the ridge – these sections may widen at different rates and times. This leads to frictional stresses building up, with shallow-focus earthquakes (those that occur at a depth of less than 70 km) releasing the tension (Figure 1.11).

Mid-ocean ridges can rise up to 4000 m above the ocean floor. The middle of the ridges are marked by deep rift valleys in all but the most rapidly separating plate margins found in the east Pacific. Over centuries the rift valleys are widened by magma rising from the asthenosphere, which cools and solidifies to form new crust.

Volcanic eruptions along the ridges can build **submarine volcanoes**. Over time these may grow to rise above sea level, creating volcanic islands – see Surtsey on the next page.

Rift valleys

Continental divergence forms massive rift valleys. These valleys are formed when the lithosphere stretches, causing it to fracture into sets of parallel faults. The land between these faults then collapses into deep, wide valleys which are separated by upright blocks of land called horsts.

Figure 1.11 *Transform faults along a mid-ocean ridge* ▼

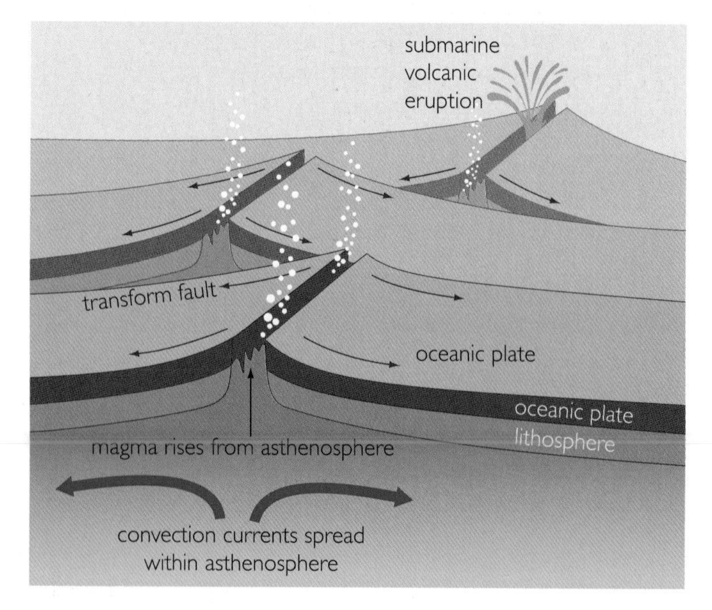

Surtsey – The Fire Giant

14 November 1963	The crew of an Icelandic vessel fishing about 5 ½ km off the south-west coast of Iceland, witness events from a science fiction film! An enormous column of black smoke and hissing steam rises from a boiling sea. Vivid flashes of lightning shoot through the smoke. Beneath the surface, the water glows a dull red. From this **submarine eruption**, the 'Fire Giant' of Surtsey is born. The eruption continues for the next 3 ½ years.
1964	Lava pours out from a new crater. When cooled, it forms a hard skin protecting the otherwise fragile island of **tephra** from wave erosion. Later, the first scientists set foot on this new island.
1965	Surtsey is declared a Nature Reserve with access restricted to 6 to 10 scientists a year.
1967	The eruption ends. Surtsey has now grown to almost 3 km^2 and a height of 154 m. But 90% of the island lies 120 m below the surface.
1970	The first seabirds start nesting. Lichens are found on the cooled lava.
2008	Added to UNESCO World Heritage List
Today	At least 60 plant, 89 bird and 335 species of invertebrates are present.

Figure 1.12 *The history of Surtsey* ▲

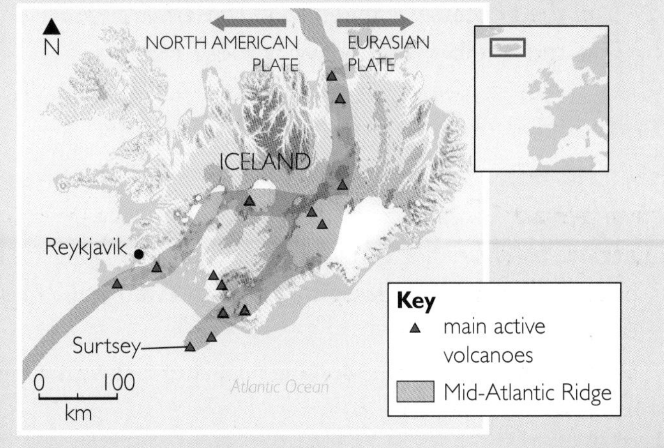

Figure 1.13 *Surtsey is located off the coast of Iceland along the Mid-Atlantic Ridge* ▲

Figure 1.14 *Birth of an island* ▶

The Great African Rift Valley

A rift is taking place right now – and it is tearing the eastern side of Africa apart. Could this be the birth of new oceanic plates and the beginnings of an ocean?

Uncertainty remains as to the exact causes of the Great African Rift Valley, but it is widely accepted that it is a result of plumes of magma rising from the asthenosphere. This magma heats the overlying plates, causing them to expand and bulge to create horsts, such as the Ethiopian Highlands. Volcanic activity may also be present, as with the famous Mount Kilimanjaro in Tanzania.

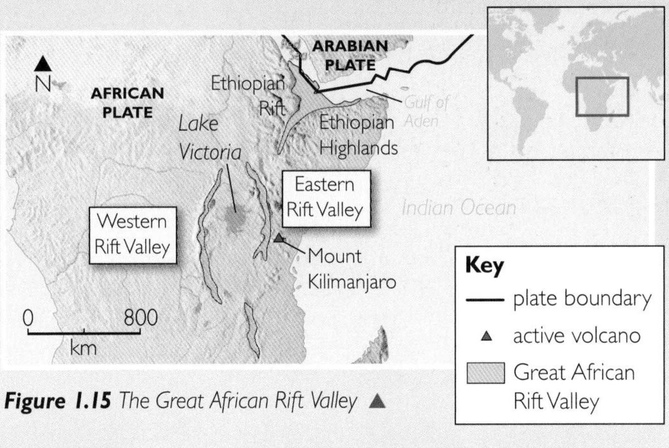

Figure 1.15 *The Great African Rift Valley* ▲

As the heated plate is stretched, it fractures along fault lines. This leads to fallen blocks of lowland called **grabens**. The resulting horst and graben landscape is one of vast interconnecting valleys running thousands of kilometres through Tanzania, Uganda, Kenya and Ethiopia.

If the stretching of the plate continues, then the thinned basaltic and continental rocks will eventually drop below sea level. This may mark the formation of a new ocean as eastern Africa splits away from the rest of the continent.

ACTIVITIES

1. Study Figure 1.11. Explain why volcanoes are formed at constructive plate margins.
2. Suggest reasons why scientists are so keen to study the island of Surtsey.
3. Do you agree that Surtsey should have such severe restrictions on human access? Justify your answer.
4. Study Figure 1.15. Can scientists use knowledge gained by studying the East African Rift Valley to help them understand more about the Mid-Atlantic Ridge?

Destructive plate margins

In this section you will learn about:

- different types of destructive plate margins
- the processes happening at these margins
- the landforms found along these margins

Destructive (convergent) plate margins

When two plates collide (converge) they form a destructive plate margin. Three types of convergence are possible:

- oceanic plate meeting continental plate, such as along the Pacific coast of South America
- oceanic plate meeting oceanic plate, such as along the Marianas Trench in the western Pacific
- continental plate meeting continental plate, such as the Himalayas.

Oceanic plate meets continental plate

The colliding of an oceanic plate with a continental plate is associated with an important process called **subduction**. This involves one plate diving beneath the other and being destroyed by melting.

Oceanic plate is denser than the lighter continental plate and so subducts underneath it. The exact point of collision is marked by the bending of the oceanic plate to form a **deep ocean trench** such as the Peru-Chile trench along the Pacific coast of South America. As the two plates converge, the continental land mass is uplifted, compressed, buckled and folded into chains of **fold mountains** such as the Andes (Figure 1.16).

The descending oceanic plate starts to melt at depths beyond 100 km and is completely destroyed by 700 km. This zone of melting is called the **Benioff Zone**. Melting is caused by both increasing heat at depth and friction. This friction may also lead to tension (stresses) building up which, depending on their depth, may eventually be suddenly released as intermediate and deep-focus earthquakes. The melted oceanic plate creates magma, which is less dense than the surrounding asthenosphere. As a result, it rises in great **plumes**. Passing through cracks (faults) in the buckled continental plate, the magma may eventually reach the surface to form explosive volcanic eruptions (See Andes on page 21).

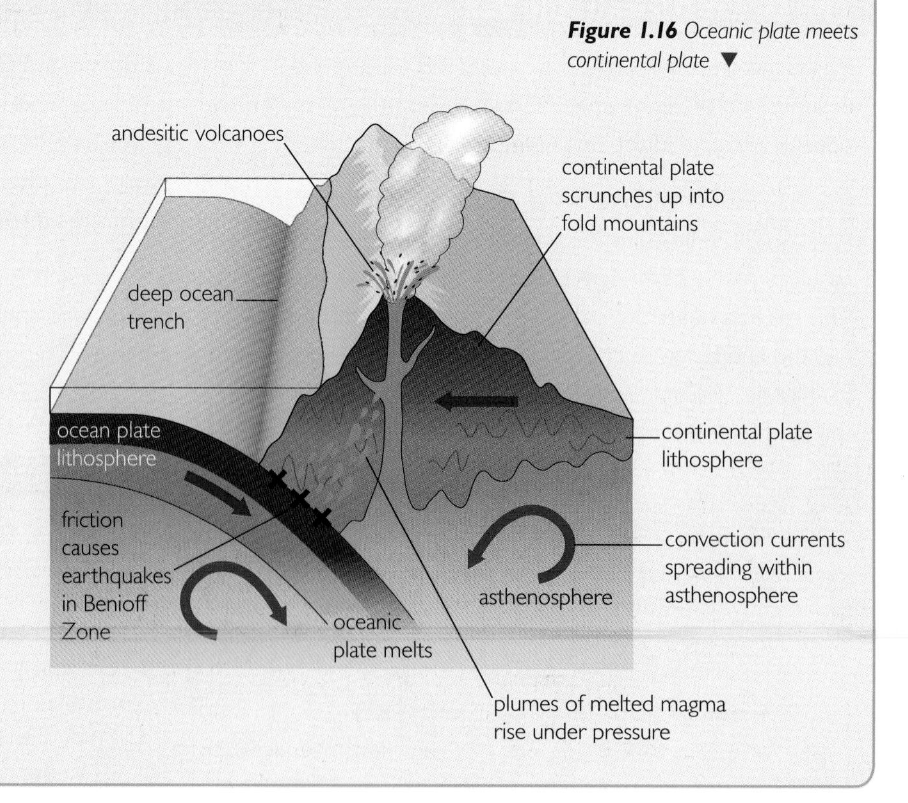

Figure 1.16 *Oceanic plate meets continental plate* ▼

Oceanic plate meets oceanic plate

When two oceanic plates collide, one plate (the faster or denser) subducts beneath the other. This leads to the formation of a deep ocean trench and melting, as described previously. The resulting rising magma from the Benioff Zone forms crescents of submarine volcanoes along the plate margins which may grow to form **island arcs** (Figure 1.17. Also see Montserrat later in this chapter). The Marianas Trench and the associated Marianas Islands in the Western Pacific illustrate this particularly well. Here, the Pacific plate is being subducted beneath the smaller Philippine plate.

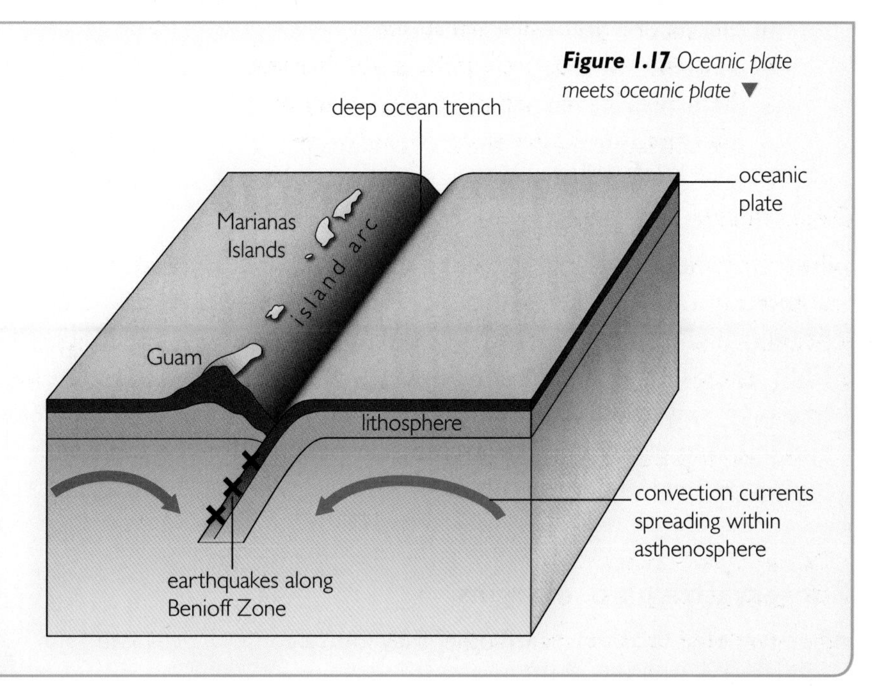

Figure 1.17 *Oceanic plate meets oceanic plate* ▼

Continental plate meets continental plate

Continental plates are of lower density than the asthenosphere beneath them. This means that subduction does not occur. The colliding plates, and any sediments deposited between them, simply become uplifted and buckle to form high fold mountains such as the Himalayas (Figure 1.18). Volcanic activity does not occur at these margins because there is no subduction; however, shallow-focus earthquakes can be triggered.

Figure 1.18 *Continental plate meets continental plate* ▼

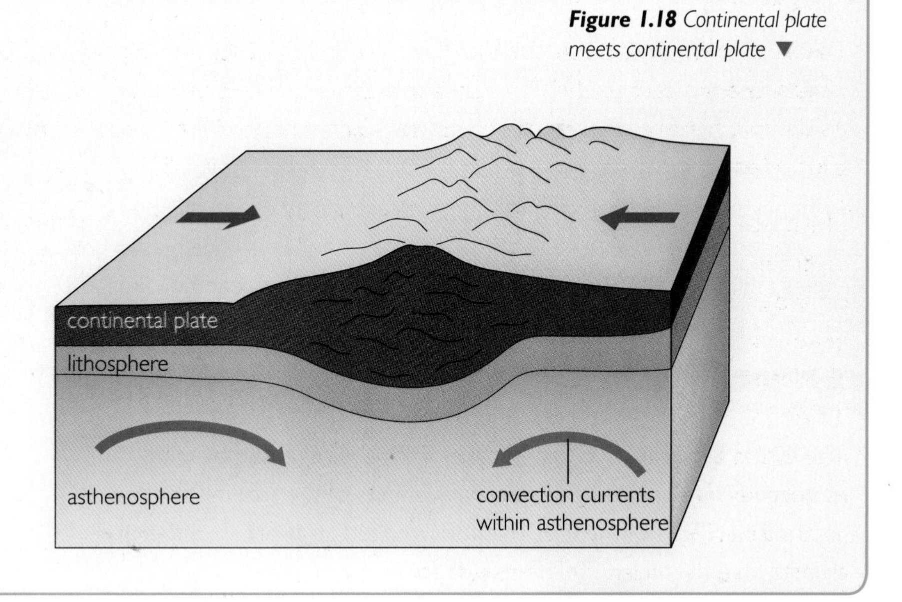

ACTIVITIES

1 Study the information about the different types of destructive plate margins.

- **a** What is meant by subduction and why is it an important process at destructive plate margins?
- **b** Why is there no volcanic activity at a continental collision margin?
- **c** Describe and explain the pattern of earthquakes at the different destructive margins. Use simple sketches to support your answer.
- **d** Imagine that you discovered a small group of islands in the Pacific Ocean. What evidence would you look for to confirm that they are an island arc and, therefore, represent a destructive plate margin?
- **e** What tectonic hazards are associated with destructive plate margins?

Did you know?

Ongoing compression of fold mountains makes them continually grow. So 5 mm growth of the Himalayas every year means that each new climber to summit Everest really has climbed higher than any other person before!

1.5 Conservative plate margins

In this section you will learn about:

- conservative plate margins
- the processes happening at these margins
- the consequences of these processes
- the San Andreas fault system

A deadly magnitude 6.6 earthquake – the strongest in modern Los Angeles history – ripped through the pre-dawn darkness Monday, awakening Southern California with a violent convulsion that flattened freeways, sandwiched buildings, ruptured pipelines and left emergency crews searching desperately for bodies trapped under the rubble.

Triggered by a fault that squeezed the northern San Fernando Valley between two mountain ranges like a vice, the 4:31 a.m. earthquake swamped hospitals with hundreds of injured people and left thousands more homeless as fires, floods and landslides dotted a landscape that has been visited by destruction with disturbing regularity in the last two years.

— *Los Angeles Times, 18 January 1994*

Conservative plate margins

When two plates slide past each other they form a conservative plate margin (Figure 1.19). Along these margins crust is not being destroyed by subduction. There is no melting of rock and, therefore, no volcanic activity or formation of new crust. Despite the absence of volcanic activity, these margins are tectonically extremely active and are associated with powerful earthquakes.

Friction between the two moving plates leads to stresses building up whenever any 'sticking' occurs. These stresses may eventually be released suddenly as powerful shallow-focus earthquakes – such as in Los Angeles (1994) and San Francisco (1989 and 1906). These earthquakes occurred along California's infamous San Andreas fault system.

CASE STUDY

The Great San Francisco Earthquake of 1906 is regarded by many as particularly significant. The primary effects of the shaking and the secondary effects of the fires that followed left large areas of the city destroyed and at least 700 people dead.

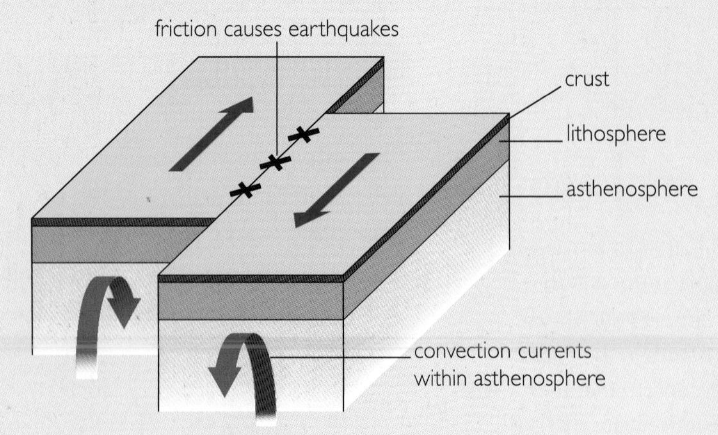

Figure 1.19 *Conservative plate margin* ▲

The San Andreas fault system

Try snapping a pencil in half. By applying force to opposite ends of the pencil you should be able to break the pencil into two pieces. Given sufficient pressure, breaking is inevitable. But is it possible to predict the exact moment when the snapping occurs? Seismologists have long attempted to answer these questions of earthquakes.

As recently as the L'Aquila earthquake in 2009 (see later in the chapter), what was later proven as a close prediction was ignored as a false alarm only days before. But in California, scientists may be close to achieving the first dependable earthquake prediction. The stakes could not be higher.

The San Andreas fault system runs for 1300 km but mostly cuts through California, the 6th richest economy in the world. The fault marks the boundary between the Pacific plate to the west and the North American plate to the east. As the plates slowly grind north-westwards against each other, powerful earthquakes are triggered.

One long-term effect of the earthquake involved the rebuilding and re-planning of the city. The 1906 earthquake helped inform scientists about the need to avoid building on soft mud and sand (materials found in the San Francisco Bay area) which become particularly unstable during an earthquake.

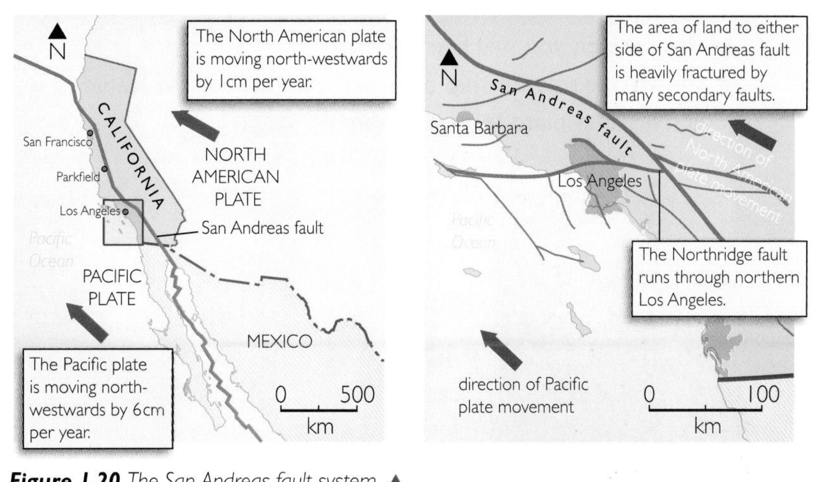

***Figure 1.20** The San Andreas fault system* ▲

Earthquakes affect all who live in California regardless of race, age or wealth. In Los Angeles, the 6.6 magnitude Northridge earthquake of 17 January 1994 left an immediate death toll of 57 and thousands injured. More than 24 hours after the earthquake, 82 000 homes and businesses were still without electricity and 50 000 without water. Overpasses collapsed as the concrete columns supporting upper road sections failed. The busiest road in the USA, the Santa Monica Freeway, and the major route northwards, the Golden State Freeway, were both closed. The economic cost of the 'quake approached $30 billion – more than four times the cost of the 1989 San Francisco Bay Area earthquake. Ironically, as the **epicentre** was in the north of the city, some of the poorest areas of Los Angeles, such as inner city Watts, suffered less damage, while affluent Santa Monica and Beverly Hills had hundreds of buildings declared unsafe.

***Figure 1.21** Damage to freeways caused by the 1994 Los Angeles (Northridge) earthquake* ▲

The San Andreas Fault Observatory at Depth (SAFOD) is an attempt to finally understand what triggers earthquakes, even to observe the earthquake as it happens several kilometres below the ground. SAFOD began in the summer of 2002 when a 2.2 km 'L' shaped hole was drilled across the San Andreas fault at Parkfield, a rural site situated halfway between Los Angeles and San Francisco. **Seismometers**, thermometers and other sensitive recording equipment have been placed in the cement-filled borehole to monitor future earthquakes. Scientists hope that such information may help them to predict earthquakes in the future.

***Figure 1.22** The San Andreas fault* ▼

ACTIVITIES

1. Why do earthquakes occur along conservative plate boundaries?
2. Use the text and photos in this section to describe some of the short-term and longer-term impacts of earthquakes that have occurred along the San Andreas fault system.
3. Why might trying to predict earthquakes be a near-impossible goal? Use information about the San Andreas fault system to support your answer.

1.6 Seismicity and volcanicity at plate margins

In this section you will learn about:

- the strong relationship between earthquakes, volcanic activity and plate tectonic theory

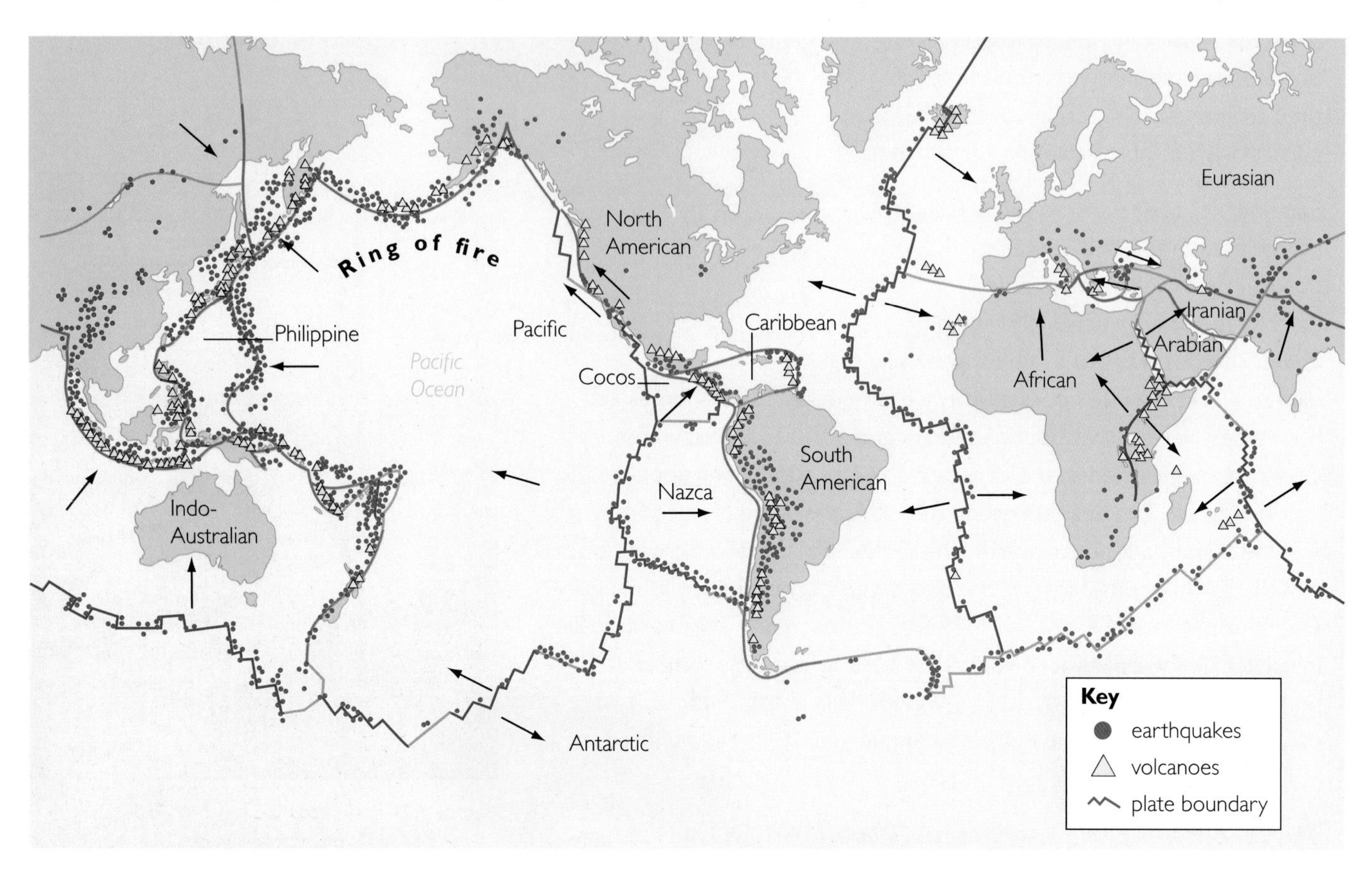

Figure 1.23 *The relationship between tectonic plate margins, earthquakes and volcanoes* ▲

Figure 1.24 *The explosive eruption of an acid volcano – Mount Pinatubo in 1991* ▼

Earth shaking (**seismicity**) and volcanic activity (**volcanicity**) are strongly associated with plate tectonic theory. In fact, 95% of the world's earthquakes and most volcanoes are located along plate margins (Figure 1.23). Even seemingly notable exceptions, such as the volcanic Hawaiian Islands, are explained in part by plate movements (see Hot spots in the next section).

Did you know?

The Indian Ocean volcano Krakatoa's apocalyptic eruption in 1883 was heard in central Australia and described as 'the loudest noise on Earth'. For three years the dust spread globally, creating wonderful sunsets and lowering world temperatures!

The relationship between seismicity, volcanicity and plate tectonics

Type of plate margin	Processes	Geographical features	Examples
Constructive margins	The plates diverge – move away from each other	• New oceanic crust is formed by basaltic magma rising from the asthenosphere • New basaltic rocks • Mid-ocean ridges broken up by transform faults • Shallow-focus earthquakes (0 – 70 km deep) • Submarine, basic volcanoes • Volcanic islands • Continental rift valleys	• Mid-Atlantic Ridge and East Pacific Rise • Ascension Island, Tristan da Cunha and Surtsey (page 15) • The Great African Rift Valley (page 15)
Destructive margins • Subduction zones • Collision zones	The plates converge – move towards each other	*Oceanic v. oceanic* • Oceanic crust is destroyed by subduction and melting at depth • Deep ocean trenches • Island arcs • Shallow, intermediate (70 – 300 km deep) and deep-focus (300 – 700 km deep) earthquakes • Explosive, acid volcanoes *Oceanic v. continental* • Oceanic crust is destroyed by subduction and melting at depth • Deep ocean trenches • Continental land mass is uplifted, compressed and buckled into fold mountains • Intermediate and deep-focus earthquakes • Explosive, acid volcanoes *Continental v. continental* • Colliding plates, and any sediments between them, uplift and concertina into particularly high fold mountains • Shallow-focus earthquakes	• Marianas Trench • West Indies, Aleutian and Marianas Islands • Krakatoa, Pinatubo and Montserrat (page 28) • Peru-Chile trench • Andes • Cotopaxi • Himalayas
Conservative margins	The plates move sideways past each other	• Shallow-focus earthquakes	• The San Andreas fault system (page 18)
Exceptions to plate margins	Hot spots near the centre of a plate	• Basic volcanoes	• Hawaiian Islands (page 22) • Yellowstone (page 34)

Figure 1.25 *Comparing the different plate margins* ▲

ACTIVITIES

1 Study Figure 1.26.

  a Using Figure 1.23 and your own knowledge, label the diagram with the names of specific tectonic plates. Begin by locating New Zealand and the Andes on the map.

  c Finish your diagram by adding annotations to the suggested boxes.

2 Explain the relationship between earthquakes, volcanic activity and plate margins. Be sure to use examples in your answer.

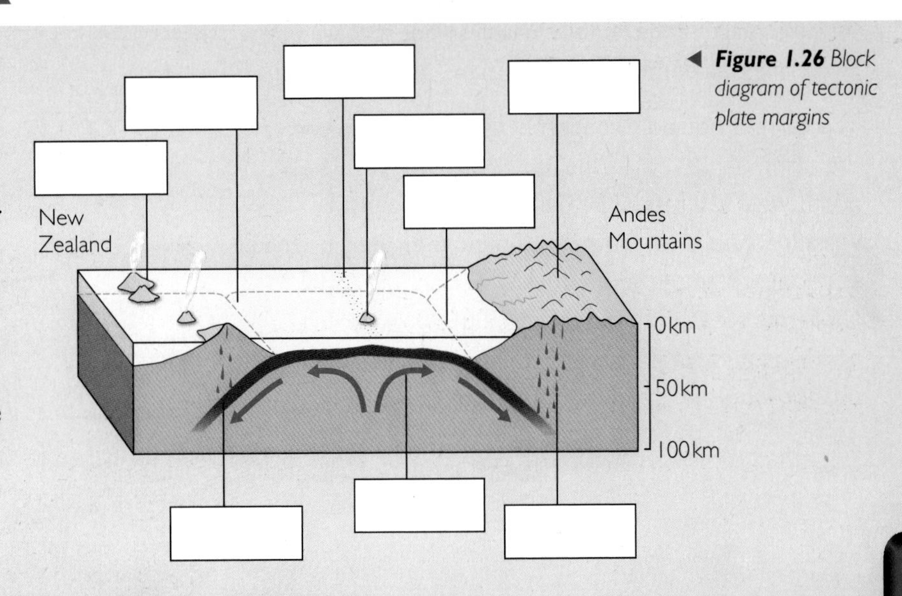

◀ ***Figure 1.26*** *Block diagram of tectonic plate margins*

Hot spots

In this section you will learn about:

- the processes that cause hot spots
- where hot spots are located
- a hot spot case study: the Hawaiian Islands

Hot spots

Whilst most of the world's volcanoes are located along plate margins, there are notable exceptions. As already described, radioactive decay within the Earth's core generates very hot temperatures. If the decay is concentrated, then hot spots will form around the core. These hot spots heat the lower mantle, creating localised thermal currents where plumes of magma rise vertically.

Whilst these plumes are usually found close to plate margins, they may occasionally rise within the centre of plates. The plumes then 'burn' through the lithosphere to create volcanic activity on the surface. As the hot spot remains stationary, the movement of the overlying plate results in the formation of a chain of active and extinct volcanoes. The Hawaiian Islands, near the centre of the Pacific Plate, are a classic example of this.

CASE STUDY

An example of a hot spot: Hawaii

In December 2009, scientists were at last able to end a forty year debate. The existence of volcanoes in the middle of plates, such as the Hawaiian Islands and Emperor Seamount chains (Figure 1.27), had proven to be difficult to fit into the newly-developing plate tectonic theory. Now, networks of seismometers on land and on the sea floor around Hawaii have provided the first seismic images of this missing jigsaw piece of plate tectonic theory – a mantle plume or hot spot.

As the less dense magma pushes through the lithosphere, the movement is picked up by the delicate seismometers. Following the analysis of two years of data from the wonderfully named PLUME (Plume-Lithosphere Undersea Melt Experiment) project, the first image of the famous Hawaiian hot spot – stretching to a depth of over 1500 km – was published.

The current hotspot is underneath the southern end of the island of Hawaii. As the Pacific plate moves slowly north-west at a rate of 5-10 cm per year, the next volcano in the chain, Lo'ihi Seamount, is forming on the sea floor just to the south of Hawaii. It should appear above the sea in around another 200 000 years! To the north-west of Hawaii, extinct volcanoes, battered by erosion, cool and slowly sink into the ocean.

Figure 1.27 *The Hawaiian and Emperor Seamount volcanic chains and hot spot* ▼

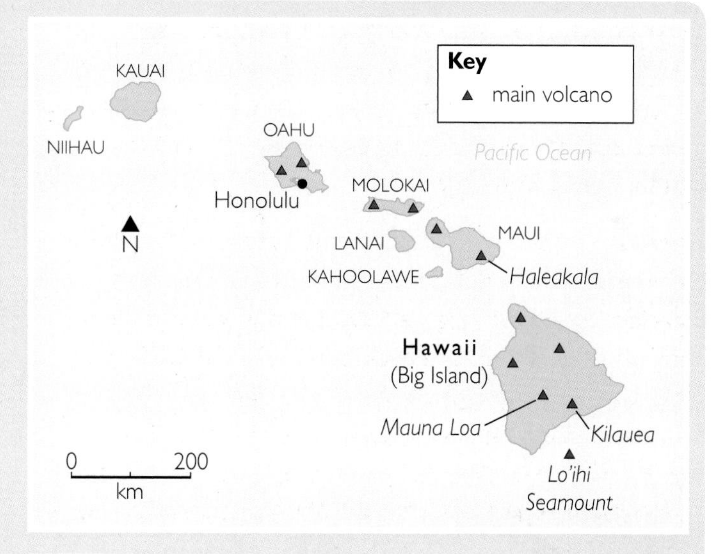

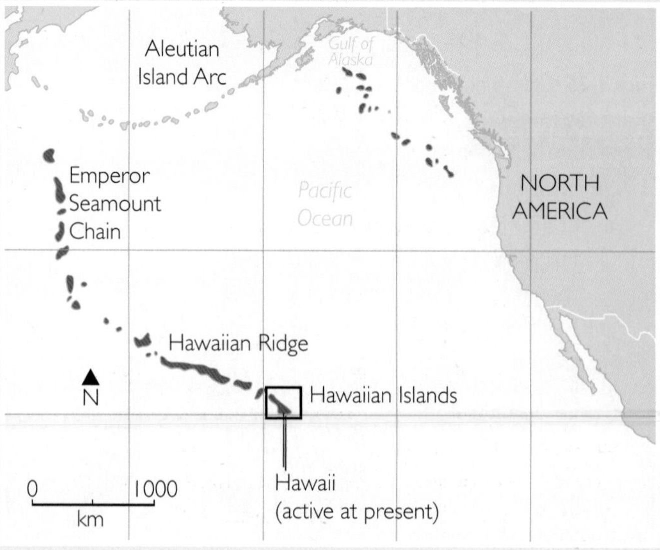

Bends in the island chain, such as between the Hawaiian Ridge and Emperor Seamounts, are a result of changes in the direction of movement of the Pacific plate. Today, the plate is pulling the whole island chain towards the Aleutian island arc, where it will be subducted and disappear forever.

New shield volcanoes above the hot spot erupt large volumes of runny lava which builds wide and broad shaped mountains. Surface flows of ropy (**pahoehoe**) and blocky (**aa**) lava are common (Figure 1.28). For example, Big Island is home to two world record volcanoes! Mauna Loa is the largest volcano on Earth and Kilauea is arguably the world's most active in terms of the volume of lava erupted in a year.

As the volcano ages and moves away from the hot spot, the lava becomes more alkaline, and cinder and **spatter cones** may appear, for example, Haleakala on the island of Maui. The final phase is associated with more violent eruptions. The lavas are now strongly alkaline and steep sided cones may erupt **pyroclastic flows**. An example of this is Hanauma Bay on the island of Oahu.

Figure 1.28 *Ropy pahoehoe lava* ▲

Volcanic tropical islands may appear a strange environment in which to study glacial melting, but the ancient coral reefs (Figure 1.29) that ring many of the islands have proven invaluable sources of dating. Coral reefs grow around the islands and, providing sea level change is not rapid, continue to flourish as the island sinks. But if sea level rises quickly, for example during an interglacial period when rapid melting of ice may occur, the coral drowns and dies. By examining and dating layers of fossil coral, it is possible to analyse past environments (or paleoenvironments). For example, the reefs off Lanai, Oahu, and Niihau islands record the glacial cycles up to several million years ago.

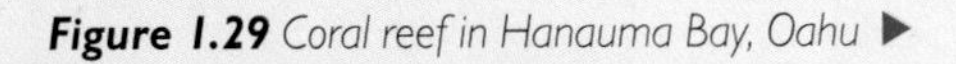

Figure 1.29 *Coral reef in Hanauma Bay, Oahu* ▶

ACTIVITIES

1. Using a simple labelled diagram, explain how a deep plume of magma at a hot spot was responsible for the formation of the chain of Hawaiian Islands.
2. Explain why hot spot volcanoes presented a problem when the original theory of plate tectonics was developed.

Did you know?

'Aa' and 'pahoehoe' are both Hawaiian words. 'Aa' is pronounced ah-ah – the sound made when walking barefoot over this jagged lava!

1.8 Volcanicity

In this section you will learn about:

- the hazards and benefits associated with volcanicity
- different forms of volcanic activity
- intrusive and extrusive volcanic features
- the materials ejected from volcanic eruptions
- different types of lava
- volcanic landforms
- how volcanoes are classified

Volcanicity

Volcanicity is most commonly associated with natural hazards. But the Earth's crust, water and atmosphere all have volcanic origins. Volcanicity should, therefore, always be thought of in terms of both hazards and benefits (Figure 1.30).

Volcanicity refers to all volcanic activities related to **magma** being forced into the crust. Usually at plate margins, although not always, magma at high temperatures and pressures exploits any weaknesses in the crust and may even erupt onto the surface as **lava**. Volcanic activity is, therefore, classified as **intrusive** or **extrusive** depending on whether the magma breaks the surface or not.

Did you know?

In 1943, in a Mexican cornfield, the volcano Paricutin started to grow. Within a year it was 335 m high. Nine years of eruption left a volcano covering a village and over 26 square kilometres!

Figure 1.30 *The hazards and benefits of volcanicity* ▼

Hazards	Benefits
Lava flows burn and bury crops.	Lava flows can create new land, for example, as we saw in the last section in Hawaii.
Submarine, coastal or island eruptions can cause tsunamis.	Hot rocks allow development of **geothermal power**, such as in Iceland, Italy and New Zealand.
Ash falls ruin crops and machinery, disrupt transport, pollute the air and cause breathing difficulties.	**Igneous rocks** contain valuable mineral deposits, including gold, silver, copper and diamonds.
Dust emissions endanger air transport and seed torrential rainstorms resulting in dangerous wet ash and mud lahars (mudflows that flow off the side of a volcano).	Volcanic sulphur is an important raw material used in many manufacturing processes, including the pharmaceutical and agrochemical industries.
Flooding results from lava flows and volcanic debris blocking and diverting rivers. For example, the water level at Spirit Lake rose by 80 metres after the eruption of Mount St. Helens in 1980.	Igneous rocks make excellent building materials. In Aberdeen, for example, most buildings are made from granite.
Very violent eruptions of **pyroclastic flows** (glowing clouds of superheated gases, ash and pumice called nuées ardentes) destroy life and property.	Historically, extinct volcanoes have made excellent defensive sites. A well-known example of this is Edinburgh Castle.
Volcanic melting of snow creates dangerous lahars.	Lava and ashes weather quickly into fertile soils.
Volcanic dust absorbs solar energy and so lowers atmospheric temperatures.	Volcanoes can be great tourist attractions generating huge revenues. Likewise geysers, fumaroles, hot springs and boiling mud make geothermal tourism big business.

Intrusive volcanic features

Magma in intrusive volcanic features cools, crystallises and solidifies into **igneous rocks** below the surface. Slow cooling results in large crystals forming – typical of rocks such as granite and dolerite. The resulting features may then only become part of the landscape once later erosion removes the overlying rocks (Figure 1.31). There are many examples of intrusive volcanicity in Scotland, Northern Ireland and northern England.

Dykes form where the magma solidifies in a vertical crack (fissure). They often form in groups called swarms, such as on the Scottish islands of Arran, Mull and Skye. Normally, the dyke material is more resistant than the surrounding (country) rock, leaving prominent wall-like features on an eroded landscape. Erosion of less resistant dyke material leaves ditch-like features.

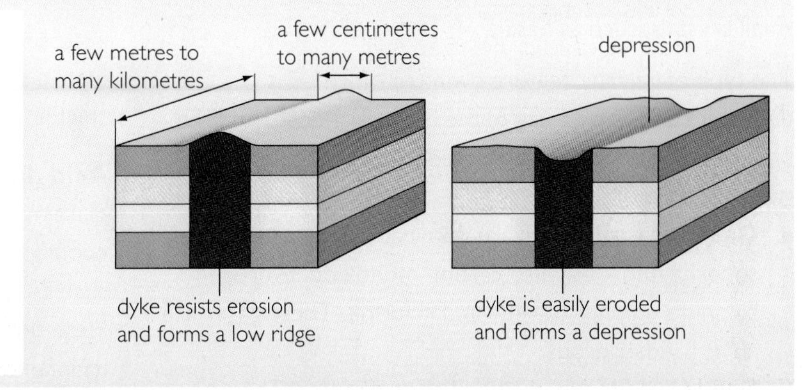

Sills form when magma solidifies into horizontal or inclined sheets in-between layers of pre-existing rock. For example, the Great Whin Sill in Northumberland forms steep north-facing cliffs of coarse-grained dolerite where it breaks the surface. Hadrian's Wall uses the natural defensive advantage of this landform (Figure 1.32).

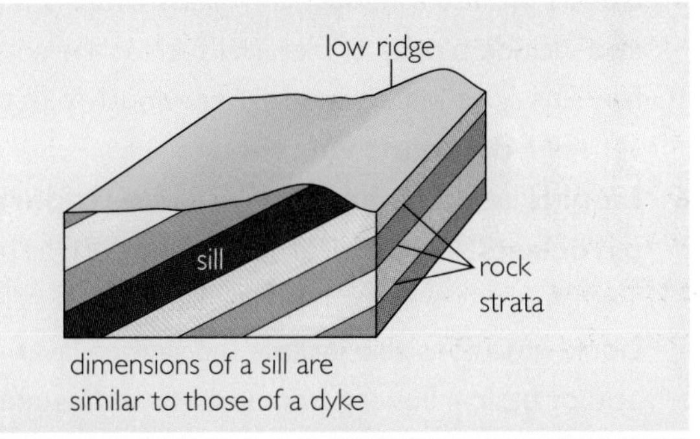

Laccoliths occur where viscous magma (resistant to flow) forces the overlying rock strata or layers to arch into a dome.

Batholiths are much larger-scale – they are massive in size and depth. Very often, dykes, sills and laccoliths will feed off the domed granite batholith before it solidifies. Surrounding the batholith, extreme heat and/or pressure alters the adjacent country rock in a process called metamorphism, to form a **metamorphic aureole**.

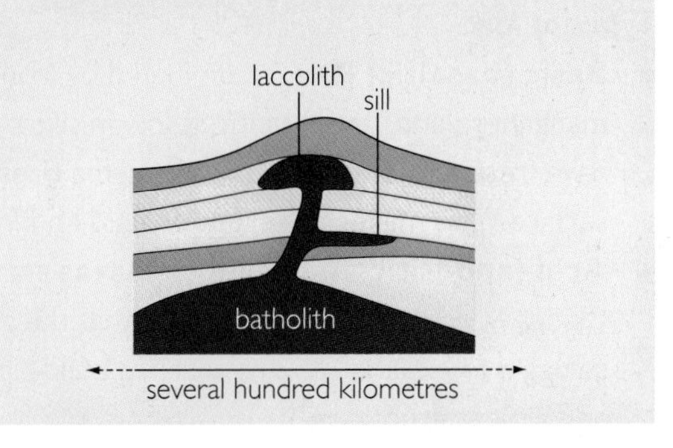

Figure 1.31 *Intrusive volcanic features* ▲

◄ ***Figure 1.32*** *Hadrian's Wall built on the Great Whin Sill in Northumberland*

Did you know?

The Romans named Vulcano after their 'blacksmith of the gods' – Vulcan. The Mediterranean island was believed to be the chimney of his forge – spitting out dust, clouds and bolts of lightning!

Extrusive volcanic features

Magma in extrusive volcanic features cools, crystallises and solidifies from surface lavas. In contact with the air, lava cools far quicker then magma still underground – and cooling in the sea is even more rapid! As a consequence, the resulting igneous rocks tend to be finer grained, with small crystals, such as basalt.

The type of volcanic feature eventually formed is determined by the nature of the material ejected and the type of eruption. Material ejected from an eruption may be gaseous, solid or liquid:

- **Gaseous emissions** are dominated by steam – often superheated – but also carbon monoxide, hydrogen sulphide, sulphur dioxide and chlorine. These gases are all highly dangerous.
- **Solids** include ash, dust and glassy cinders. They may also include blocks of material such as the shattered remains of solidified lava that previously plugged the vent of a **dormant volcano**.
- **Liquids** include lava bombs (known as **tephra** or **pyroclasts**) which solidify in mid-air, such as pumice. Tephra in its finest form of hair-like trails is called **lapilli**. Liquid emissions also include the surface lavas – either acid or basic – flowing from **vents** and **fissures**.

Types of lava

- **Basic (basaltic) lava** is dominated by iron and manganese and, importantly, is low in silica. Gas bubbles have freedom to expand as the magma rises to the surface. This makes the eruption fluid and free-flowing.
- **Acid (andesitic and rhyolitic) lavas** are rich in silica. Critically, acid lavas are so viscous that the gas bubbles struggle to expand. This results in a build-up of pressure and violent eruptions.

Landforms

Lava can erupt from long cracks called fissures or localised vents. The nature of the opening(s) from which the magma emerges to form lava will determine much about the resulting landform.

Fissure eruptions of basic lava, such as blocky (aa) or smooth ropy (pahoehoe), can create extensive **lava plateaus**. Hollows in the existing landscape are filled to create flat, featureless basalt plains.

Vent eruptions usually create cone-shaped landforms, often the classic volcano image, complete with summit crater (Figure 1.33).

Classification of volcanoes

Volcanoes are usually classified according to the violence of the eruption. This is determined by the pressure and amount of gas in the magma (Figure 1.34).

Basic shield volcanoes are associated with mid-ocean ridges, rift valleys and hot spots where magma has direct access to the surface. Repeated eruptions of runny lava from a central vent form gently sloping volcanic cones and shields, such as those forming the Hawaiian Islands.

Acid dome volcanoes are associated with destructive plate margins. The viscous lava cannot flow far before cooling, resulting in steep-sided convex volcanic cones, such as Mount Pelée on the Caribbean island of Martinique. Here the rhyolitic lava is so thick that it is unable to flow! Its infamous eruptions are therefore deadly, explosive **pyroclastic flows** of superheated, glowing (incandescent) gases, ash and pumice, called **nuées ardentes**.

- **Dormant volcano**: A volcano which has not erupted within historic times is called dormant. If it has not erupted for at least 10 000 years we say it is **extinct**.

Figure 1.33 *Mount Fuji, Japan – a composite volcanic cone* ▼

Composite cones are formed from alternating eruptions of ash, tephra (including lapilli) and lava building up the volcano in layers (Figure 1.33). The classic conical shape of Mount Fuji is relatively rare compared to many irregular-shaped composite volcanoes covered by numerous **secondary (parasitic) cones** and **fissures**, such as Mount Etna in Sicily.

Ash and **cinder cones** are formed from ash, cinders and tephra erupted from a central crater. They are usually symmetrical, steep-sided and concave in shape.

Calderas result from violent eruptions blowing off a volcano's summit. This empties the **magma chamber** and allows the sides of the volcano to collapse inwards. A vast pit crater up to many kilometres in diameter is left to be flooded by the sea or fill as a lake.

Did you know?

In April 2010, ash from the Eyjafjallajökull volcano in Iceland caused the cancellation of all flights from UK airports – the biggest shutdown in UK aviation history to date.

ACTIVITIES

1. As a three-column table, summarise the nature of gases, liquids, and solids ejected from the different types of volcanic eruption.
2. 'The thicker the lava, the more violent the eruption.' In no more than 100 words, give reasons for this general rule.
3. Find out the literal translation of 'nuée ardente'. This will help you remember the nature of the phenomenon very clearly.

Name	Type of magma	Characteristics of eruption
Icelandic	basaltic	Lava flows gently from fissures.
Hawaiian	basaltic	Lava flows gently from a central vent.
Strombolian	(thicker) basaltic	Frequent, explosive eruptions of tephra and steam. Occasional, short lava flows.
Vulcanian	(thicker) basaltic, andesitic and rhyolitic	Less frequent, but more violent eruptions of gases, ash and tephra (including lapilli).
Vesuvian	(thicker) basaltic, andesitic and rhyolitic	Following long periods of inactivity, very violent gas explosions blast ash high into the sky.
Peléean	andesitic and rhyolitic	Very violent eruptions of nuées ardentes.
Plinian	rhyolitic	Exceptionally violent eruptions of gases, ash and pumice. Torrential rainstorms cause devastating lahars.

Figure 1.34 *A classification of volcanic eruptions* ▲

Internet research

Use the Internet to research why volcanoes vary in shape. Is there any link between shape and the type of tectonic plate margin a volcano is located near? Support your answers with examples using any three of the volcanic eruptions classified in Figure 1.34.

Two contrasting volcanoes

In this section you will learn about:

- the causes of a volcanic eruption in a low-income and a high-income country
- the effects of this volcanic activity
- responses to the volcanic eruptions
- management of the associated hazards

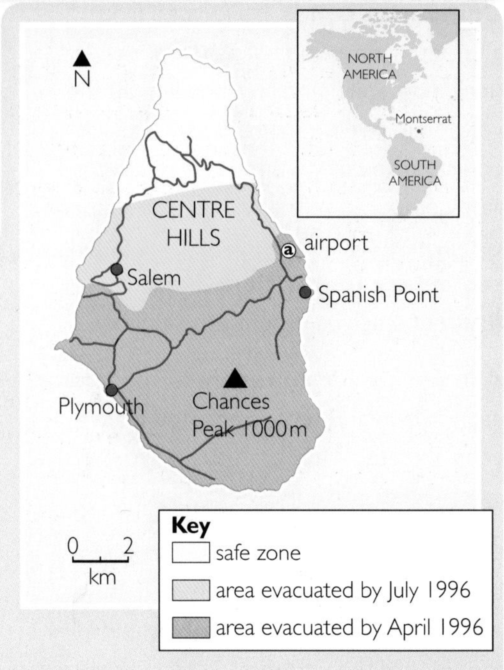

Figure 1.35 *Montserrat in the Lesser Antilles island arc* ▲

CASE STUDY

Montserrat – a low-income country (LIC) case study

How much would you pay to live in the shadow of an active volcano? Until the eruption of the Chances Peak volcano in July 1995, the Caribbean island of Montserrat was a playground for the rich and famous. Pop stars including Paul McCartney and The Rolling Stones visited and recorded music there, staying in luxurious holiday villas or hotels. Today, similar villas in the north of the island may be purchased at around a third of their original value.

Location

The British colony of Montserrat is situated within the northern part of the Lesser Antilles (Figure 1.35). This is an island arc formed where the Atlantic tectonic plate subducts beneath the Caribbean tectonic plate. Most of the islands are composite volcanoes formed as a result of violent eruptions.

History

Montserrat is only 16 km long and 10 km wide, and is built almost entirely of volcanic rock. Throughout the island's history, lava domes have been created as a result of thick sticky (andesitic) lava building up at the top of the volcano. When the lava eventually becomes too heavy, the domes collapse, resulting in andesitic lava and pyroclastic flows. These volcanic flows have left rich volcanic soil supporting an abundance of tropical vegetation and **cash crops**, including the soft, locally woven Sea Island cotton.

Figure 1.36 *Ash-covered Plymouth after the eruption* ▼

Eruption

On 18 July 1995, the youngest of three volcanoes on Montserrat, the Chances Peak volcano in the south of the island, began erupting ash and dust. Scientists began to monitor gases, microquakes and changes in the volcano's shape, discovering that it was only dormant, and not extinct as previously thought.

One month later, the evacuation of the south of the island began, with residents being moved to churches and halls in the north of the island. In April 1996, the entire population was forced to leave the capital city of Plymouth (Figure 1.36).

On 25 June 1997, Chances Peak volcano catastrophically erupted. The dome of the volcano collapsed – sending 5 million cubic metres of hot rock and gases down the sides of the Soufrière Hills. The south of the island was covered by these pyroclastic flows of hot ash, mud and rock.

Effects

The immediate effects of the eruption included the death of 19 people killed in fires associated with the pyroclastic flows (Figure 1.37). There were a number of burn and inhalation injuries and two-thirds of all houses were either buried by ash or flattened by rocks. Similarly, farmland, vegetation and three-quarters of the infrastructure were destroyed. More than half of the 11 000 population was evacuated to Antigua, the USA and the UK.

This active volcano continues to create headlines. On 11 February 2010, Chances Peak explosively erupted and ejected a 15 km column of ash. 10% of the lava dome collapsed, leading to pyroclastic flows and ash fall.

Figure 1.37 *Pyroclastic flows continue on Montserrat* ▲

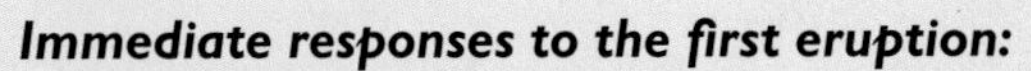

Immediate responses to the first eruption:

- The Montserrat Volcano Observatory was set up in 1995 to closely monitor changes to the volcano and successfully predicted its eruption on 25 July 1997.
- **NGO**s, such as the Red Cross, set up temporary schools and provided medical support and food.
- Warning systems were set up to alert inhabitants – sirens, speaker systems and via the media.
- Troops from the USA and the British Navy came to aid the evacuation process.
- £17 million in UK government emergency aid was given to pay for items such as temporary buildings and water purification systems.

Did you know?

The largest pyroclastic flows can move at rates of 200 metres per second and have been known to cover ridges more than 1000m high!

Long-term responses:

- A 3-year redevelopment programme for houses, schools, medical services, infrastructure and agriculture was funded by the UK.
- A top-heavy population pyramid was created as younger people did not see an economic future on the island and moved elsewhere.
- By 2005, many people had moved back, but the south of the island, including Plymouth, remains an exclusion zone.
- Vegetation is slowly re-growing as the ash and lava break down. Fertile soil will mean that land can again be used for crops and farming.
- The potential exists for rebuilding tourism, with the volcano itself as an attraction. In 2010, Montserrat was selected by *National Geographic Adventure Magazine* as one of the "Top 25 New Trips of 2010". However, despite a new airport, hotel and dive shop, it remains questionable as to whether tourists will return in numbers.

ACTIVITIES

1 Study Figure 1.35 and the satellite image of Montserrat on the right.
   - **a** Identify the man-made feature marked as X on the satellite image.
   - **b** List evidence to suggest that the volcano is active.
   - **c** Approximately what area of the island is considered 'safe'?
   - **d** Explain why the 'safe zone' is several kilometres from the site of the volcano.

2 Volcanic disasters can sometimes be avoided if scientists and local people recognise the early warning signs of a catastrophic eruption. Suggest how the Chances Peak volcano may have given such 'early warning signs' and allowed for an organised evacuation of the island.

Satellite image of the active Mount Soufrière Hills volcano in Montserrat ▲

CASE STUDY

Mount Etna ('Mongibello') – a high-income country (HIC) case study

A Russian doll is a rather clever toy. At first glance, there only appears to be a simple wooden figure. But on removing the top half of the doll, there is a slightly smaller second doll neatly nested inside. This action can be repeated until an extended family of dolls is removed from the original, each one slightly smaller in size.

At 3350 m tall and covering an area of 1250 km², Mount Etna dominates the landscape of northeastern Sicily, Italy (Figure 1.38). Although it is the largest active volcano in Europe, Etna is in fact a 'Russian doll' of nested composite volcanoes with summit **calderas** (large collapsed craters). This complex geology results in a wide variety of eruptive styles. Interestingly, Mount Etna has the longest documented record of eruptions of any volcano in the world.

In simple terms, Etna is a result of the collision of the African and Eurasian continental plates. While there is no agreement on the exact cause of each eruption, most theories suggest it is linked to rifting, normally associated with constructive plate margins. As seen in the Great African Rift Valley (see page 15), this pulling apart of the Earth's crust is thought to be separating eastern Sicily from the rest of Italy.

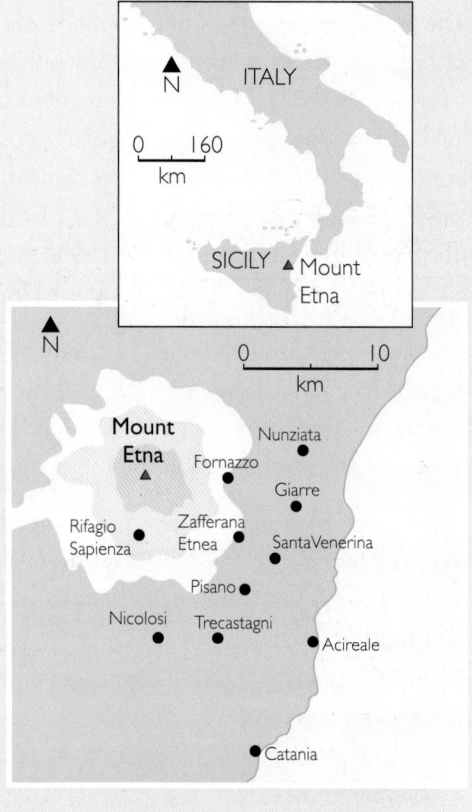

Figure 1.38 *Mount Etna, Sicily, Italy* ▲

Volcanologists classify the eruptions of Etna as mostly effusive (lava flows) and occasionally mild Strombolian. The Strombolian-type eruptions, with short lava flows, tend to occur on the summit (Figure 1.39). Less frequent, but more effusive, are the eruptions from fissures on the sides of Etna.

Figure 1.39 *Strombolian eruption of Mount Etna* ▼

At the summit, volcanic cones mark the locations of the active Northeast (1911) and Southeast Craters (1971). The remaining summit craters, the Voragine (1945) and the Bocca Nuova (1968), actually lay in the pre-existing 250 m wide Central Crater.

Secondary (parasitic) volcanic cones and jagged fissures combine to create a rich and varied landscape on the slopes of Etna. A catastrophic collapse of Etna's eastern slopes resulted in the impressive 7 km long horseshoe shaped Valle del Bove (Valley of the Oxen). This natural funnel is layered with ancient and recent lava flows. By drilling through these tree-ring like layers, scientists have been able to uncover a detailed geological record of past eruptions. In contrast, ridges such as the active Northeast Rift stand upright and may be identified by eruptive cones separated by deep fissures.

Today, 900 000 people continue to be at risk from the volcanic hazards of Etna. Figure 1.40 below gives some information about Etna's most recent eruptions – and the consequences.

Date	Nature of eruption	Impact	Responses and management
28 March 1983	Lava effusion from a 750m fissure lasting for 131 days.	Whilst slow moving, the lava flows threatened to destroy property.	Huge (750 000 cubic metres) earthworks used to redirect lava flows away from inhabited areas.
15 December 1991 – 31 March 1993	Effusive lava flows and fountaining from a fissure on the eastern side of the Southeast Crater. The eruption lasts for 473 days.	Town of Zafferana Etnea threatened by the largest volume of lava in hundreds of years.	1992 • January – earth dam built to temporarily hold lava flow. • April – blasted rock and concrete blocks dropped to 'plug' the lava channel. • May 27–29 – explosives used to divert the lava flow into a new man-made channel.
17 July – 9 August 2001	Seven fissures erupt effusively on the south and northeast flank.	Significant damage to tourist facilities. The town of Nicolosi is threatened.	Catania airport intermittently forced to close as result of ash fall.
26–27 October 2002	Strombolian, Hawaiian fountaining and phreatomagmatic (where magma is mixed with water) eruptions.	Most explosive flank eruption in the last 150 years! Lava flows threaten the mountain village of Rifugio Sapienza.	
8 November 2009	Strombolian eruptions in Southeast Crater.	4.4 magnitude earthquake beneath southwest flank.	

Figure 1.40 *A history of some of Mount Etna's most recent eruptions* ▲

- **Effusive activity**: Lava flows out onto the ground surface rather than exploding into the air.

Did you know?

Footage from the 2002–03 eruption of Mount Etna was used in Star Wars Episode III: Revenge of the Sith.

ACTIVITIES

1 Tectonic activities are defined as hazardous when they have the potential to cause the loss of life or property. This means that people living in areas affected by tectonic hazards are at risk – particularly when an area's vulnerability and the natural hazard overlap (see the Risk of Disaster diagram to the right).

Answer the following questions in **pairs**. One recommended approach is for each person to study one of the volcanoes mentioned below.

a For either Montserrat or Mount Etna, answer the questions that are presented in the Venn diagram. You may wish to keep the same headings of Natural Hazard and Vulnerability.

b Rate the risk of disaster on a sliding scale. For example, 1 = highly likely to 5 = highly unlikely.

c Share and justify your results with your partner.

d Which volcano poses the greater risk of disaster? As a pair, write a concluding paragraph to explain your answer.

Natural hazard
- When did it last occur?
- What is the probability that it will occur again?
- How quickly did it start?
- How big was the event?
- How long did it last?
- How large was the affected area?

Risk of disaster

Vulnerability
- What is the population?
- How developed is the economy?
- What is the land used for?
- Are emergency services prepared?
- Are buildings, roads and wider infrastructures hazard proof?
- How are vulnerable (e.g., elderly) people and buildings (e.g., historic) protected?

Risk = Hazard × Vulnerability

The risk of disaster – when vulnerability and a natural hazard collide ▲

Minor forms of extrusive volcanic activity

In this section you will learn about:

- geysers
- fumaroles
- hot springs and boiling mud
- geothermal tourism and geothermal power
- Yellowstone National Park

Figure 1.41 *Geyser erupting in New Zealand* ▲

Geysers

Great Geyser is a hot spring in Iceland. In the 1770s, every half hour it shot a jet of scalding water and steam over 60 m into the air! This, and other spouting hot springs in the centre of Iceland, gave us the name **geyser**, which in Icelandic means 'to erupt'. This name is now used throughout the world, including in Indonesia, Japan, New Zealand and the USA. Many of these popular tourist attractions are closely associated with volcanoes or young volcanic rocks.

The water in geysers is thought to boil from contact with hot rocks or volcanic gases building pressure until an explosive eruption takes place (Figure 1.41). The vent then refills with water – and the cycle is repeated. In geological terms they are short-lived features because groundwater conditions change, vents collapse and new geysers appear elsewhere.

Fumaroles

Much less spectacular, but far more widespread are the low-pressure outlets of steam and gas called **fumaroles**. These gas vents are associated with both active and dormant volcanoes and usually persist for thousands of years after a volcano has become extinct. Fumarole gases are mainly steam and carbon dioxide, but carbon monoxide, hydrochloric acid and sulphur gases amongst others can make them highly unpleasant to be around!

If the gas is sulphurous, fumaroles are called **solfatara**, named after the coneless volcano on the outskirts of Naples in Italy (Figure 1.42). As you can see in the photograph, solfatara leaves distinctive yellow deposits of sulphur on the surrounding rocks.

> **Did you know?**
>
> *The word geyser derives from the Icelandic word gjósa – literally meaning 'to erupt'.*

Hot springs and boiling mud

Hot thermal springs are not under pressure and so do not explode from the surface like geysers. They are common in Iceland and other volcanic areas, but they are not all volcanic. Some hot springs are acidic, others alkaline and, on occasions, may even be radioactive (having absorbed radon gas). Closely associated with fumaroles, the groundwater heated deep below the surface may contain minerals dissolved from contact with the hot rocks. These minerals build on the surface as colourful **boiling mud** deposits – spattering to form **mud volcanoes**, which may bubble quietly or erupt like geysers.

Figure 1.42 *Solfatara Crater, near Naples, Italy* ▼

Geothermal tourism

Tourism is hot business! Whilst volcano and geothermal tourism is not new (for example, Vesuvius AD 79), the numbers globally are now significant (Figure 1.43). Volcanic and geothermal environments offer exciting tourist destinations, yet there are concerns too that this particular brand of **geotourism** may be both hazardous (there are no international visitor safety guidelines) and unsustainable. Kenya, for instance, has struggled to manage the tourist pressure of big game safaris – yet increasing numbers of adventure tour companies now offer experiences of the volcanic landscape of the Great African Rift Valley (see page 15).

- **Geotourism**: Sustainable tourism that enhances the geographical character of an environment and its people. **Ecotourism** is an example of this.

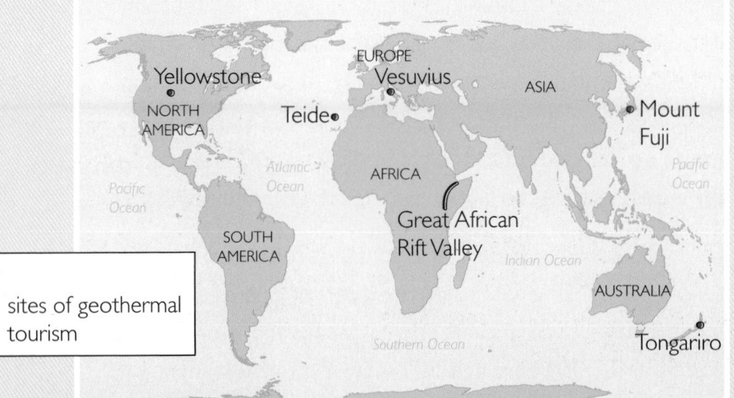

Figure 1.43 *Geothermal tourism – Mount Fuji in Japan attracts over 100 million visitors every year whilst the geothermal landscapes of national parks in, for example, Yellowstone, Teide in Spain and Tongariro in New Zealand, attract between one and four million tourists a year* ▲

Geothermal power in New Zealand: an environmentally harmful energy source?

In North Island, New Zealand, native Maoris have used hot spring water for cooking, bathing and heating for centuries, particularly within the major geothermal regions of Taupo and Rotorua. But it is the more recent human activity of geothermal power stations that is causing environmental damage. Geothermal power stations in the Taupo geothermal region generate electricity by tapping into naturally occurring **hydrothermal convection systems**. In practice, this means drilling bore holes into underlying aquifers, or reservoirs, then storing the hot water. Steam from the pressurised water is then captured and used to drive turbines and generate electricity.

New Zealand's first geothermal power station was built in 1958 at Wairakei, Taupo (Figure 1.44). Shortly afterwards, the geysers at Geyser Valley and Taupo Spa ran dry and disappeared. In the Rotorua region more than 900 shallow wells have been drilled to provide local hot water. Following significant changes to surface features, in 1988 the government closed wells within 1.5km of Pohutu Geyser. It resulted in a partial recovery of the Rotorua geothermal reservoir, but no extinct geysers recovered.

Regional councils now have responsibility for monitoring and preserving geothermal features. With growing pressure on sites of geothermal significance for sustainable energy and geotourism, the New Zealand example illustrates the importance of planning any development from the outset.

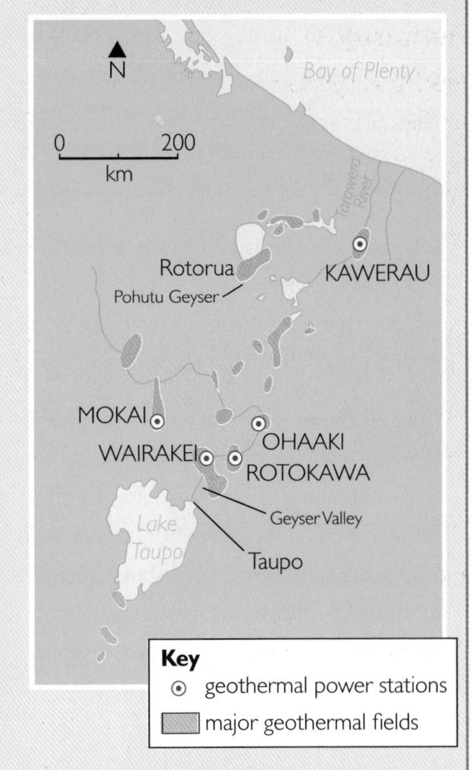

Figure 1.44 *Taupo geothermal region, North Island, New Zealand* ▲

CASE STUDY

Yellowstone National Park: supervolcano!

How safe do you feel? There is a one in 14 million chance of being struck by lightning and every 7000 years or so one of us is likely to be squashed by a meteorite! Yet statistically, the occurrence of a **supervolcanic** eruption at Yellowstone, USA is far more likely. If such an eruption occurred, it would release 1000 cubic km of magma and create a giant caldera as the ground collapsed. The erupted ash would change the global climate for years afterwards.

The Yellowstone Caldera is a constant reminder that Yellowstone National Park sits on top of one of the world's largest volcanoes. The caldera was formed around 640 000 years ago following a supervolcanic eruption and is 48 km by 72 km wide. The hotspot magma chamber lying beneath the caldera has existed for at least 16 million years. As the North American plate has moved southwest above this hot spot, it has left a trail of extinct volcanoes and calderas stretching into the nearby state of Nevada (Figure 1.46). The magma continues to push against the surface and helps to make Yellowstone unrivalled for its richness of hydrothermal resources. With half of the world's geothermal features, Yellowstone holds two-thirds of all geysers and more than 10 000 hot springs, mud volcanoes, and fumaroles.

Yellowstone National Park is arguably a geologist's sweet box. Such is the variety and richness of geothermal activity that there is an abundance of pick-'n'-mix landforms of different colours and sizes.

New landforms appear and disappear as the ingredients of water, heat and rock are shaken by seismic activity. Many are named, for example, 'Old Faithful' geyser was named in 1870 for its frequent and generally predictable eruptions of between 33 m and 56 m. With intervals of around 45 – 90 minutes between eruptions, it continues to be a tourist **honeypot**.

Did you know?

'Tickling' or 'soaping' geysers for tourists is now strictly forbidden. Soap flakes or detergent thrown down the vent to cause frothing lowered the surface water tension and made them erupt!

Figure 1.45 *Yellowstone National Park* ▼

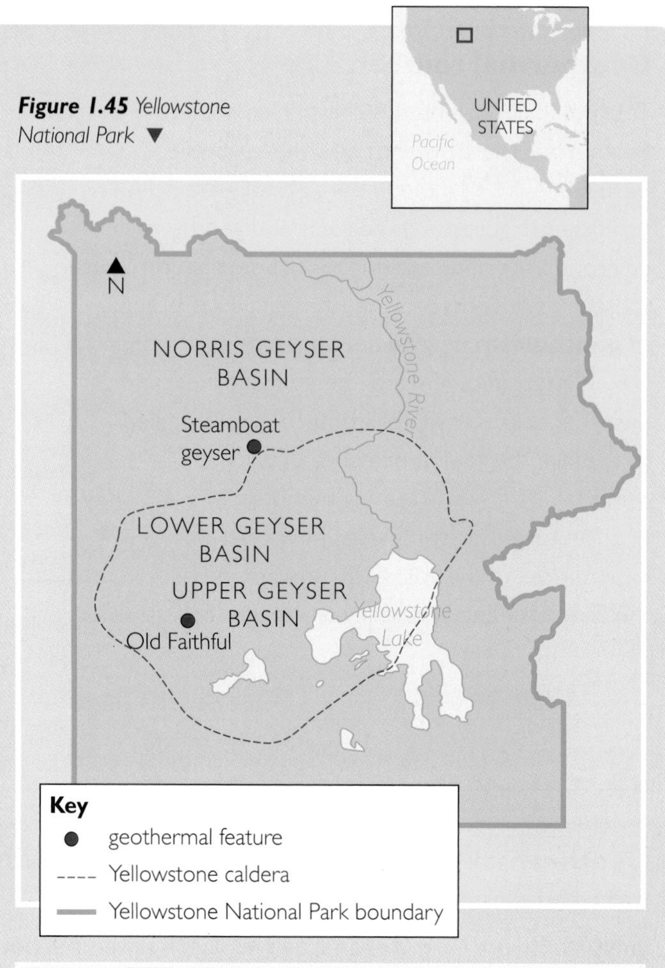

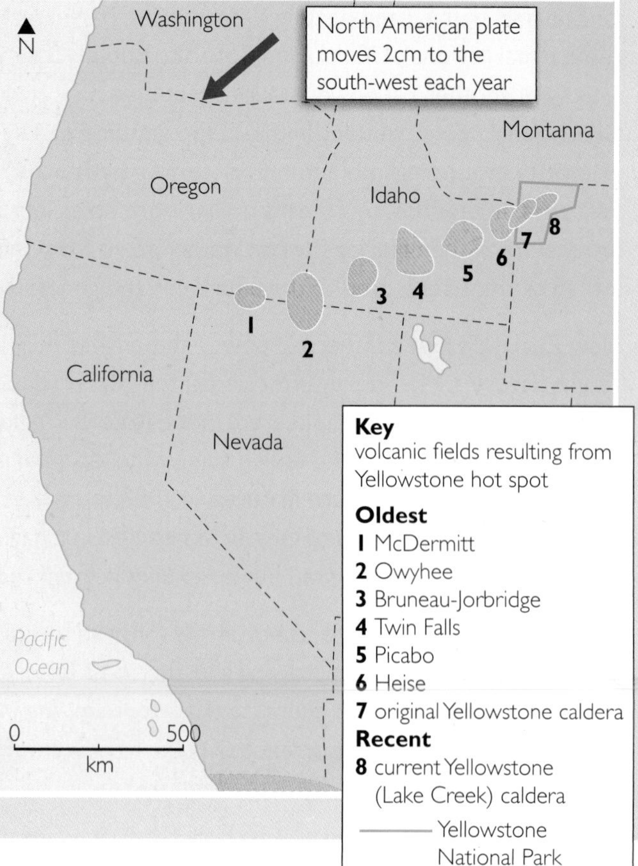

Figure 1.46 *The trail of volcanic fields resulting from the southwest movement of the North American plate* ▶

The crossing of two major fault lines in the area around Norris Geyser Basin has fractured the surface and allows precipitation to percolate to depths of 3000 m and temperatures of 200°C. This superheated and pressurised water rises again to the surface to create geysers, fumaroles, hot springs and boiling mud.

Despite the name, Porkchop Geyser is now a gently bubbling hot spring. The pressure beneath the geyser reached exploding point in 1989 and the rocks surrounding the vent were dramatically blasted outwards more than 66 m. With the release of pressure, the geyser went back to its earlier state of four years before.

Each hot spring within Norris Geyser Basin is unique. Many carry dissolved sulphur from the igneous rocks through which the groundwater has travelled. The sulphur is then deposited as crusty layers on the surface. When sunlight reflects off the water and sulphur, brilliant colours result, such as the green of Emerald Spring (Figure 1.47).

Black Growler Steam Vent (fumarole) is found on a hillside, above the water supply to Norris Basin. As it cannot be easily cooled, it is one of the hottest features – with temperatures of 138°C. Boiling mud is common on the flats of Norris Geyser Basin, yet in many locations the features appear and then disappear without warning. For example, for most of the year, Congress Pool is a pale blue colour, yet overnight may suddenly change to violently bubbling hot mud. It is not fully understood why such disturbances happen or why the majority remain only for a few days or weeks.

How safe is Yellowstone? Minute Geyser once erupted every 60 seconds, but is now clogged with rocks thrown from the nearby road. It now only erupts irregularly, showing that we remain the biggest threat to any sustainable management of Yellowstone.

Did you know?

Microscopic life thrives in the mineral-rich hot water of Yellowstone. Investigations of these tiny life forms have been used for AIDS research and DNA fingerprinting.

ACTIVITIES

1. To what extent do you agree that geothermal power in New Zealand is environmentally harmful?
2. Use the Internet and the information in this section to complete a short study of Yellowstone's supervolcano. Use diagrams to describe its features and formation. To what extent does it form a global threat to people on Earth?

Figure 1.47 *The aquamarine green of Emerald Spring* ▼

1.11 Causes and measurement of earthquakes

In this section you will learn about:
- the causes of earthquakes
- different types of seismic shockwaves
- the measurement of earthquakes

Seismicity

Earth shaking (**seismicity**) is strongly associated with plate tectonic theory. Only 5% of the thousands of earthquakes every day are *not* located along plate margins. Very few earthquakes are big enough to be noticed by people, with only three or four every year powerful enough to be considered a major hazard.

Plate movements produce energy of unimaginable proportions – but their progress is not smooth. Friction along plate margins builds stresses in the lithosphere. When the strength of the rocks under stress is suddenly overcome, they fracture along cracks called **faults**, sending a series of seismic shockwaves to the surface. The breaking point is called the **focus** (**hypocentre**) of the earthquake. The **epicentre** is a point on the surface directly above the focus. It commonly experiences the most intense ground shaking (Figure 1.48). The shaking then becomes progressively less severe with distance from the epicentre, like ripples spreading outwards in a pond. Earthquake tremors usually last for less than a minute followed by several weeks of **aftershocks** as the crust settles.

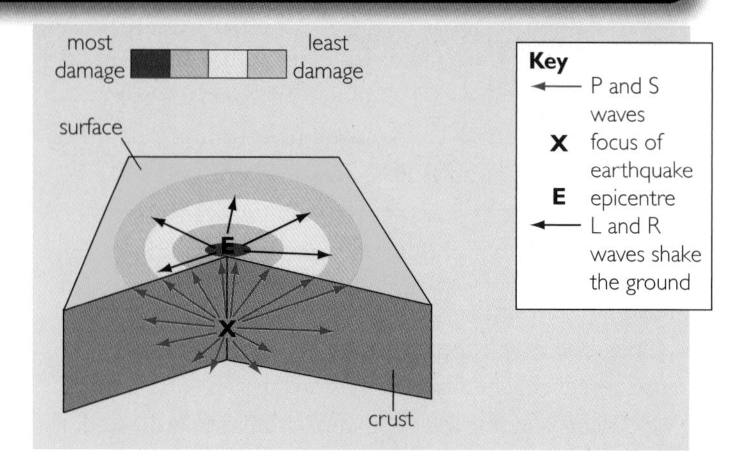

Figure 1.48 *The focus (hypocentre) and epicentre of an earthquake* ▲

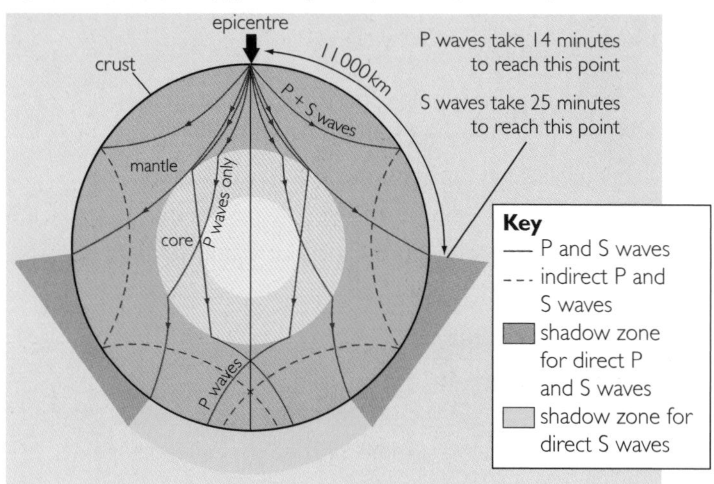

Figure 1.49 *The paths of P and S waves through the Earth's interior* ▲

Seismic shockwaves

Seismic waves are measured by sensitive instruments called **seismometers**. They tell us not just about earthquakes, but also the internal structure of the Earth (see Figures 1.48–1.50).

Scientists have identified several different types of seismic waves:

- ***Primary or pressure (P) waves*** are the fastest and reach the surface first (Figures 1.48–1.50). P waves are like sound waves – high-frequency and pushing like balls in a line. They travel through both the mantle and core to the opposite side of the Earth.
- ***Secondary or shear (S) waves*** are half as fast and reach the surface next (Figures 1.48–1.50). Like P waves, they are high-frequency but shake like a skipping rope. They can travel through the mantle, but not through the core, so cannot be measured at a point opposite the focus or epicentre.
- ***Surface Love (L) waves*** are the slowest waves and cause most of the damage (Figure 1.50).
- ***Rayleigh waves (R)*** radiate from the epicentre in complicated low-frequency rolling motions (Figure 1.50).

Figure 1.50 *Primary, Secondary, Love and Rayleigh waves* ▼

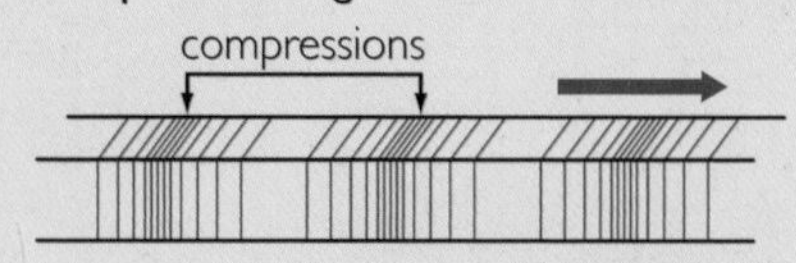

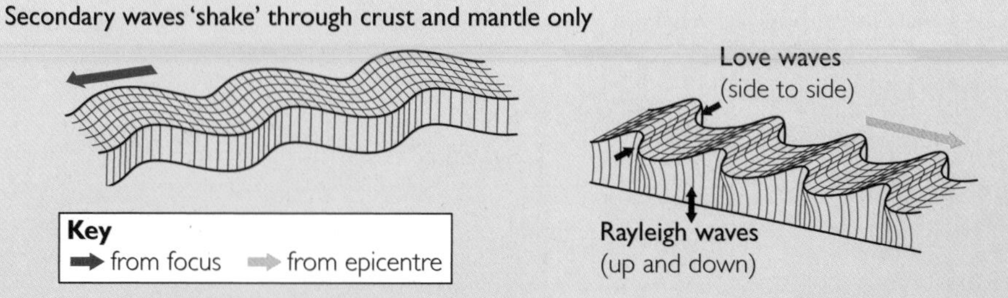

Measuring earthquakes

The **Richter scale** measures the **magnitude** of an earthquake. The scale is logarithmic – each number is ten times the magnitude of the one before it. The scale starts at 1 and could go on forever, but in practice, earthquakes rarely exceed Richter 9. The highest Richter value ever recorded of 9.5 was in Chile in 1960. Destructive earthquakes tend to have values in excess of 6.

In effect, the logarithmic jumps in this scale mean that slight differences in Richter values have an enormous effect on the ground.

The actual **intensity of damage** caused is measured on the **Modified Mercalli scale**. This scale uses observations on the ground of the actual impact of the earthquake. The twelve-point scale ranges in single units from I (imperceptible) to XII (catastrophic). For example:

- I – 'imperceptible' is measured only by seismometers
- IV – 'moderate' is likely to rattle doors and windows
- VIII – 'destructive' will cause chimneys to fall
- XII – 'catastrophic' results in complete destruction.

Can earthquakes be predicted?

The generally chaotic nature of an earthquake makes it impossible to predict either when or exactly where a disaster will strike. But an understanding of plate tectonics does allow geologists to know which areas are most at risk. Microquakes experienced minutes before a large earthquake are a sign of the immense pressure building up, but they are virtually useless in terms of preparing a population for the main event. In effect, we can only make vague predictions – such as the US Geological Survey stating that there is a 67% chance of another 'serious' earthquake striking San Francisco in the next 30 years.

The best defence remains:

- identifying areas of high risk, generally along certain sections of active plate margins
- limiting development wherever possible to earthquake-proof designs
- contingency planning to cope with the aftermath of an earthquake.

For example, in both China's Sichuan earthquake in May 2008 and the Tangshan earthquake, near Beijing in July 1976, most of the thousands of fatalities were a result of poorly built brick buildings or unreinforced concrete structures collapsing. In the reconstruction of both areas, Chinese planners have now attempted to create buffer zones within which building type and construction are restricted (Figure 1.51).

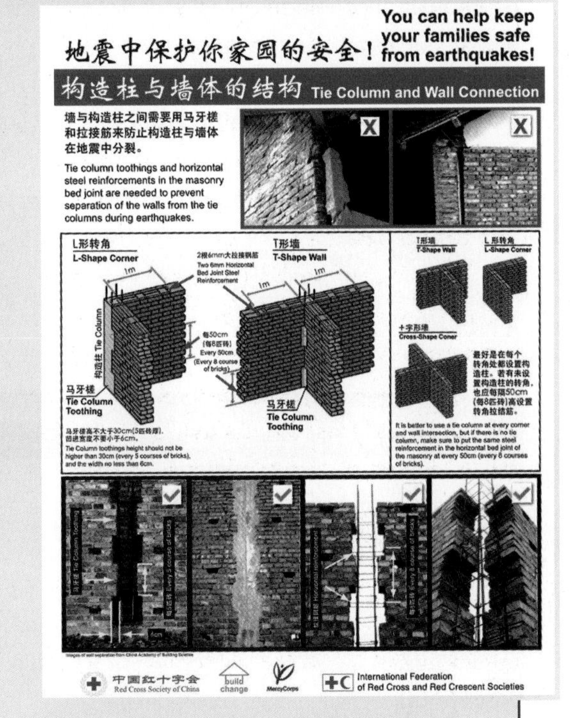

Figure 1.51 *In China, posters like this show 'good practice' construction techniques to help local people construct buildings better able to withstand earthquakes* ▲

ACTIVITIES

1. Explain how the different types of information gathered from the Richter scale and the Modified Mercalli scale might be used.
2. To what degree can the level of earthquake prediction available today help governments who are faced with potential future earthquakes? Explain your answer.

Did you know?

The UK has more than 300 earthquakes every year – most are too small to notice. But tremors from the 5.2 magnitude Market Rasen 'quake in Lincolnshire in 2008 were felt for hundreds of kilometres.

Internet research

Use the Internet to research efforts by the Chinese government to reduce the impact of earthquakes on its citizens.

1.12 Earthquakes and human activities

In this section you will learn about:

- direct and indirect effects of earthquakes
- immediate and long-term consequences of earthquakes
- planning for earthquakes

Direct and indirect effects

Earthquakes are major natural hazards, causing death and destruction particularly in urban areas. People are buried by collapsing buildings, with many more dying in the fires, chaos and confusion that follows, especially as smaller, crust-settling aftershocks destroy weakened structures. The consequences of an earthquake will depend on:

- the magnitude and depth of the earthquake
- geological conditions
- the distance from the epicentre
- population density, preparation and education
- the design and strength of buildings
- the time of day
- the impact of indirect hazards, such as fires, landslides and tsunamis.

Earthquakes have immediate and often well-publicised consequences. The long-term effects are less well reported and may cost an economy billions (US$) to put right (see the table below).

- **NGOs** – non-governmental organisations such as the Red Cross, Oxfam and Greenpeace.

Figure 1.52 *Some effects of earthquakes* ▼

Primary effects – immediate impact	Secondary effects – result as a direct consequence	Long-term effects
Ground shaking will cause: • buildings and bridges to collapse • windows to shatter • power lines to collapse • road and railway damage • water mains, gas mains and sewers to fracture.	Fires caused by broken gas pipes and power lines are difficult to put out. Emergency services are hindered. Diseases spread from contaminated water.	Higher unemployment as not all businesses recover. Repair and reconstruction of buildings. and **infrastructure** may take months or years. Longer-term illness and/or reduced life-expectancy caused by immediate suffering.
Schools, colleges and universities destroyed.	Education suspended for immediate future.	Long-term 'lost generation' to develop local/regional economy.
Immediate deaths and injuries from crushing, falling glass, fire and transport accidents.	Bodies not buried or cremated spread diseases such as cholera. Injuries may result in long-term disability or death if not treated promptly.	Trauma and grief may take months or years from which to recover. Disability and reduced life-expectancy.
Shocked, hungry people forced to sleep outside.	**NGOs** provide tents, water and food.	Emergency pre-fabricated homes may become permanent fixtures.
Slope failures setting off landslides and avalanches.	Further deaths and injuries. Flooding from blocked rivers creating 'quake lakes'.	Loss of farmland and food production. Permanent disruption to natural drainage patterns.
Liquefaction of saturated soils.	Building foundations subside, resulting in more collapses, deaths and injuries.	Repairs to buildings are difficult and reconstruction expensive.
Damage to power stations.	Power cuts restrict emergency services including immediate medical care.	Reconstruction of power stations is very expensive.
Panic, fear and hunger.	Civil disorder, looting and direct intervention by civil authorities, such as police and army.	Problem restoring trust in neighbours and civil authorities.

Despite the damage and destruction shown in the table opposite, lessons are often learned for future **contingency planning**.

Planning for earthquakes

Although earthquakes cannot be accurately predicted, scientists have identified a number of events that can occur before an earthquake strikes. These include:

- microquakes before the main tremor
- bulging of the ground
- electrical and magnetic changes within local rocks
- increased argon gas content in the soil
- curious animal behaviour.

Even in areas where earthquakes are expected, issuing warnings based on such weak evidence could result in panic and chaos so could never be risked. But **risk assessment**, contingency planning and **earthquake engineering** are possible.

New buildings, bridges, roads, pipelines and power lines can be designed to withstand ground shaking (Figure 1.53). **GIS** can be used to prepare hazard maps which show the areas at greatest risk. Growth and development can be controlled to reflect these maps. Public education may be as simple as providing earthquake preparation checklists to practising evacuation drills in schools, offices and public buildings. In Japan, these are as common as practice fire drills.

Did you know?

A Richter 5 magnitude earthquake releases as much energy as 32 kilotons of TNT - the equivalent of the atomic bomb that exploded over Nagasaki in 1945.

- **Contingency planning** by individuals, organisations, local authorities, governments and international aid agencies prepares for hazards in advance. It aids the management of hazards through prediction, prevention (if possible) and protection. It might include special building designs, evacuation plans and practice drills, as well as the stockpiling of water, food and medical supplies.

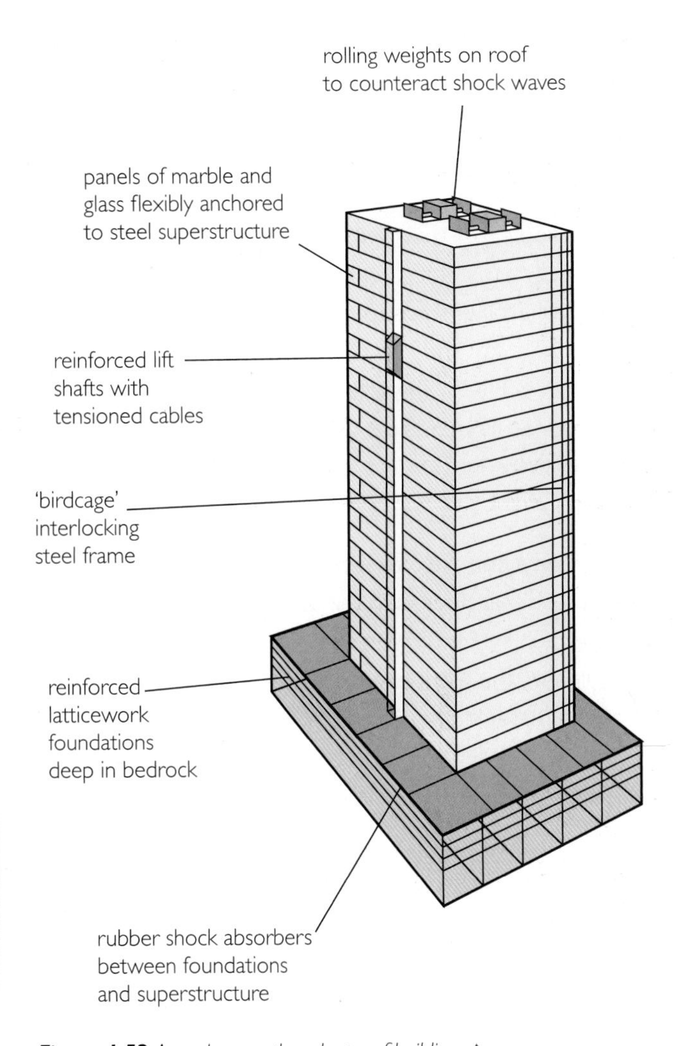

Figure 1.53 *A modern earthquake-proof building* ▲

ACTIVITIES

1. Study Figure 1.52. Severe earthquakes will affect countries such as Japan and China every few years. Work in pairs to consider the contingency planning needed to help these countries cope with primary, secondary and long-term effects of an earthquake. What needs to be in place to enable an effective response to take place?
2. Study Figure 1.53.
   - **a** Describe how buildings can be designed to withstand earthquakes.
   - **b** Do you agree that the best approach to dealing with earthquakes is to learn to live with rather than try to predict them? Explain your answer.

1.13 Two contrasting earthquakes

In this section you will learn about:

- the causes of recent earthquakes in Haiti, an LIC, and Italy, an HIC
- the effects of these earthquakes
- the responses to these earthquakes
- the management of the related hazards

CASE STUDY

Port-au-Prince earthquake, Haiti 12 January 2010

What is the price of human life? The 2010 Haiti earthquake resulted in the loss of up to 250 000 lives. The 1989 San Francisco earthquake was the same magnitude as Haiti, (about 7 on the Richter scale) but only 63 people lost their lives! Frequently, in seismically active LICs such as in Haiti, where over 70% of the population live on less than $2 a day, the loss of life is much higher than in HICs. This is largely due to inadequate quality controls on buildings in poor areas.

Causes

Haiti sits on a complex **strike-slip fault** that shares similarities with the San Andreas fault system (Figure 1.54). At this fault line, the Caribbean and North American plates slide past one another in an east-west direction. Friction builds up until one plate suddenly 'pings' back up.

In 2008, scientists made an alarming discovery: the plates, which had been moving at an average of 7mm a year since the earthquake of 1751, had become stuck. On 12 January 2010 at 16.53 (21.53 GMT), the stress of the surrounding rocks was finally overcome resulting in a magnitude 7 earthquake.

The epicentre was 24 km southwest from the capital of Haiti (Port-au-Prince) and had a shallow focus of 13 km (Figure 1.55). The equivalent energy of an atomic bomb was transmitted outwards – violently shaking the entire country.

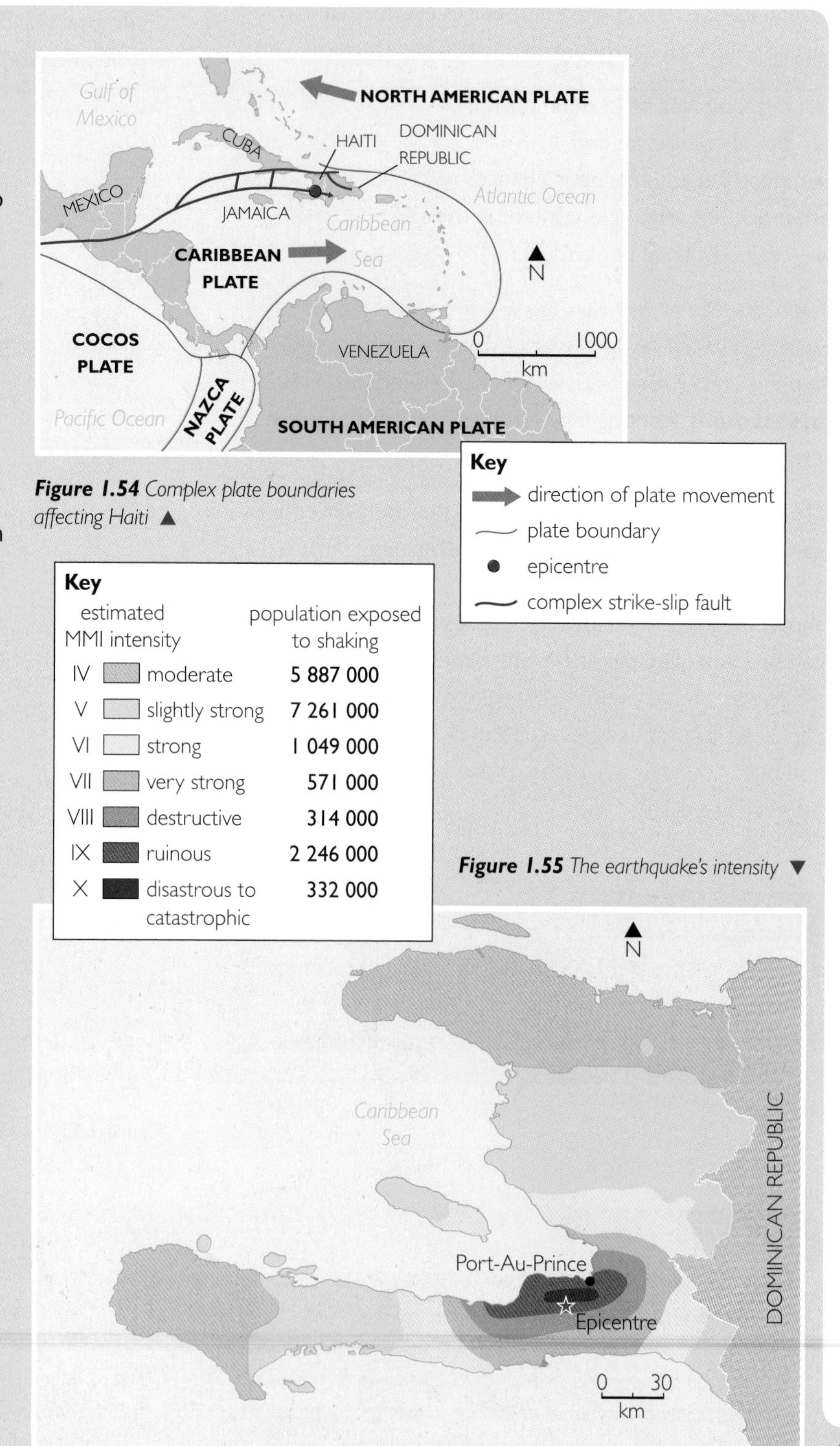

Figure 1.54 *Complex plate boundaries affecting Haiti* ▲

Figure 1.55 *The earthquake's intensity* ▼

Primary effects

- Between 230 000 and 250 000 lives were lost, with many people killed by collapsing buildings (Figure 1.56).
- Port-au-Prince, home to 2 million people, was flattened in less than 60 seconds.
- Lateral (sideways) spreading resulted in the ground slumping or falling away.
- 50% of buildings collapsed, including key government buildings such as the police headquarters and the Parliament.
- **Liquefaction** on looser sediments caused building foundations to subside.
- Infrastructure was brought down – the main port in the capital subsided and became unusable, roads were cracked and blocked by building debris.
- 1.5 million people became homeless.
- Damage was localised – for example, buildings built on hard bedrock near the epicentre suffered less damage.
- A small localised tsunami killed 7 people.
- The landscape was permanently changed – corals were pushed upwards to the north of the fault line whilst farmland collapsed into the sea in the south.

- **Amplification**: As seismic waves pass through softer sediment, such as sand, the wave amplitude or height increases. This results in more intense shaking of the surface.

Secondary effects

- Strong aftershocks – including a 6.1 magnitude earthquake on 20 January.
- Possible trigger for sequence of much larger earthquakes.
- With the loss of hundreds of civil servants and the destruction of ministries, the Haiti government was crippled.
- Local food prices at markets became too expensive for the majority of people.
- With the main prison destroyed in Port-au-Prince and the police force crippled, the city became lawless.
- By the first anniversary of the earthquake, cholera had killed over 1500 and 1.5 million people were still homeless.

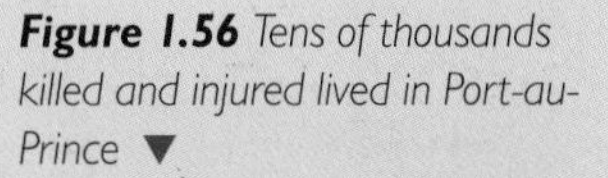

Figure 1.56 *Tens of thousands killed and injured lived in Port-au-Prince* ▼

Immediate responses

- Rescue efforts – international search teams struggled within the dense and congested urban environment. Local people employed by the UNDP (United Nations Development Project) pulled survivors out from the debris and cleared roads.
- Infrastructure – in Port-au-Prince, the US military took control of the airport to speed up the distribution of aid and re-opened one of the two piers in the port.
- Security – 16 000 UN troops and police restored law and order, coordinated by a new UN/US Joint Operations Tasking Centre.
- Food – in absence of local food markets, the UN World Food Programme provided basic food necessities. Farmers were given immediate support before the spring planting season.
- Water – bottled water and purification tablets were distributed.
- Health – emergency surgeries were established to perform life-saving operations.
- Shelter – 692 000 survivors lived in 591 make-shift camps. 235 000 took advantage of free transportation to cities in the north and southwest.
- Buildings – rapid structural assessment was made of buildings.

Longer-term responses

- Aid – a single Haiti Fund manages an $11.5bn reconstruction package (Figures 1.57 and 1.58), with controls in place to prevent corruption.
- Food – the farming sector was reformed to encourage greater self-sufficiency and less reliance on food imports.
- Health – a shift was made to focus on follow-up care, including mental health care.
- Buildings – hospitals, schools and government buildings were rebuilt to new **life safe** building codes (Figure 1.59). Local people were employed as construction workers. Slums were demolished and new settlements built away from high-risk areas, such as unstable hillsides. New homes are more affordable, safe and sustainable.
- Economy – some economic activities were moved away from Port-au-Prince to less earthquake-prone areas. A UN strategy was developed to create new jobs in clothing manufacture, tourism and agriculture, and also to reduce effects of uncontrolled urbanisation.

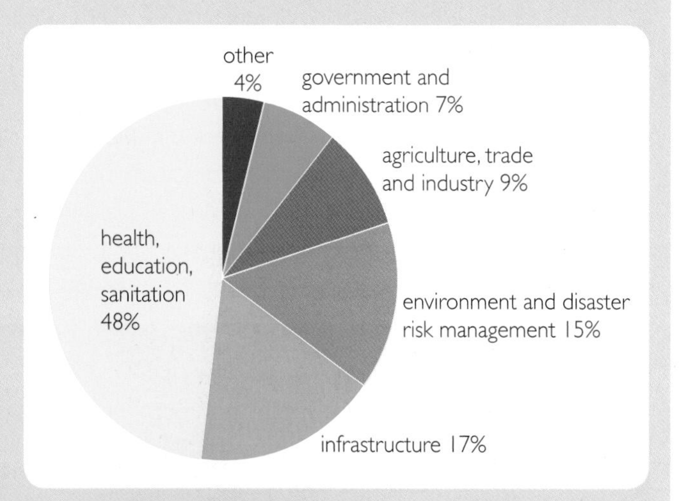

Figure 1.57 *The $11.5bn reconstruction package* ▲

◀ ***Figure 1.58*** *Huge international relief efforts on the scale of the 2004 Boxing Day tsunami*

Will it happen again?

Geologists studying the Haiti earthquake have identified a worrying pattern of earthquake activity along the strike-slip fault. An earthquake releases local pressure but may result in stress building up further along the fault line, leading to the whole system 'unzipping'.

In 1751, an initial earthquake struck Haiti only to be followed by a second bigger 'quake 19 years later. Seismic activity continued moving westwards along the fault line until 1907 when Kingston, Jamaica was flattened. The January 2010 earthquake did not significantly fracture the surface and, in seismic terms, a magnitude 7 earthquake is not unusually large. The largest earthquake ever recorded was over 300 times more powerful than Haiti. Worst may yet be delivered to Haiti.

Conclusion – Lessons to learn?

8 out of the 10 most populous cities, the majority in LICs, lie on active plate boundaries or in active earthquake zones. Rapid population growth in these urban centres will result in increasingly marginal land, such as unstable slopes or softer flood plains, being used to house people. If lessons are to be learnt from Haiti then it must be to spend both money and time on ensuring buildings are made life safe. The question is not when the next earthquake will strike, but where – and at what cost to human life?

Did you know?

Over the last 100 years, 2 million people have been killed or injured by earthquakes; but 600 000 have been killed or injured within the last ten years.

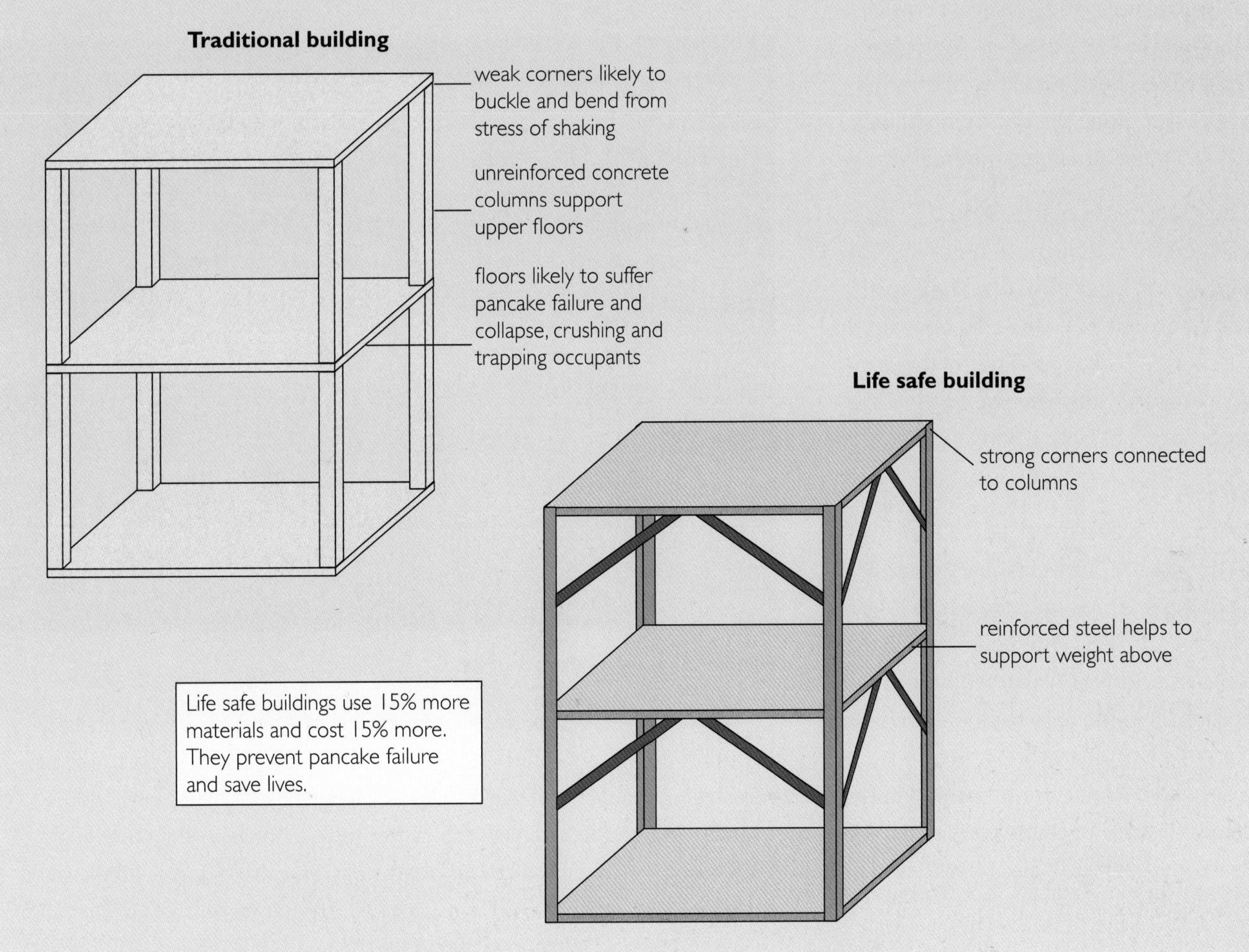

Figure 1.59 *The differences between traditional building and life safe building construction in LICs* ▲

CASE STUDY

L'Aquila earthquake, Italy 6 April 2009

Imagine living in a temporary campsite, surrounded by freezing snow-capped mountains and coping with the after-effects of a 6.3 magnitude earthquake. The Prime Minister then comments that living in the emergency accommodation compares to a 'camping weekend'! In April 2009, this is precisely the situation that faced thousands of survivors of the L'Aquila earthquake, Italy following the visit of Silvio Berlusconi, the Italian Prime Minister (Figures 1.60 and 1.61).

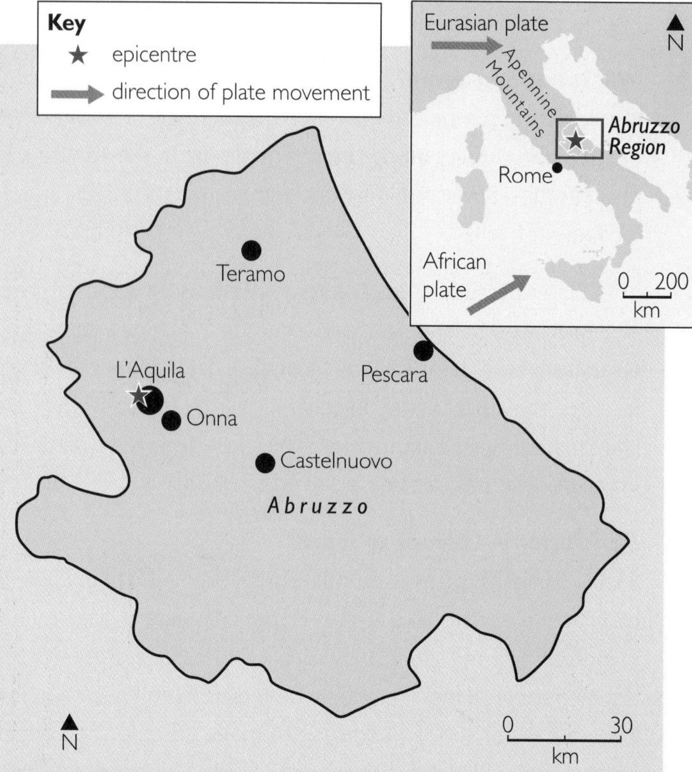

Figure 1.60 *The L'Aquila earthquake, Abruzzo, Italy* ▲

Causes

Italy is one of the most seismically active countries in Europe. The African plate is pushing northwards towards the Eurasian plate at a rate of around 3 cm per year. The Alps, a young fold mountain range, are a result of this squeezing of continental crust. As these rigid continental plates collide and the surface is subjected to immense pressure and stress, the crust cracks along lines of weakness or geological faults. The linear fault-lines criss-cross all the way along the backbone of Italy, following the Apennine mountain range.

The exact cause of the L'Aquila earthquake is thought to be a result of slippage on a fault-line that runs north-south for tens of kilometres beneath the Apennine Mountains. The earthquake was especially destructive, as the focus was shallow at only 10 km.

Figure 1.61 *'A camping weekend'?* ▲

Effects

The earthquake struck at 3.32 am on 6 April 2009. The epicentre was at Tornimparte, 7 km northwest of L'Aquila. Many of the 305 who died did so in their sleep as poorly constructed homes collapsed. Immediate rescue efforts were hampered by nearly 300 aftershocks. Sporadic looting forced some reluctant home and business owners to remain in their unstable properties. Many thousands more were left homeless – with as many as 34 000 living in temporary tents.

L'Aquila, the historic capital of Abruzzo, was particularly badly affected. Some of the 10 000 buildings damaged were of particular architectural significance. These included the San Massimo cathedral in L'Aquila, and further afield the third-century Baths of Caracalla in Rome. Medieval mountain villages close to the epicentre, such as Onna and Castelnuovo, also suffered heavy damage.

Responses and management

Italy is a member of the world's richest group of nations (G8). But it was again criticised for failing to meet architectural and safety standards typical in other seismically affected G8 countries, such as Japan or the USA. Modern buildings constructed within the last 25 years were badly damaged or collapsed. These included San Salvatore hospital, completed in 2000, and a university dormitory – both apparently seismically engineered (Figure 1.62).

In 2002, a lesser earthquake registering 5.5 on the Richter Scale caused a primary school in the southern Molise region to collapse, killing the teacher and 27 children. This was despite the school being built according to national 'earthquake-resistant' standards imposed after 2500 people lost their lives in the town of Irpinia in November 1980. Ironically, in an attempt to boost the regional economy, the July 2009 G8 summit was moved to L'Aquila.

An official government enquiry was immediately launched to identify whether human error was to blame for the loss of life in L'Aquila. However, the momentum of public support has pushed for a longer-term strategy, rather than 'the fatalistic attitude' and short-termism of political leaders who do not see money spent on seismic safety standards as a vote winner.

Following a declaration of a state of emergency, Silvio Berlusconi was quick to announce a rebuilding programme. He promised that L'Aquila would receive the first of a series of 'new towns' planned for provincial capitals within two years, and the old town would also be rebuilt. But the tent dwellers of L'Aquila have reason to be pessimistic about an early return to new homes.

Figure 1.62 *Seismically engineered buildings such as the San Salvatore hospital above failed to withstand the shaking* ▲

The Italian government has gained an unenviable reputation for unmatched promises of reconstruction following earthquake damage. In Sicily, home of the mafia and associated corruption, people are still living in emergency huts built for victims of earlier earthquakes. Slow recovery from the economic recession of 2008 is further pressure on the costs of reconstruction, estimated at more than $16 billion.

Did you know?

20 million of Italy's 57 million residents are at risk from earthquakes.

ACTIVITIES

1 Study the table on the right.
  - **a** List the differences between the three earthquakes.
  - **b** Give possible reasons for these differences.

2 'People living in poorer countries suffer more from the effects of earthquakes than people living in wealthier ones.' Use evidence from the table to support or refute this statement.

	Haiti 2010	China 2008	Italy 2009
Magnitude	7.0	7.9	6.3
Deaths	up to 300000	87476	295
Injuries	300000	360000	2000
Homeless	1200000	5000000	24000
Survival rate	1 in every 15 affected died	1 in every 595 affected died	1 in every 190 affected died
Rescue rate	1 in every 16588 affected rescued	1 in every 690 affected rescued	1 in every 373 affected rescued
Economic cost	estimated more than $10 billion	$85 billion	$2.5 billion
GDP per capita	$1300 (2009 est.)	$6600 (2009 est.)	$30300 (2009 est.)
LIC/ HIC / LMC*	LIC	LMC	HIC

* lower- middle- income country

1.14 Tsunamis – causes and characteristics

In this section you will learn about:

- how tsunamis are formed
- how they differ from normal ocean waves
- their destructive nature
- the 2004 Boxing Day Indian Ocean tsunamis
- the March 2011 Japan earthquake and tsunami

Did you know?

Most dictionaries say the plural of 'tsunami' can be 'tsunami' or 'tsunamis'.

Shortly before 8am on the morning of December 26, 2004, Ida Sabri, 42, was tidying her kitchen when she heard neighbours shouting hysterically outside. "The sea is coming," they said. "Leave now!" When she left her house, the sky had turned dark, and then she saw a wave that gives her nightmares to this day. "It was higher than a telegraph pole," she says, "and in the shape of a dragon that was rushing towards me." As Sabri ran to escape, she turned to see "the dragon lower its head, then crash all around, breathing in houses, people and trees."

— adapted from an article in *The Times*, 29 December 2009

Causes

Tsunamis are sometimes called 'tidal' waves. This is extremely misleading because they have nothing to do with tides, or even the wind like normal ocean waves. Tsunamis are usually generated by seismic activity caused by ocean floor earthquakes or submarine volcanic eruptions (Figure 1.63). Massive landslides into the sea and even meteor or asteroid strikes can also cause them!

Characteristics

Tsunamis differ from normal ocean waves in many respects:

- The wave height is very low (less than 1 m), but on reaching the shore can rise to over 25 m.
- The wavelength (distance between crests) is long – very, very long. In open water the wavelength can be anything from 100 – 1000 km.
- They travel very quickly – at speeds of 640 – 960 km per hour.
- They usually consist of a series of waves, with the first not always being the biggest.
- The time between each wave (the wave period) is long – between 10 and 60 minutes.
- On approaching the coast – especially if funnelled into an inlet – they will slow down and pile up as a massive wall of water before breaking.

Up to 90% of tsunamis occur in the Pacific Ocean and are associated with seismicity along plate boundaries surrounding the Pacific basin (**Ring of Fire**). The effects of tsunamis will vary according to the factors above but are also influenced by population density, the coastal relief and the land use of the coastal region.

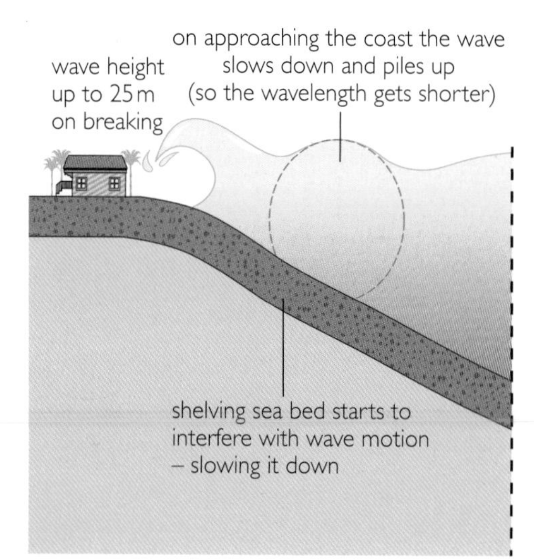

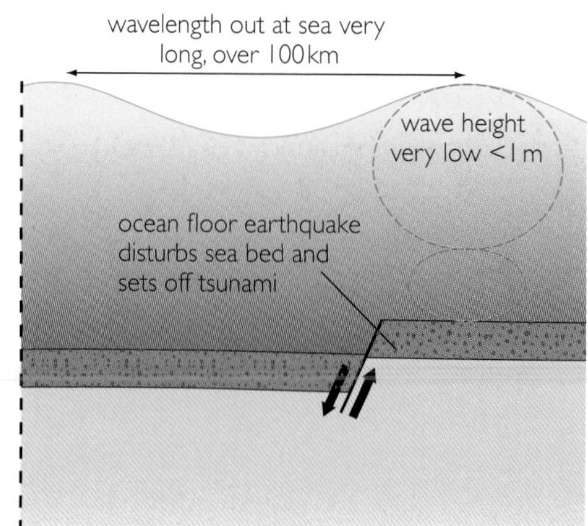

***Figure 1.63** Characteristics of a tsunami* ▲

Did you know?

The Japanese term tsunami comes form tsu – meaning harbour and nami – meaning wave.

Did you know?

Tsunamis generated by the eruption of Krakatoa in 1883 killed 36 000. The waves caused catastrophic flooding and were estimated to be over 40 m high!

The effectiveness of warnings can also have a significant effect on peoples' vulnerability to tsunamis. Warning systems do exist, for example, the Pacific Tsunami Warning System based in Hawaii. These systems give many hours' warning of approaching waves following 'important', 'major' and 'serious' seismic events. But with no such warning, the first sign – the apparent draining away of the sea in front of the tsunami – will be too late. A tsunami will sweep away its victims, uproot vegetation, destroy wooden buildings and bridges and wash boats hundreds of metres inland. The destructive power of tsunamis is legendary, especially if caused by ocean floor earthquakes.

Country affected	Dead and missing
Indonesia	236 169
Sri Lanka	31 147
India	16513
Thailand	5395
Somalia	150
Burma	61
Maldives	82
Malaysia	68
Tanzania	10
Seychelles	3
Bangladesh	2
Kenya	1
Total	**289 601**

Figure 1.64 The distribution of the dead and missing, by country ▲

CASE STUDY

The Indian Ocean disaster (2004)

Figure 1.65 Countries affected by the 2004 Indian Ocean tsunamis ▼

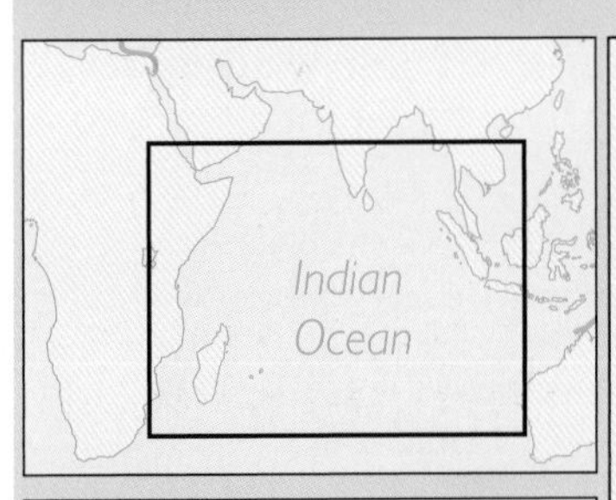

The Indo-Australian plate is pushing north, diving under the Eurasian plate. The ocean floor here is under huge pressure.

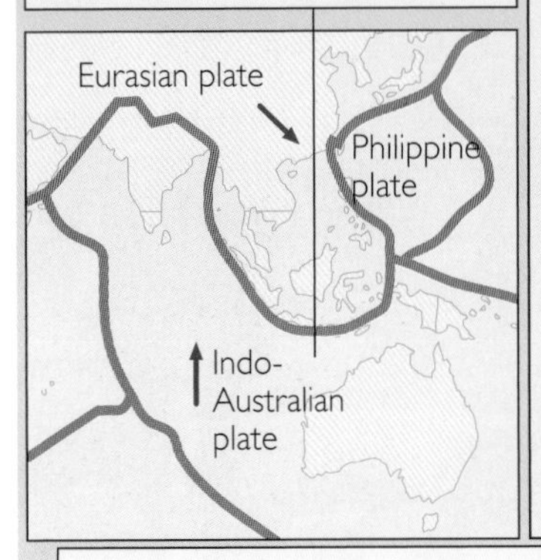

1 Time zero: the waves begin.
2 +15 minutes, tsunami hits **Sumatra**. This is the worst affected region, with 700 000 left homeles.
3 +60 minutes, in **Malaysia**, the island of Penang is worst hit as tourists are swept away from beaches.
4 +90 minutes, **Thailand** is hit. 1000 tonnes of water crashes down on each metre of Thailand's beaches. 1700 tourists from 36 countries, many staying in Phuket, are included in the death toll.
5 +2 hours, **Sri Lanka** is hit. Shallow water causes wave refraction – waves bend around the island and rush ashore in otherwise safe areas. City of Galle is destroyed at a cost of 4000 lives. 1 million people are left homeless.
6 +2 hours, **India** is hit. Tsunami reaches up to 3km inland. Isolated Andaman and Nicobar islands suffered severe damage. 376 000 people are left homeless.
7 +3 hours 30 minutes, the **Maldives** are hit. 20 of 199 inhabited islands are totally destroyed. Low-lying islands are saved further losses because:
- surrounding coral reefs break up the wave's energy
- an absence of a continental shelf keeps the wave height down

8 + 7 hours, the **Seychelles** are hit.
9 + 8 hours, **Madagascar** is hit.
10 + 8 hours, in **Somalia** 2000 structures are damaged.
11 + over 9 hours, the east coast of Africa is hit. In **Kenya**, warnings issued in time allow beaches to be evacuated and reduces the loss of human life.
12 + over 11 hours, **South Africa** is hit.

Think of a big number. Now change that total into any number of people, young and old, male and female, familiar and unknown. On 26 December 2004, at 00.59hrs local time, the second biggest earthquake in history, a 9.3 magnitude 'quake, 240 km off the coast of northwest Sumatra tore apart the ocean floor. It caused massive tsunamis that travelled thousands of kilometres across the Indian Ocean. It also cost the lives of 290 000 people across two continents.

Figure 1.66 *Debris carried by the powerful waves along the southern Indian coastline* ▲

The cause of the tsunamis was the release of stress along a giant 1000 km thrust fault on the India and Sunda subducting plate boundary. The India plate suddenly moved downwards by around 15 m. This made the ocean floor above the thrust fault punch upwards by several metres and begin the series of devastating tsunamis.

Such was the extent of the initial movement that at least eight aftershocks followed. As the focus of some of these earthquakes was shallow, they added to the already significant loss of life – and hampered relief efforts, particularly in the area around the Andaman Islands.

Scientists knew in advance that southern Asia was going to be hit by a tsunami. But warnings were too late – radio and television alerts were not broadcast in Thailand until nearly an hour after the first waves hit. Thirty minutes after the initial earthquake, waves of up to 20 m began to crash into the coastline of Sumatra, Indonesia. It was the first of many communities to be devastated.

Figure 1.67 *A devastated community in Ampara, Sri Lanka* ▲

Primary effects

- Up to 290 000 people died with thousands of bodies still missing.
- Vegetation and top soil was removed up to 800 m inland.
- Infrastructure was destroyed. For example, Andaman and Nicobar Islands were all but cut off as jetties washed away.
- Coastal settlements were devastated. For example, the city of Banda Aceh in Sumatra was obliterated.

Secondary effects

- Widespread homelessness. 500 000 were forced into refugee camps in the worst hit region of Aceh Province, Indonesia (Figure 1.68).
- Economies were devastated, including fishing, agriculture and tourism sectors. For example, 44% of the population living in Aceh Province, Indonesia lost their livelihoods. Furthermore, in Thailand the cost to the fishing industry was £226m – far more than the widely reported impact on tourism.
- Negative **multiplier effects** weakened economies further. In Sri Lanka, the loss of deep sea trawlers resulted in fewer catches. These catches were then less likely to reach market because of fewer refrigerated trucks and bridges.
- Water supplies and soils are now contaminated by salt water.
- The gap between the rich and poor has increased.

> **Did you know?**
>
> *The earthquake which started the 2004 Indian Ocean tsunamis released the energy of 23 000 Hiroshima-type atomic bombs! The tsunamis' waves travelled at the speed of a jumbo jet.*

Immediate responses

- Massive international relief efforts were established involving more than 160 aid organisations and UN agencies.
- Foreign military troops provided assistance. For example, the Australian air force improved air traffic control at Banda Aceh airport, Indonesia.

Longer-term responses and their effectiveness

- Large-scale programmes of reconstruction were implemented, but still many thousands were left in tents one year on.
- Political barriers slowed aid distribution. In Sri Lanka, aid was delayed to areas held by rebel Tamil Tigers.
- Existing government prejudices were highlighted. For example, in India, the Dalits, an underclass, were ignored by the government.
- Tourist resorts were quickly rebuilt. But elsewhere some native coastal communities were forced out by new developments.
- A UN group set up a tsunami warning system for the Indian Ocean. But as individual governments are responsible for sending out their own alerts, there are huge contrasts in their possible effectiveness. For example, sirens in tourist beaches of Thailand contrast with radio warnings in isolated and poor rural communities.
- Education on tsunami awareness began in schools.
- Practice drills and evacuation plans were established.
- Coastal zones were mapped out to identify areas most at risk.

Figure 1.68 *Refugee camp in Aceh province, Indonesia* ▲

ACTIVITIES

1. Discuss some of the ways people and organisations try to manage the tsunami hazard and its effects.
2. Survivors of natural disasters may suffer mental health problems related to grief and the sight of so many deaths and injuries.
   - **a** Are such mental health problems a primary or secondary effect?
   - **b** What actions by individuals, civil authorities, and NGOs would be necessary to respond to this? Refer to both the immediate and longer-term.

CASE STUDY

The March 2011 Japan earthquake and tsunami

What happens when a 9.0 magnitude earthquake and 10 m tsunami hits one of the wealthiest, most sophisticated countries on Earth?

At 2.46 pm Tokyo time, on 11 March 2011, the fourth-largest earthquake ever recorded occurred under the Pacific Ocean, 100 km due east of Sendai on northern Honshu's eastern coast (Figure 1.70).

A 400 to 500 km long segment of the North American plate, which was being dragged down by the subducting Pacific plate, suddenly slipped upwards between 5 and 10 m. The resulting seawater displacement caused a tsunami to spread in all directions – at hundreds of kilometres an hour.

Japan's tsunami warning system kicked in, but people along a 3000 km long stretch of coastline had just minutes to escape. The first wave hit the north-east coast about 30 minutes after the earthquake.

There were ten waves, each about 1 km apart as they reached the shallower coastal water. Here they slowed and piled up, reaching a staggering 10 m in places. They overwhelmed tsunami defence walls and surged inland up to 10 km.

Figure 1.69 *The destroyed town of Minami-Sanriku* ▲

Did you know?

Mobile phones and news helicopters mean it was the best-recorded tsunami in history. Amateur footage of the disaster spread worldwide via YouTube and Facebook.

Figure 1.70 *The epicentre of the earthquake, and the area hit by the tsunami* ▼

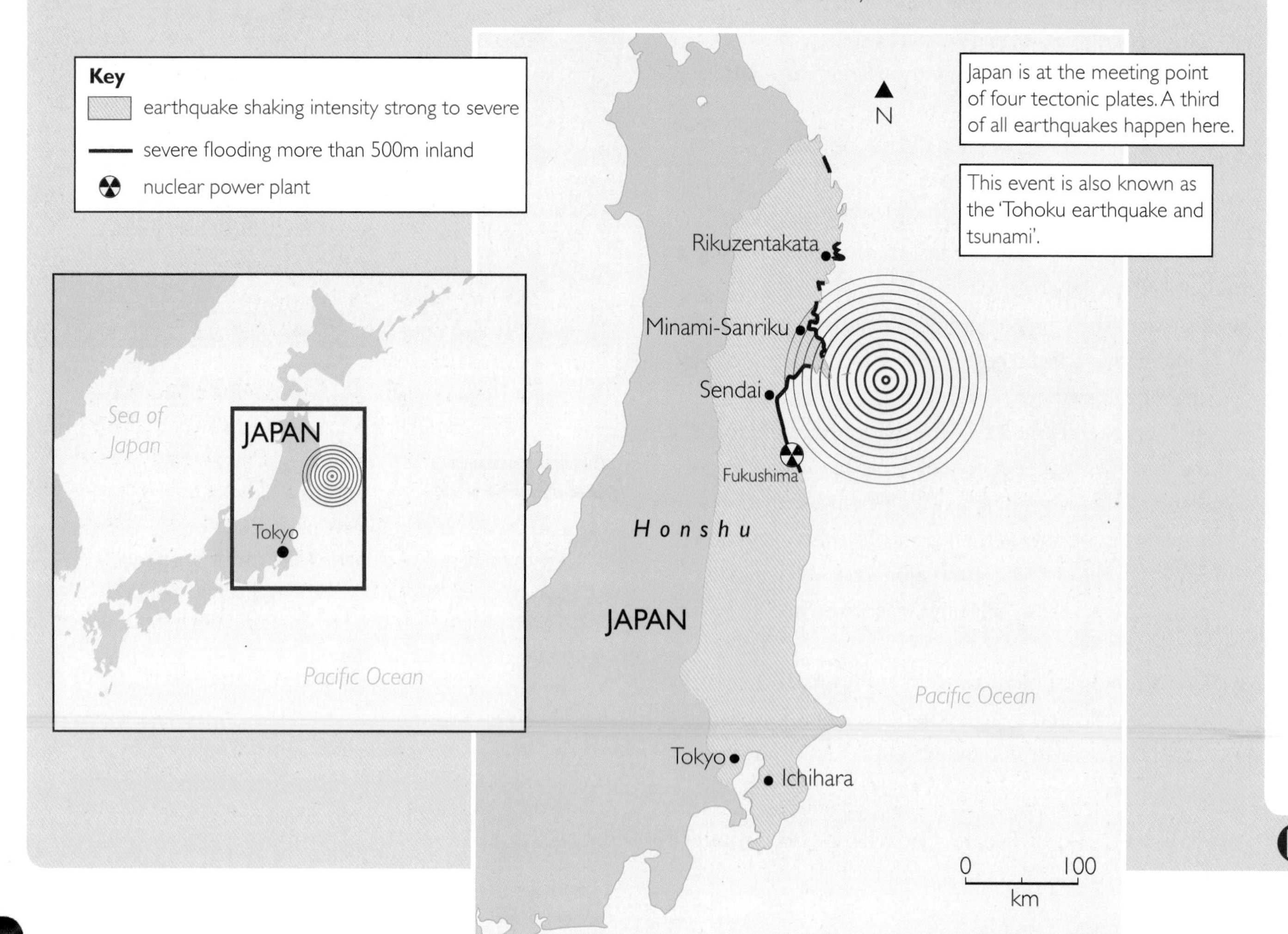

Primary effects

- Part of Japan's east coast was moved up to 4 m out into the Pacific Ocean. And parts sunk by more than a metre.
- Ground shaking caused buildings to collapse. Some were set ablaze by broken gas and petrol pipes.
- The tsunami swept inland, mainly along the north-east coast, swallowing boats, homes, vehicles, trees, and everything else.
- It flooded nearly 500 square kilometres.
- When the waters receded, whole cities were in ruins. Trains had vanished. Ships and cars lay tossed like toys.
- In Tokyo, skyscrapers had 'started shaking like trees'. But their earthquake-proof design meant damage was limited. In Ichihara, a commuter town of Tokyo, an oil refinery was engulfed in flames as fuel tanks exploded.
- In Sendai, areas near the sea were badly damaged, but just inland the city centre was largely unscathed. Rikuzentakata was almost completely submerged, and suffered near-total destruction. In Minami-Sanriku only half the population of 17 000 escaped alive, and few buildings were left standing (Figure 1.69).
- Over 25 000 dead or missing, mainly due to the tsunami. But Japan's tsunami warning system saved many lives.

Immediate responses

- In freezing temperatures, survivors huddled in shelters and hoarded supplies as rescue workers searched the mangled coastline of submerged homes.
- Helicopter crews plucked survivors from rooftops and flooded farmland.
- 100 000 soldiers were mobilised to establish order, organise rescue work, and distribute blankets, bottled water, food, and petrol.
- Offers of aid poured in from other countries, including the USA and China.
- The UK sent 63 fire service search and rescue specialists, two rescue dogs, and a medical support team. People were rescued after being trapped for several days.
- An exclusion zone was set up around the Fukushima nuclear plant. Homes were evacuated and iodine tablets, to prevent radiation sickness, were distributed.
- There were no reports of looting or violence.

Secondary effects

- Half a million people were homeless. For weeks, 150 000 people lived in temporary shelters.
- At least 1 million homes were left without running water. Electricity was cut off in almost 6 million homes. Cars queued for rationed petrol.
- There were shortages of food and water, and of medical supplies.
- Aftershocks terrified people – in the two weeks after the earthquake, there were more than 700.
- Explosions and radiation leaks at the Fukushima nuclear power plant in the days after the earthquake spread fear around the world. The earthquake severed the power supply to the cooling system and then the tsunami destroyed the back-up generators. Workers struggled to prevent a meltdown.
- Fears of a nuclear disaster caused panic selling across global stock markets.

Longer-term responses?

- Japan coped well with the earthquake. But the tsunami defences were inadequate against the previously unimaginable height and force of the water. Future contingency planning will have to consider the level of defence. Should tsunami defences be built to defend the coast against a similar high-magnitude, low-frequency event? Should settlements be re-built in areas devastated by the tsunami?
- Japan will pay a big price for this disaster. The country was already the most heavily indebted in the industrialised world and the repair bill will have to be raised by more government borrowing. Private companies also faced heavy costs. But Japan is a developed, rich country and will recover.

ACTIVITIES

1. The nature of tsunami damage varies considerably between areas. With reference to both the 2004 Boxing Day Indian Ocean and 2011 Japan tsunamis, discuss the human and physical factors which may be responsible for this variation.
2. For both HICs and LICs, which should come first – prediction or contingency planning? Explain your answer with reference to tsunamis or any other tectonic hazard.

aftershock A minor earthquake or tremor following a larger and more powerful earthquake

asthenosphere The partial liquid layer – part of the mantle – on which the tectonic plates 'float'

Benioff Zone The zone of melting at a destructive (convergent) margin as one plate dives (subducts) beneath another

core The central 'core' of the Earth rich in iron and nickel comprising a solid inner core (1200km thick) and a liquid outer core (2300km thick)

crust The outermost 'skin' of the Earth that varies in thickness between 12 and 120 km

earthquake engineering Mitigation approaches involving engineering solutions to reduce damage associated with ground shaking during an earthquake, such as strengthening bridge supports and the foundations of buildings

epicentre The point on the ground surface immediately above the focus of an earthquake

faults Lines of weakness in the form of cracks in rocks along which relative movement (displacement) takes place. Sudden movement along faults is often associated with earthquakes

focus The point within the Earth's crust where an earthquake originates – put simply, it is the origin of an earthquake

fold mountains Mountain chains such as the Alps and the Andes formed by enormous compressional forces associated with a destructive (collision) plate margin

igneous rocks Rocks formed by the cooling of a molten magma, either underground (intrusive) or on the ground surface (extrusive)

liquefaction The jelly-like state of silts and clays resulting from intense ground shaking that may result in the subsidence and collapse of buildings following an earthquake

lithosphere Resting on the asthenosphere, this is the outermost solid layer of the Earth comprising the crust and the outer mantle. Approximately 100 km thick, it is divided into several large plates and many smaller ones

magma Molten rock, gases, and liquids occupying vast magma chambers at great pressures deep within the mantle. On reaching the ground surface, magma is known as lava

mantle The concentric layer of the Earth between the crust and the outer core. The mantle is some 2870km thick

metamorphic aureole The zone of rock adjacent to an igneous body, such as a magma chamber, that has been affected (heated) by metamorphism and where rocks have changed in their mineralogy and/or texture

ocean trenches Deep gashes in the sea-floor up to 10 971 m deep (the Marianas Trench) marking destructive plate margins

plumes Rising currents of heat (part of convection cells) usually associated with rising magma

'ring of fire' The name given to the spatial pattern (ring) of volcanoes (and earthquakes) marking plate margins around the edge of the Pacific Ocean

seismicity Activity associated with the transmission of seismic (shock) waves radiating outwards from the focus of an earthquake

seismometer Instrument used to record the shockwaves (seismic waves) associated with an earthquake. Seismometers are also used to monitor earthquakes in volcanic areas to help predict eruptions

spatter cone A somewhat indistinct volcanic cone forming a mound up to 12 m high resulting from the eruption of 'globs' of very fluid lava, typically found in Hawaii

stratovolcano Otherwise known as a composite volcano, a stratovolcano is made up of alternating layers of lava and ash representing a series of eruptions over many years

strike-slip fault A crustal crack along which movement takes place in a horizontal plane (sideways) rather than vertical (up and down). Earthquakes are often associated with strike-slip faults, for example the San Andreas Fault in California, USA, and the North Anatolian Fault, Turkey

submarine volcanoes Volcanoes formed beneath the sea that can either be single vent volcanoes or fissure volcanoes, where lava is emitted along a crack in the Earth's crust (e.g. the Mid-Atlantic Ridge)

supervolcano A huge volcano that often takes the form of a caldera (collapsed volcano cone) and is associated with massive eruptions capable of having a global impact on people. Examples of supervolcanoes include Yellowstone in the USA and Taupo in New Zealand

tephra Fragmental material emitted during a volcanic eruption, often resulting from the destruction of the volcanic cone. Tephra can vary from very fine ash to volcanic bombs (over 64 mm in size)

vesicular lava A type of basic (rich in iron and magnesium) lava containing holes (vesicles) that represent former gas bubbles

volcanicity Generic term used to describe a range of volcanic activity often associated with a place or region including volcanic eruptions, geysers, and hot springs

EXAMINER'S TIPS

Section A

1 (a) Study Figure 1.37 (page 29). Outline the characteristics and impacts of the volcanic eruption shown in the photograph. *(7 marks)*

(a) Make sure that you focus on the characteristics and impacts shown in the photograph only. Use the correct geographical terminology.

(b) Explain the processes responsible for the formation of volcanoes. *(8 marks)*

(b) Consider the processes leading to formation of magma and subsequent eruptions. Refer to plate margins and different types of volcano. Use the correct terminology.

(c) With reference to examples, discuss the success of management strategies in reducing the hazards associated with either volcanoes or earthquakes. *(10 marks)*

(c) Decide whether to study earthquakes or volcanoes – don't do both! Focus your answer on the management strategies and be sure to consider their success.

2 (a) Study Figure 1.2 (page 8). How does the satellite image provide evidence to support the theory of plate tectonics? *(7 marks)*

(a) Focus on the physical landforms shown in the satellite image. Identify what they are and where they are found. Suggest why these landforms support the theory of plate tectonics.

(b) Explain the reasons for the distinctive tectonic landforms found at constructive plate margins. *(8 marks)*

(b) Here you need to focus on constructive margins only. Identify the features and explain why they are distinctive. Use annotated diagrams to support your answer.

(c) With reference to examples, explain the causes of earthquakes. *(10 marks)*

(c) Make sure you consider a range of earthquake causes and use examples. You should certainly make use of simple diagrams to support your answer.

Section C

1 'The extent of damage caused by an earthquake is more to do with levels of development than the power of the earthquake.' Discuss. *(40 marks)*

(1) Briefly describe the impacts of an earthquake. With reference to examples, consider the significance of levels of development (rich and poor countries, and regions within countries) and earthquake magnitude. Contrast and discuss the importance of these two factors using words such as 'whereas', and don't forget to conclude.

2 To what extent do successful monitoring and prediction make volcanoes less of a hazard today than in the past? *(40 marks)*

(2) Briefly consider the hazards associated with volcanoes and describe the methods used to monitor and predict eruptions. Use case studies to support your points. Consider whether modern techniques have reduced the volcanic hazard, and remember to focus on 'to what extent'.

2 Weather and climate and associated hazards

Winter wonderland

In which country was this photo taken?
How much snow has fallen on the telephone box?
What impact will this snow have on people in the local area?
Why is so much snow falling if the world is supposed to be warming?
What impact does heavy snow have on the economy?

Introduction

Many of us are fascinated by the weather. In the USA, one of the most popular television channels is the 'Weather Channel', which features up-to-the minute reports from around the country tracking tornadoes, hurricanes, and snowstorms.

The weather affects the daily lives and businesses of people throughout the world – from farmers to retailers, fishermen to financiers. Whilst we are all affected by the daily changes in the weather, many of us are increasingly concerned about the longer-term trends associated with climate change.

In this chapter you will learn about the processes and mechanisms responsible for weather patterns in the UK and elsewhere in the world. You will learn about the patterns and causes of the world's climates. And you will investigate the issues surrounding climate change.

Books, music, and films

Books to read

Our Choice by Al Gore
How Can I Stop Climate Change? by Helen Burley & Chris Haslam
There Is No Such Thing as a Natural Disaster by Chester Hartman & Gregory D. Squires (about the New Orleans flood 2005)
The Global Climate System by Howard A. Bridgman and John E. Oliver
The Climate Files by Fred Pearce

Music to listen to

'Reckoner' by Radiohead
'Last Good Day of the Year' by Cousteau

Films to see

The Age of Stupid
Perfect Storm
Flood
The Day After Tomorrow
When the Levees Broke: A Requiem in Four Acts (a documentary about New Orleans after Hurricane Katrina)

About the specification

'Weather and climate and associated hazards' is one of three Physical Geography options in Unit 3 Contemporary Geographical Issues – you have to study at least one.

This is what you have to study:

Major climate controls

- The structure of the atmosphere, the atmospheric heat budget, the general atmospheric circulation, planetary surface winds, latitude, oceanic circulation, and altitude.

The climate of the British Isles

- Basic characteristics: temperature, precipitation, and wind.
- Air masses affecting the British Isles.
- Depressions: origin and nature. The weather changes associated with the passage of a depression.
- Anticyclones: origin and nature. The weather conditions associated with anticyclones in winter and summer.
- Storm events: occurrence, impact, and responses to them.
- One case study of a storm event.

The climate of one tropical region

(Tropical wet/dry savanna, monsoon, or equatorial.)

- Basic characteristics: temperature, precipitation, and wind.
- The role of sub-tropical anticyclones and the ITCZ.
- Tropical revolving storms: their occurrence, their impact, the management of the hazard, and responses to the event.
- Two case studies of recent tropical revolving storms (from within the last 30 years) from contrasting areas of the world.

Climate on a local scale: urban climates

- Temperatures: the urban heat island effect.
- Precipitation: frequency and intensity, fogs, and thunderstorms – and relationship to urban form and processes.
- Air quality: particulate pollution, photochemical smog, and pollution reduction policies.
- Winds: the effects of urban structures and layout on variations in wind speed, direction, and frequency.

Global climate change

- Evidence for climatic change over the last 20 000 years.
- Global warming: possible causes.
- Global warming: possible effects – on a global scale, on the chosen tropical region (see above), and on the British Isles.
- Responses to global warming: international, national, and local.

The structure of the atmosphere

In this section you will learn about:

- the importance of weather and climate
- the structure of the atmosphere
- the concept of the atmospheric heat budget
- the effect of latitude and altitude on temperature

Temperatures in Big-Freeze Britain on Par with South Pole

The Day the Sea Froze: Temperatures Drop so Low Seas in UK Freezing

In the winter of 2009-10 headlines such as these filled newspapers and television news reports as the UK suffered the worst weather conditions for 30 years. For several weeks much of the nation was gripped by a prolonged Arctic blast that brought heavy snow, extremely low temperatures (below -20°C in parts of Scotland) and widespread ice.

The effects of this extreme weather were widespread. Schools and businesses were closed; in some parts of the country 1 in 5 people could not get to work. Hospitals were overwhelmed as people suffered injuries from falling on icy roads and pavements. Transport systems ground to a halt for days at a time: trains were cancelled or delayed as tracks froze; motorists were stuck for hours, some overnight, as snow and ice blocked roads and motorways (Figure 2.1).

The disruption caused by the harsh winter weather had a considerable impact on the UK economy, costing an estimated £1.2 billion. In its 2010 budget, the national government set aside £100 million to repair the over 1 million potholes caused by the unusual winter freeze.

The importance of weather and climate

There is an important difference between the terms weather and climate. The **weather** is the day-to-day condition of the atmosphere (the air above our heads). When we hear a weather forecast we are told about the likely temperature and whether it will rain or stay dry. **Climate** is the average weather calculated over a 30-year period.

The UK has a temperate climate, which means that it does not suffer from extremes of temperature. Regions of the world experience different climates, including desert (hot and dry) and equatorial (warm and wet).

While weather affects our day-to-day life, whether we ride a bike to work or plan a weekend barbecue, climate affects our way of life. Consider how in the UK we have adapted to a climate that has distinct seasons. Farming clearly shows the seasonal nature of our climate: it determines the types of crops we grow and when they are planted. Shops sell different goods in winter than they do in summer. We even alter the clocks twice a year to cope with the changing hours of darkness.

Figure 2.1 *Snow stops traffic on a busy UK road* ▼

Elsewhere in the world aspects of climate have a significant effect on the way people live. In India the monsoon rains shape lives: farmers rely on them for up to 90% of the annual rainfall, without which millions of people would starve. Yet the floods they can bring also destroy villages (Figure 2.2). The summer hurricane season affects people across the world from the Caribbean in the west to Japan in the east. In Florida, new buildings must comply with 'hurricane building laws' to help them better withstand the storms.

Figure 2.2 *The much-needed monsoon rains in India can also bring destructive floods* ▲

Climates change over time. Some 12 000 years ago ice spread south from the Arctic to cover vast areas of North America and northern Europe. This cooling resulted largely from very slight changes in the pattern of heat energy reaching the Earth from the sun. Beginning in about 1500, Europe and North America entered a cold spell called the 'Little Ice Age'. During its height, many rivers in Europe, including the River Thames in London, froze over. In parts of the northern hemisphere long, cold winters saw average temperatures fall by about 1°C causing glaciers in Europe to advance, sometimes overrunning entire villages. Sea ice spread south into the North Atlantic to the extent that in 1695 no open water flowed around Iceland.

Today there are concerns about an apparent warming trend commonly referred to as '**global warming**'. In the last 100 years average global temperatures have risen by 0.7-0.8°C. Since 1950, they have risen by 0.5°C. This rapid rise in temperature is attributed to increases in greenhouse gases, in particular carbon dioxide, mainly caused by human activities. This highly controversial topic will be studied in more detail later in the chapter.

Did you know?

Accurate instruments to measure temperature have only been available since the mid-1800s. To study earlier periods, scientists have to examine ice cores, pollen records and tree rings.

The structure of the atmosphere

The air above our head is called the **atmosphere** (Figure 2.3). It is a relatively thin but vital support system for life on Earth. The atmosphere is roughly 1000 km thick, although most of it (99% of its mass) is in the lowest 40 km above the Earth's surface. The atmosphere does not have a sharply defined outer edge; instead the air just becomes progressively thinner.

The atmosphere is made up of gases, such as oxygen, nitrogen, carbon dioxide and water vapour, together with liquid water and solid particles of dust and ice. Influenced by the Earth's rotation and surface features, the atmosphere acts like a liquid twisting and turning in highly complex interconnected waves. No wonder it is so difficult to forecast the weather!

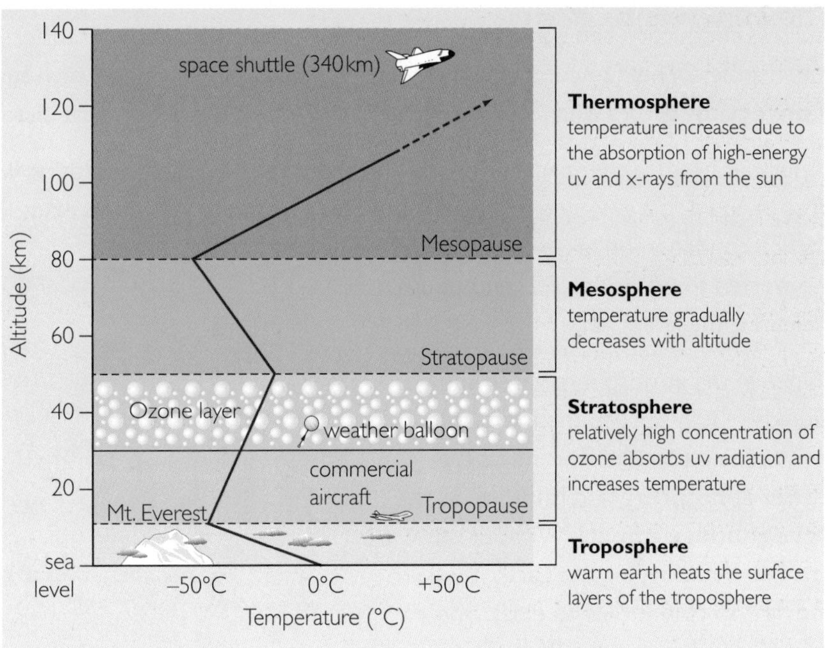

Figure 2.3 *The atmosphere can be divided into a number of layers which are determined by changes in temperature – note how the temperature changes as the altitude rises. The vast majority of the atmosphere's mass and almost all weather-related processes (cloud formation, winds, etc.) occur within the lowest layer called the* ***troposphere****. This explains why, when cruising in an aeroplane at 11 km (about 36 000 ft) almost all of the clouds are below you.* ▲

The atmospheric heat budget

The driving force behind our weather and, indeed, all life on Earth is heat energy from the sun. Look at Figure 2.4 which describes the movement of heat within the atmosphere. This is called the **atmospheric heat budget**.

The energy from the sun is called **solar radiation**. It arrives in the form of short-wave radiation. Not all of the radiation reaches Earth. Some is lost as it is scattered, absorbed and reflected by the atmosphere's gases, liquids and solids. Notice, for example, that in Figure 2.4, 20% of the incoming solar radiation is reflected back to space off the top of clouds. Have you ever noticed the dazzling upper surfaces of clouds when you have been in an aeroplane? This is caused by the solar radiation being reflected off the clouds.

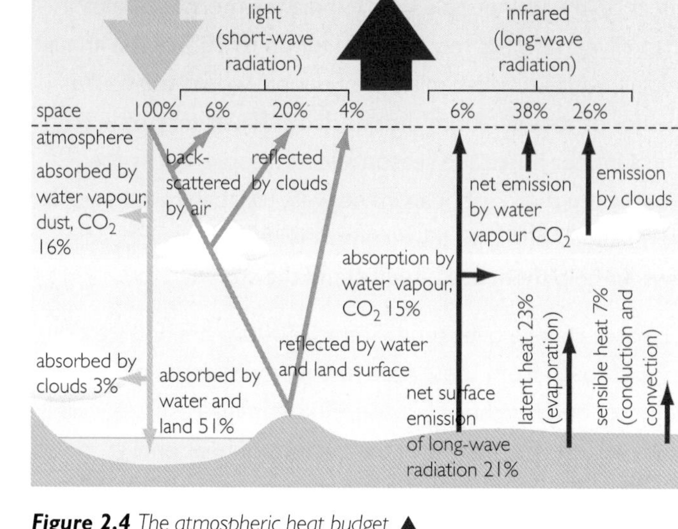

Figure 2.4 *The atmospheric heat budget* ▲

The solar radiation that does eventually reach the Earth's surface is called **insolation**. Some of this radiation is immediately reflected back into the atmosphere off white surfaces such as ice and snow. The reflectivity of a surface is called its **albedo** – the greater the reflectivity of a surface, the higher its albedo (Figure 2.5). The oceans and rainforests have low albedos, which means they reflect only a small part of the solar radiation. Deserts, ice and clouds have high albedos because they reflect more solar radiation. The remaining energy that isn't reflected back into the atmosphere is absorbed and heats the Earth's surface.

Heat is transferred back to the atmosphere from the Earth's surface as **terrestrial radiation**. This takes the form of long-wave radiation. Heat transfers may involve processes such as conduction and convection (together called **sensible heat** transfers). Conduction is the transfer of heat by contact with a substance (such as touching a warm surface) whereas **convection** occurs when warm air rises, taking with it the energy that it stores.

A further heat transfer process involves **latent heat**. During evaporation water changes from a liquid to a gas – the heat that is used during this process is stored as latent heat. When condensation takes place, the gas is converted to water droplets and clouds, releasing the latent heat.

Once in the atmosphere gases and liquids readily absorb the long-wave terrestrial radiation. This heats the lower atmosphere and is known as the **'greenhouse effect'**. Without this warm blanket surrounding the Earth it would be far too cold for life to exist. Some scientists believe that it is the increasing absorption of heat caused by emissions of gases such as carbon dioxide that lies behind the recent global warming trend.

Figure 2.5 *The albedo of different types of land and water surfaces* ▼

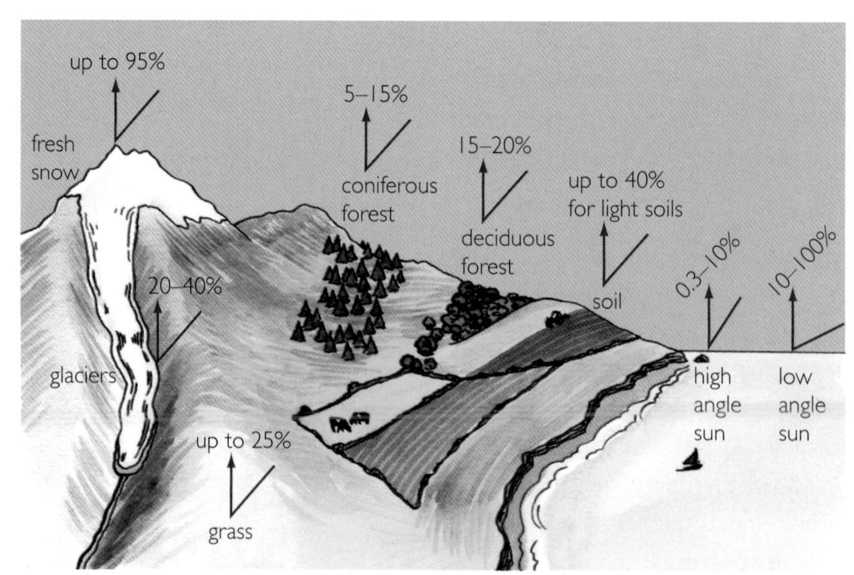

Patterns of global insolation

Despite periodic shifts in climate, the atmospheric heat budget displays a balance between incoming and outgoing radiation. So the Earth is neither getting significantly warmer nor colder.

However, insolation is not evenly distributed around the world. Look at Figure 2.6 which shows the global distribution of insolation. Notice how the highest insolation values are in the equatorial regions and how they decrease towards the poles. There are several reasons for this difference:

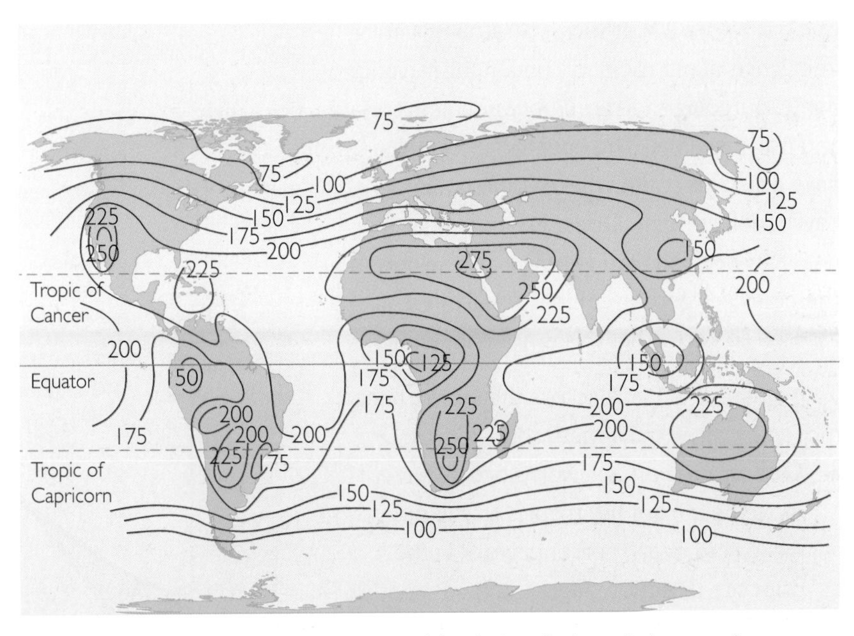

Figure 2.6 *This map shows the average annual distribution of solar radiation around the world* ▲

- **Angle of the sun** – the curvature of the Earth and its position as it rotates around the sun mean that the relative angle of the sun varies. In equatorial regions the sun's rays approach the Earth roughly at right angles (Figure 2.7). This means that the sun appears to be almost directly overhead. Rather like a torch beam hitting a flat surface at a right angle, the light, or heat energy, is very concentrated. Towards the poles, the sun's rays now approach at more of an acute angle so that the sun appears lower in the sky, reducing the concentration of heat energy. Equatorial regions, then, receive a higher concentration of energy from the sun than the poles.
- **Albedo** – the large areas of ice and snow in polar regions mean that large amounts of energy are reflected back into the atmosphere, reducing the amount of insolation.
- **Cloud cover** – look again at Figure 2.6 and notice that the highest insolation values lie close to the Tropics of Cancer and Capricorn and not at the Equator. This is because these sub-tropical regions are largely free from cloud whereas at the Equator there is much more cloud cover. So, despite there being more potential insolation at the Equator due to the high angle of the sun, the thick cloud cover reduces it considerably.
- **Land and sea** – evaporation from the oceans tends to cause more clouds to form than evaporation from land areas. This reduces the amount of solar radiation reaching the Earth's surface. Land areas also tend to be darker than the oceans so absorb more radiation, particularly when the sun is at a low angle (Figure 2.7).

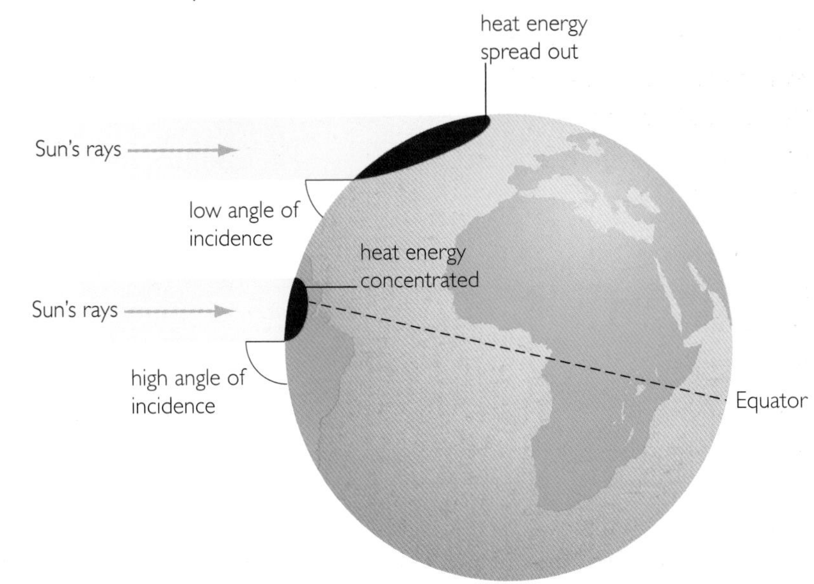

Figure 2.7 *The angle of the sun's rays as they reach Earth has an effect on the amount of insolation of different areas* ▲

Look at Figure 2.8. It shows a rather alarming imbalance of net radiation (incoming radiation minus outgoing radiation) between equatorial and polar regions. The equatorial regions have a surplus (gain) whereas the poles have a deficit (loss). Does this mean that the equatorial regions are getting hotter whereas the poles are getting colder? You'll be re-assured to know that this is not happening!

To maintain a balance across the world, heat is moved by three main methods:

- **Ocean currents** – warm ocean currents move heat from the tropics to the poles while cold ocean currents work in the opposite direction (Figure 2.9).
- **Trade winds** – these transfer large amounts of heat from the tropics to the poles.
- **Storms** – such as tropical cyclones (hurricanes) transfer large amounts of heat energy from the tropics (where they develop) to the subtropics and temperate zones.

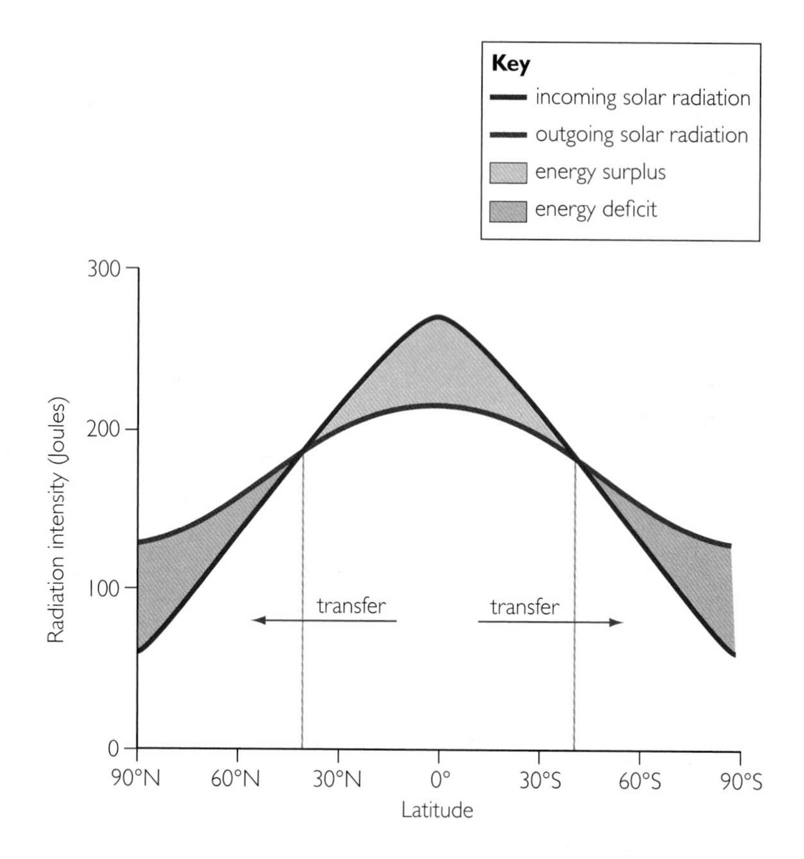

Figure 2.8 *Latitude influences the balance of incoming and outgoing energy* ▲

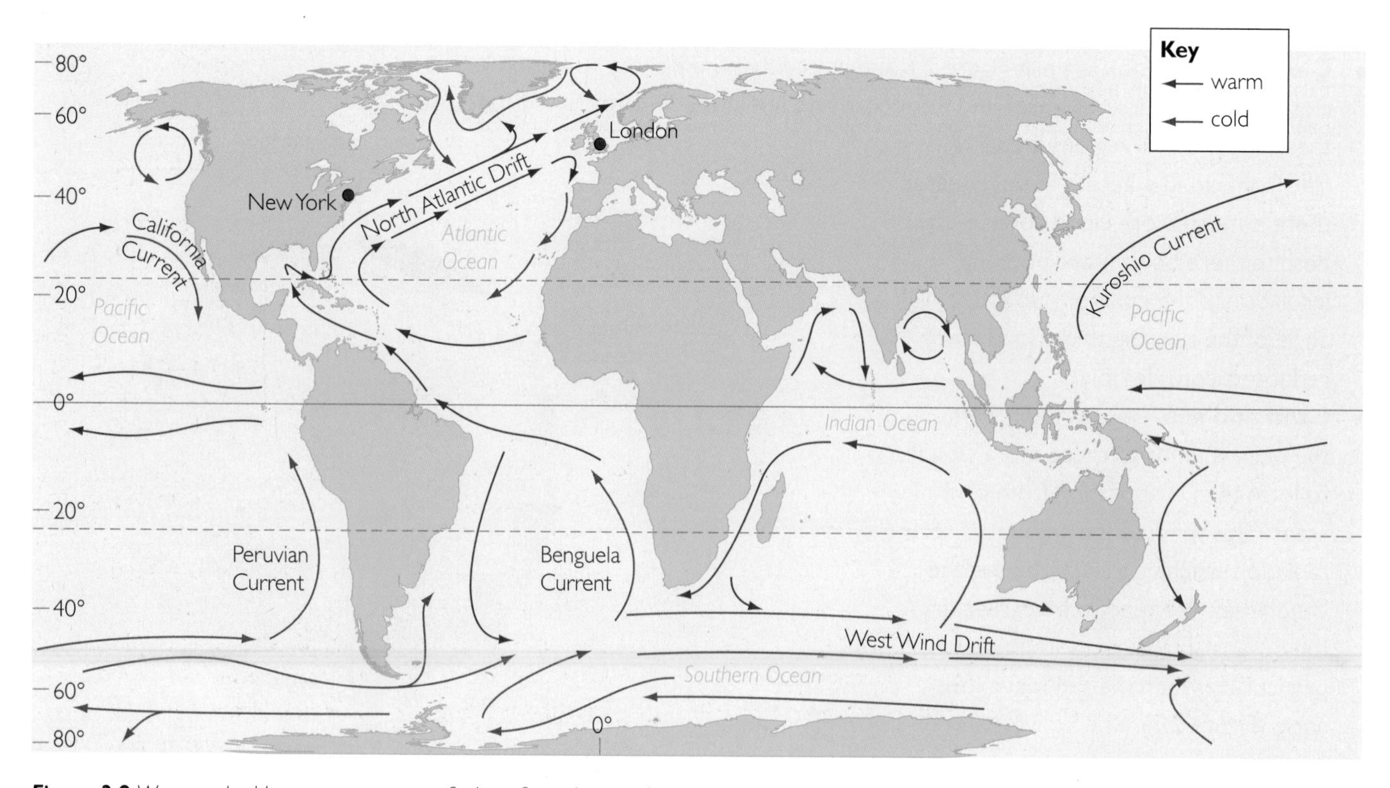

Figure 2.9 *Warm and cold ocean currents transfer heat from place to place* ▲

Patterns of global temperature

While insolation is the main overriding factor determining the temperature of an area, a number of additional factors also play a role:

- **Land and sea** – land and sea heat up and cool down at very different rates. Water has a higher heat capacity than land, which means that it takes longer to heat up. Water is transparent so radiation can spread deeper in the sea as compared to land, where the heat is concentrated on the surface. And because water moves, it is easier for the heat to be spread.
- **Seasonal factors** – there are significant differences in temperature from season to season, particularly in mid-latitude regions. For example, central Europe generally has very high temperatures in the summer but low temperatures in winter.
- **Altitude** – temperature decreases with altitude (about 1°C for every 100 metres in altitude). This is because the higher the altitude, the further we are from warmer lowland areas. Also, air pressure gets lower as altitude increases, causing temperatures to drop.
- **Ocean currents** – warm and cold ocean currents are important methods of transferring heat (Figure 2.9). For example, the North Atlantic Drift accounts for the relatively mild conditions in northwestern Europe. Without it, the UK could be as cold as parts of Canada.
- **Air masses** – large bodies of air can transfer heat from place to place. During the 2009-10 winter cold spell, an Arctic air mass swept down over the whole of the UK introducing bitterly cold conditions. (See page 56).
- **Urbanisation** – large urban areas, such as Greater London, Tokyo or Mexico City, generate heat from industry and transport, which causes them to have slightly higher temperatures than that of surrounding rural areas (Figure 2.10).

Figure 2.10 Heat generated from a refinery in Beijing, China ▲

Figure 2.11 Average sea level temperatures in January (C °) ▼

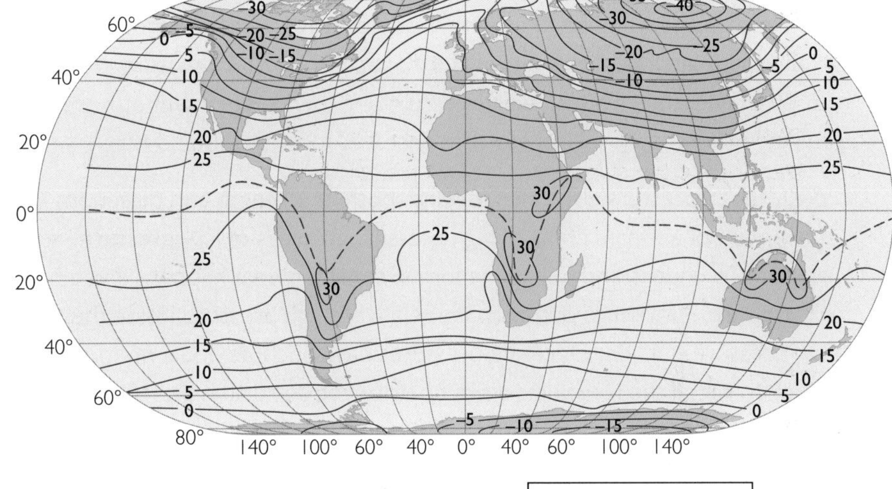

ACTIVITIES

1. Study Figure 2.4.
   a. Explain the meaning of 'sensible heat' transfer.
   b. Would the percentage of latent heat be different depending on whether the surface was land or ocean? Explain your answer.
2. Study Figure 2.6.
   a. Describe the global pattern of insolation. Refer to values and locations from the map.
   b. Explain in your own words how the angle of the sun, as shown in Figure 2.7, explains the difference in insolation between the equatorial regions and the poles.
   c. Why are the higher values of insolation mostly over land rather than the seas?
3. Study Figure 2.11 above. It shows average sea level temperatures in January. Remember, this means that it is winter in the northern hemisphere and summer in the southern hemisphere!
   a. Find the lowest temperatures on the map. Why are they in this area?
   b. Explain why there are warmer temperatures to the west of the UK and Norway.
   c. The Arctic air mass often affects the United States. Is there any evidence of this on the map?
   d. The 'thermal Equator' shows the highest temperatures. Why doesn't it form a straight line along the Equator?

General atmospheric circulation

In this section you will learn about:

- atmospheric pressure and winds
- global patterns of atmospheric pressure and winds
- the tri-cellular model of atmospheric circulation

Figure 2.12 *A television weather forecaster presenting the weather. Note the map which uses isobars to illustrate the atmospheric pressure* ▲

Atmospheric pressure and winds

The weight or mass of the atmosphere presses down on the Earth's surface – this is called **pressure**. It is measured with an instrument called a barometer in units called millibars (mb). On a weather map, areas that have equal atmospheric pressure are joined together by lines called **isobars** (Figure 2.12).

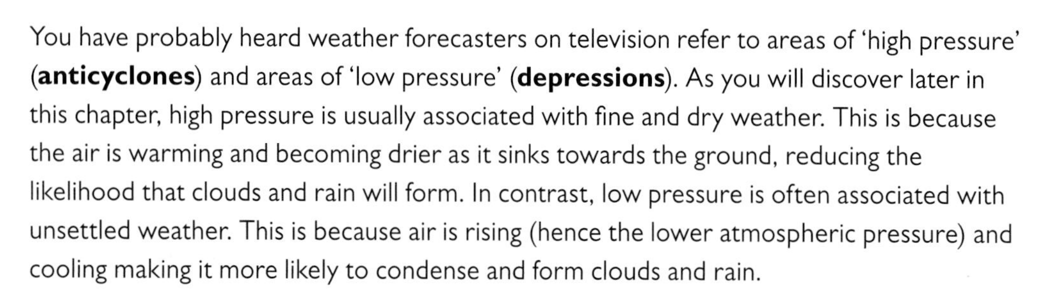

You have probably heard weather forecasters on television refer to areas of 'high pressure' (**anticyclones**) and areas of 'low pressure' (**depressions**). As you will discover later in this chapter, high pressure is usually associated with fine and dry weather. This is because the air is warming and becoming drier as it sinks towards the ground, reducing the likelihood that clouds and rain will form. In contrast, low pressure is often associated with unsettled weather. This is because air is rising (hence the lower atmospheric pressure) and cooling making it more likely to condense and form clouds and rain.

Weather forecasters also refer to wind, both its strength and direction. **Wind** is simply the movement of air from areas of high pressure to areas of low pressure – think of it as air in motion. To understand how wind forms, consider what happens if you have a puncture in a bicycle tyre. Air rushes out from the relatively high pressure inside the tyre to the relatively low pressure outside. Once the pressures have become equal no more air is released. In order to get all the remaining air out before you repair the puncture, you have to squeeze the inner tube.

- **Wind strength** – this is affected by the difference in pressure between two points known as the **pressure gradient**. The greater the difference in pressure, the faster the winds. On a weather map, this would be shown by closely spaced isobars. Look back at Figure 2.12 – can you see where it was windy on this day?
- **Wind direction** – in common with many other phenomena, air movement is affected by the rotation of the Earth – this is called the **Coriolis effect**. It causes winds to veer to the right in the northern hemisphere and to the left in the southern hemisphere (Figure 2.13). While the Coriolis effect is a major influence on wind direction, there are many other factors at the local level which also influence wind direction, for example, the relief of an area.

Figure 2.13 *A high pressure (anticyclone) and a low pressure (depression) centre with wind direction in the northern hemisphere. The circles are isobars with pressure measured in millibars (mb)* ▼

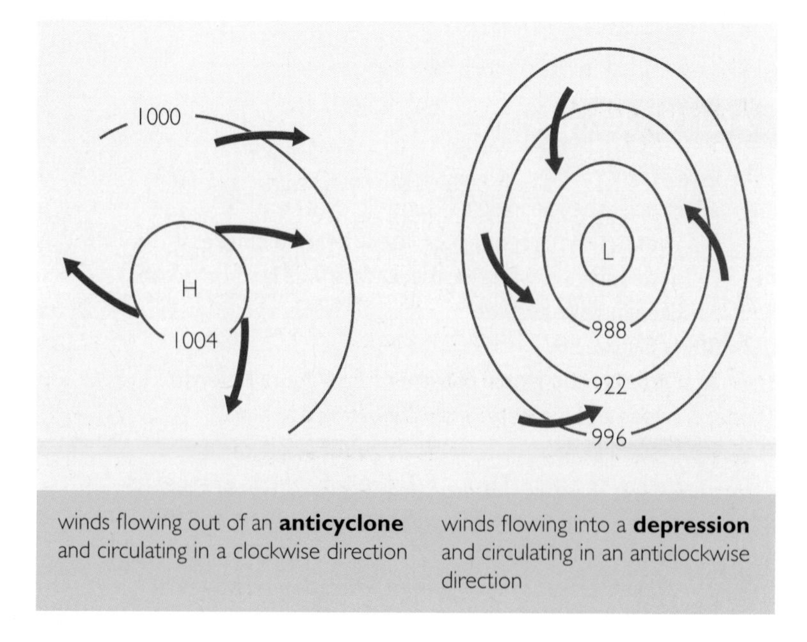

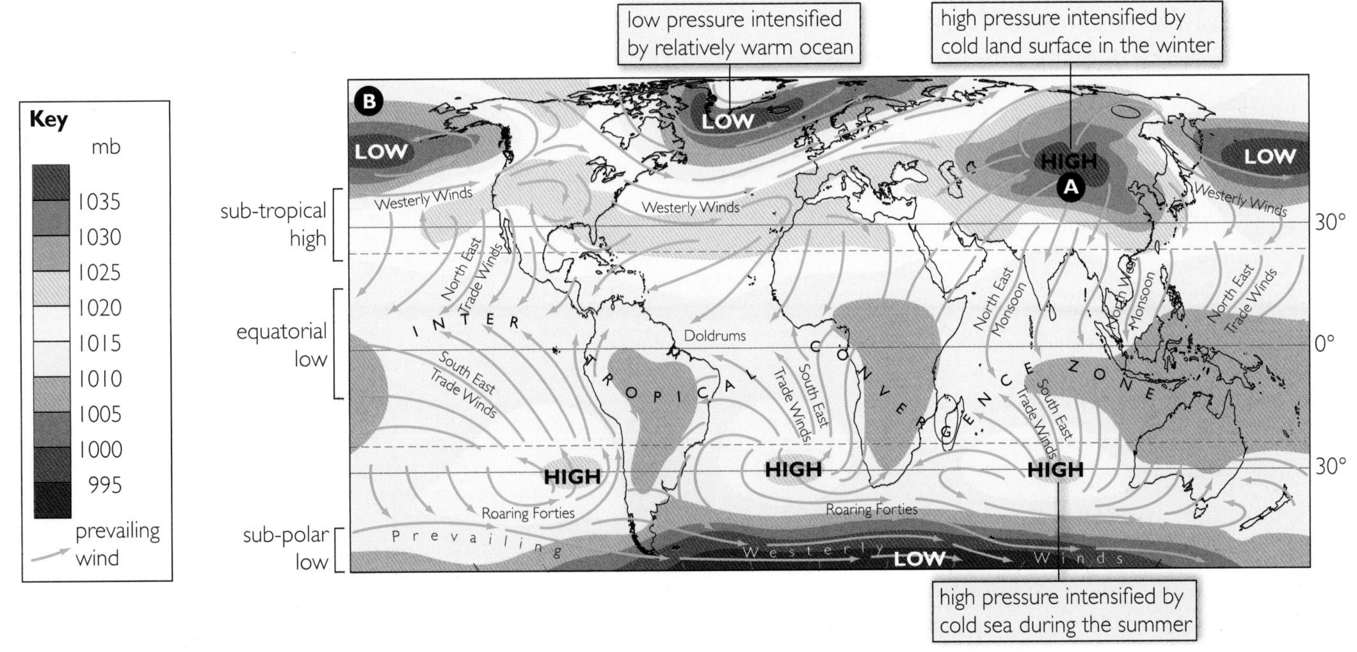

Figure 2.14 This map shows atmospheric pressure and winds across the world in a typical January ▲

Global pressure patterns

Figure 2.14 shows the global patterns of atmospheric pressure and winds for a typical January. Find the anticyclone labelled A. Notice that the winds are flowing out of the anticyclone and that, due to the Coriolis effect, they are curving towards the right so are circulating in a clockwise direction. Compare this with the depression labelled B. Here the winds are flowing towards the centre of the depression in an anticlockwise direction.

While it is possible to identify individual anticyclones and depressions on Figure 2.14, notice that a number of distinct bands or belts of pressure run across the map from west to east. These large-scale pressure belts are caused by differences in ground surface temperature and the circulation of the atmosphere.

- **Ground surface temperature** – in lower latitudes, the ground surface tends to have high temperatures which warm the air causing it to rise. This creates a belt of low pressure called the equatorial low (Figure 2.15). In contrast, the very cold temperatures at the poles cool the air, making it denser, causing it to sink. This creates a zone of relatively high pressure called the polar high.
- **Circulation of the atmosphere** – the atmosphere is in constant motion. In some places the air in the upper atmosphere converges (comes together) then sinks towards the ground surface, creating a zone of high pressure. Elsewhere, the air that converges close to the ground surface rises quickly, creating an area of low pressure.

Figure 2.15 This map shows atmospheric circulation and global wind patterns ▼

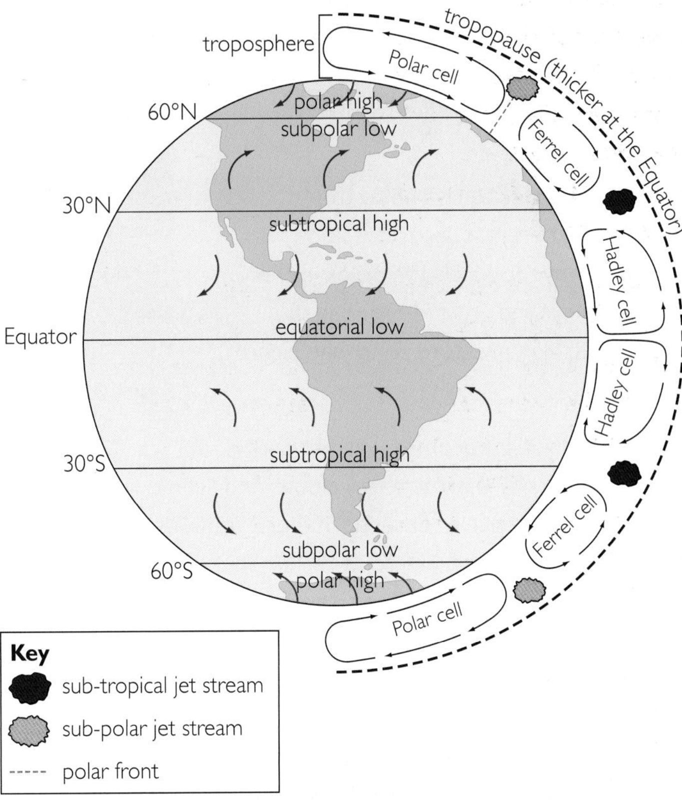

Jet streams are very fast, high-level winds that have a velocity of up to 400km per hour. They have an influence on our weather and are used by aeroplanes to speed journeys, for example, between the USA and Europe.

The global atmospheric circulation system

Look back at Figure 2.15. It shows the connection between surface pressure belts, the major surface winds and the circulation of the atmosphere. It is important to understand the connections between what is happening in the atmosphere and what is happening on the ground.

Circulation in the atmosphere involves three different circles or **cells** that connect with each other. This concept is known as the **tri-cellular model** of atmospheric circulation.

Hadley cell

The first of the cells is called the Hadley cell – it is the most distinctive of the three cells. At the Equator the ground is strongly heated by the overhead sun (Figure 2.7, page 59). This causes air to rise and create a low pressure zone on the ground surface.

As the air rises, it cools and forms thick storm clouds in a zone called the **inter-tropical convergence zone** (ITCZ) (Figure 2.16).

The air continues to rise up to the tropopause before it separates and starts to move towards the poles. It then sinks towards the ground, forming the subtropical high pressure zone (see Figure 2.15).

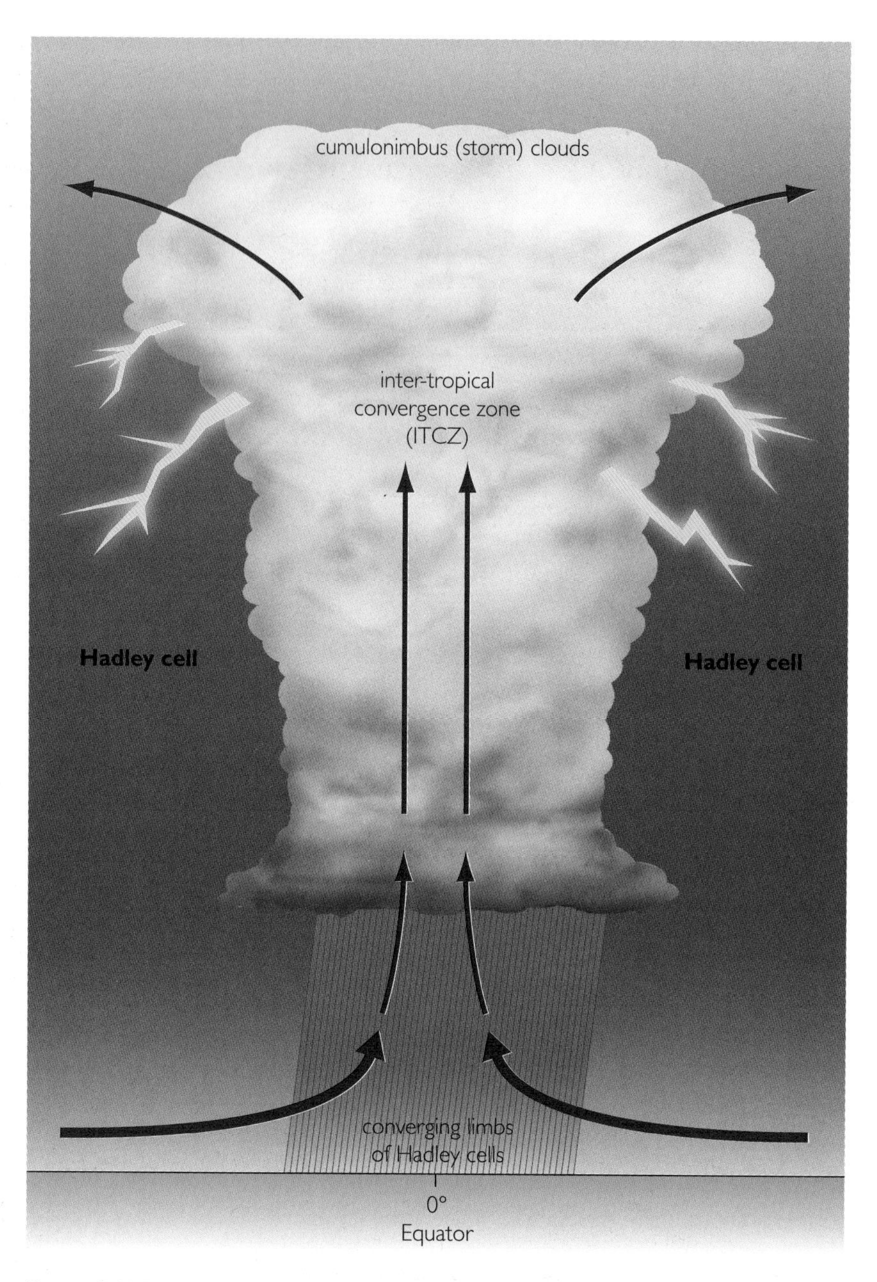

Figure 2.16 *The inter-tropical convergence zone* ▲

As the air sinks, it becomes warmer and drier – this is why there are few clouds and little rain, as well as the existence of deserts, in these latitudes. The Hadley cell is then completed as the air flows back towards the Equator – this is known as a **trade wind**.

Ferrel cell

The Ferrel cell is the weakest of the three cells. It is largely driven by **convergence** (coming together) and **divergence** (moving apart) of air associated with the other cells. On the ground, it is characterised by surface winds moving from the Equator towards the poles. Much of our weather in the UK is determined by these winds.

Did you know?

The Hadley cell takes its name from the Englishman George Hadley. A lawyer and amateur meteorologist, in 1686 Hadley was the first person to suggest that equatorial circulation cells actually existed. He came up with this idea when he was trying to explain the permanent trade winds that made it possible for ships to sail west across the Atlantic from Europe to North America. The Hadley cell together with the influence of the Coriolis effect explain the driving force behind these important winds.

Polar cell

The Polar cell is a fairly shallow cell (see Figure 2.15). At the poles, air is cooled and sinks towards the ground, forming the pressure zone called the polar high. It then flows towards the lower latitudes. At about 60 ° latitude north and south, the cold polar air mixes with warmer tropical air and rises upwards, creating a zone of low pressure called the subpolar low. The boundary between the warm and cold air is called the **polar front** – it is a very important junction and accounts for a great deal of the unstable weather seen in these latitudes (Figure 2.17).

The three cells do not always remain in the same place. As the overhead sun moves with the seasons, the cells shift north and south of the Equator. This helps explain the seasonal differences in the weather experienced in different parts of the world.

Figure 2.17 *An approaching storm on the west coast of Norway* ▲

ACTIVITIES

1 Copy Figure 2.13 but switch the positions of the anticyclone and depression.
  - **a** Write labels to describe what is happening.
  - **b** Use two separate colours to shade:
    - an area with strong winds
    - an area with light winds
  - **c** Give your diagram a title.

2 Study Figure 2.14.
  - **a** Using the key, what is the atmospheric pressure in the centre of Anticyclone A?
  - **b** Explain why high-pressure areas (anticyclones) in the northern hemisphere are more intense over land areas.
  - **c** Describe and explain the location of the low-pressure areas (depressions) in the northern hemisphere.
  - **d** Find the areas on the map labelled the 'Roaring Forties'.
    - **1** Where are they located?
    - **2** What do you think they are?
    - **3** What do you think causes them?
  - **e** Describe the location of the inter-tropical convergence zone.
  - **f** Using Figure 2.16, describe what happens within this zone.
  - **g** How in the past would sailing ships have used the winds over the Atlantic Ocean to trade between Europe and North America?

3 Figure 2.18 below shows the tri-cellular model of global circulation. Some areas of the diagram are not named.
  - **a** Make a copy of Figure 2.18.
  - **b** Using information from Figures 2.15 and 2.16, write in the name of each part of the diagram and add as many labels as you can.

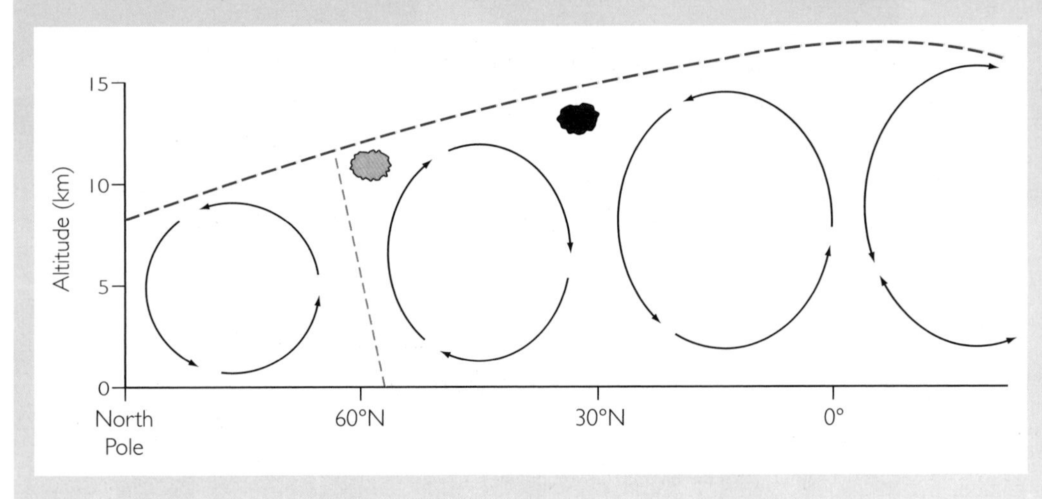

◀ ***Figure 2.18*** *The tri-cellular model of atmosphereic circulation – not quite complete*

2.3 The climate of the British Isles

In this section you will learn about:

- the climate characteristics of the British Isles
- the factors affecting climate in the British Isles
- the importance of air masses in affecting weather and climate in the British Isles

In 2010, the Met Office abandoned its long-term weather forecasts. This decision followed two inaccurate headline forecasts. As reported in *The Times*, "Britain was in for a season of mildness, the Met Office announced, shortly before the coldest winter in 31 years. That followed the prediction of a barbecue summer that went on to leave the nation with some very soggy sausages" (Figure 2.19).

The media-grabbing headlines don't tell the whole story. The Met Office predictions are made on the basis of probability. While models might suggest a most likely scenario, there is always a chance that a different set of weather conditions will occur.

The UK's temperate climate

When the Met Office announced that it was stopping its seasonal forecasts, the reason it gave was that the UK's temperate climate made it hard to predict more than a week ahead. What exactly does this mean?

To begin with, a temperate climate has two main characteristics:

- rain is experienced throughout the year, and
- temperatures are relatively mild in the winter and reasonably warm in the summer.

In other words rather bland and uninteresting! So, how does this make it hard to predict the weather?

Figure 2.19 *A typically wet day in the summer of 2010* ▼

As we saw earlier in this chapter, the **climate** is the average weather conditions over a period of 30 years. This means that weather extremes, such as storms (the 1987 'Great Storm'), droughts (1976), floods (Boscastle 2004 and Cockermouth 2009), heatwaves (2003) and severe winters (1947, 1963, 2010) will tend to be evened-out to produce a rather bland average.

The actual weather in the UK is anything but 'average' and varies enormously from day to day. This explains the difficulties weather forecasters, including the Met Office, have in making long-term forecasts.

Factors affecting the UK's climate

There are several factors that affect the climate of the UK and influence the weather we experience on a day-to-day basis (Figure 2.20).

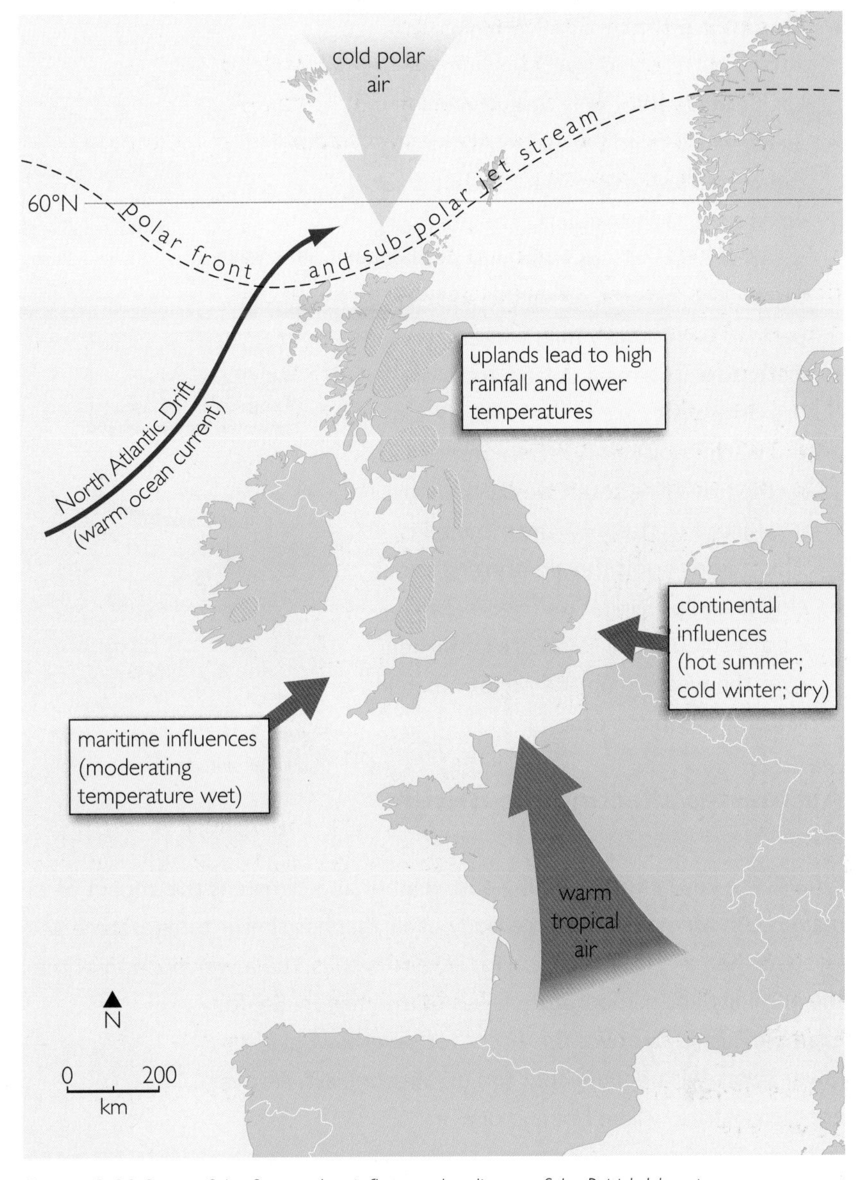

Figure 2.20 *Some of the factors that influence the climate of the British Isles* ▲

- **Global position of the UK**. The UK lies in an atmospheric 'battleground' between warm tropical air to the south and cold polar air to the north. As the two distinctly different types of air battle one another for control over the mid-latitudes, the UK experiences contrasting and changeable weather conditions.
- **Seasonal changes in global pressure patterns**. The pressure belts of the world (see Figure 2.14, page 63) are not fixed in place. As the position of the overhead sun moves between the tropics, the pressure belts shift north and south. This means that in the summer the subarctic low pressure belt, and unsettled conditions that come with the subpolar jet stream, tend to be north of the British Isles. This leads to high pressure and settled conditions building from the south. In the winter, the low pressure belts and the jet streams move further south, causing unsettled weather over the British Isles.
- **Seasonal variations**. The UK experiences significant seasonal shifts in the weather. During the summer, the sun is relatively high in the sky, making high daytime temperatures a possibility. During the winter months, the days are shorter and the angle of the sun is lower so less heat reaches the ground, causing lower temperatures.
- **Influence of land and sea**. The seas surrounding the UK influence the climate in two ways:
  - First, moisture from the sea leads to generally humid conditions and explains why the UK is often cloudy and wet.
  - Second, because the sea takes longer to heat up and cool down than land, it moderates the UK's climate, making the temperatures more even throughout the year than they might otherwise be. As such, summers rarely become very hot and winters rarely very cold.

However, because Europe is so near the UK its continental climate periodically spreads across, bringing dry periods with hot summers and cold winters.

- **Ocean currents**. The warm North Atlantic Drift significantly impacts the UK's climate. Not only does it carry warm water from the South Atlantic to the western shores of the UK, but it also warms the air it contacts. The prevailing south westerly winds then spread these warmer conditions, giving the western parts of the country mild winters.
- **Altitude**. The main upland areas of the British Isles are in the west. When the warm moist air is driven onshore by the prevailing south westerly winds it is forced to rise over the mountains. This quickly cools the air, bringing about clouds and precipitation (Figure 2.21). As a result, rainfall totals are far higher over the western upland regions than in lower areas in the east.

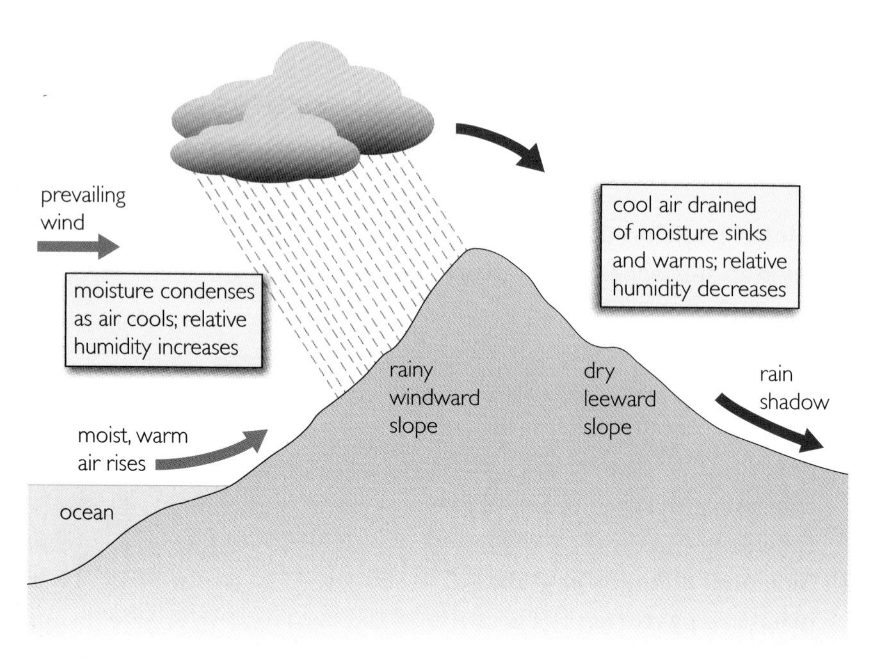

***Figure 2.21** This diagram shows why hilly and mountainous areas get more rain than areas at a lower altitude* ▲

Air masses affecting the British Isles

One of the key factors affecting our weather and climate is the impact of various air masses. An **air mass** is a large body of air that has similar temperature and humidity levels. When air sits in a region for several weeks, the lower portion of the air takes on the humidity and temperature levels of that region. So, for example, if air stays over the Arctic for several weeks, its lower layers will become cold and dry. In contrast, air over northern Africa will be hot and dry.

When the air mass eventually moves it takes the temperature and humidity levels with it. For example, when an Arctic air mass drifts south over the British Isles it introduces cold conditions. Of course, as an air mass moves its temperature and humidity levels will be changed slightly by the surface it travels over. For example, a dry air mass travelling over water will become more humid.

The British Isles is affected by several different air masses (Figure 2.22). Each brings very different weather conditions and helps explain our characteristically changeable weather.

The dominance of certain air masses at certain times of the year affects our climatic characteristics too. For much of the year, the polar maritime and tropical maritime air masses drive frontal systems over the British Isles, causing the relatively even spread of precipitation during the year and the lack of extreme temperatures.

***Figure 2.22** Air masses influence the weather of the British Isles* ▼

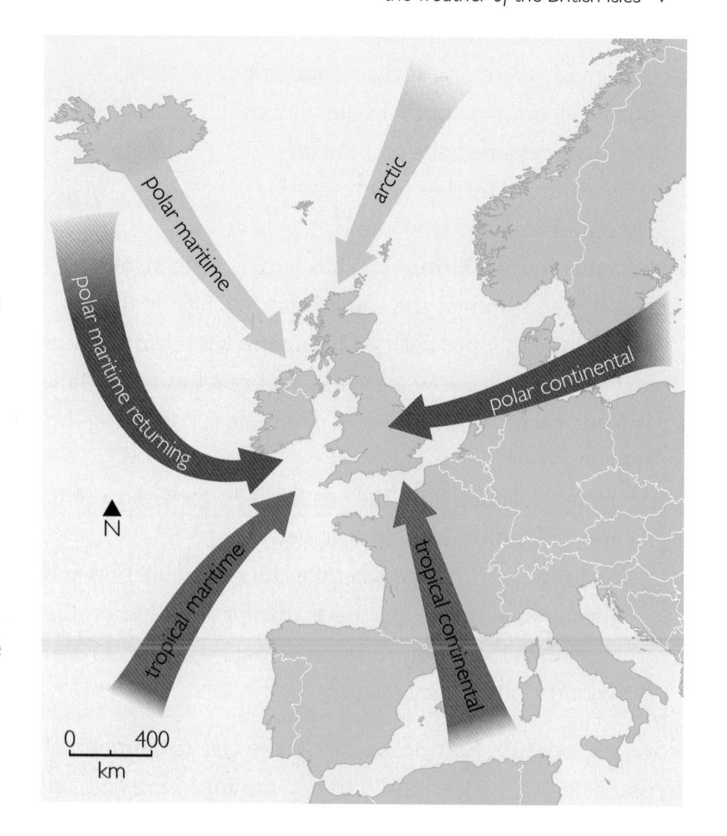

Air mass	Source	Temperature	Humidity	Modifications on route to the British Isles	Typical conditions experienced in the British Isles
arctic	Greenland, Arctic	cold	dry	becomes less cold and more humid as it travels south over the sea	• cold and showery conditions particularly in the winter • snow showers in the north
polar continental (winter)	Scandinavia and Siberia	cold	dry	can pick up moisture over the North Sea	• very cold and frosty conditions • can bring snow showers to the east
tropical continental (summer)	Southern Europe/ North Africa	hot	dry	can pick up moisture over the Mediterranean Sea	• hot and humid • possible thunderstorms
tropical maritime (typically forms the warm sector in a frontal system)	South Atlantic	warm	wet	becomes slightly cooler and even more humid having travelled over the Atlantic Ocean	• unsettled wet and often windy conditions • mild
polar maritime (typically forms the colder air behind a cold front)	North Atlantic	cold	wet	becomes slightly less cold as it moves towards the British Isles	• cool and unsettled conditions
polar maritime returning	North Atlantic	cold	wet	takes a longer route south so becomes warmer than a polar maritime air mass	• cool/mild and unsettled

ACTIVITIES

1 Study Figure 2.23 below.

a Describe the similarities and differences between the climates for London and Valencia (an island off the western coast of Ireland).

b Using the information from this section together with additional Internet research, suggest reasons for the differences in average monthly temperature and average monthly rainfall in the two locations.

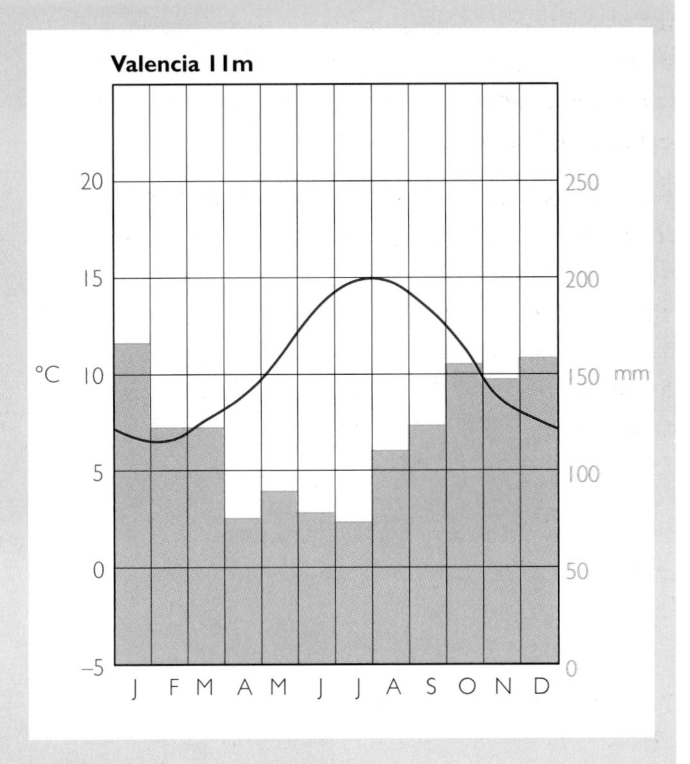

Figure 2.23 *These graphs show the average monthly climate for London and Valencia, Ireland* ▲

Mid-latitude depressions

In this section you will learn about:
- the characteristics of a depression
- the formation and development of a depression
- the weather associated with a depression over the UK

And the cause of the situation shown in the cartoon? No, not bad weather forecasters, but rather a weak weather system called a **depression**. A depression is an area of low atmospheric pressure. It is associated with unsettled weather, often in the form of clouds and precipitation. This is because in the centre of the depression air on the ground is converging and rising. As it rises, the air cools to form cloud and rain. Look back to Figure 2.14 (page 63) to remind yourself of the relative size of depressions on a global scale. Each year, about fifty depressions of varying sizes pass over or close to the UK bringing cloud, rain, snow or strong winds. In fact, you could say that depressions are responsible for giving the British Isles its characteristic changeable weather.

Look at Figure 2.24. It is a satellite photograph showing the swirls of cloud associated with a mid-latitude depression. Notice the band of cloud stretching down through central and western England and Scotland – it's bringing heavy rain to these areas. Ahead of it, however, people living in eastern England are still enjoying clear skies and sunshine.

Figure 2.24 *A mid-latitude depression over the British Isles as seen from space* ▼

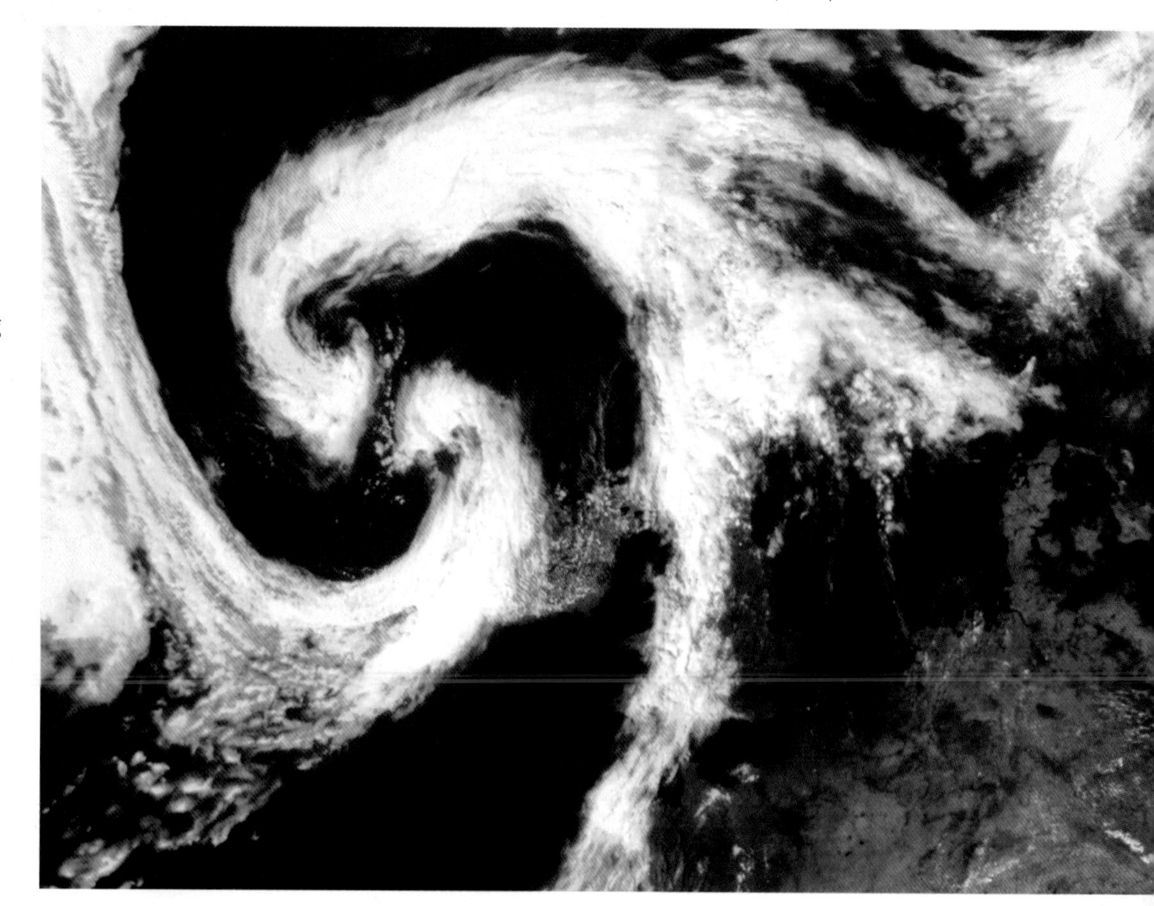

The development of a depression

Depressions are said to have a life cycle: they are born, live for three to five days, and then die away. Look at Figure 2.25 – it shows the three stages, or the life cycle, of a depression.

A Early stage

Most depressions form over the mid-Atlantic at a point where cold polar air and warm tropical air meet – this area is called the **polar front** (see Figure 2.20 page 67). The polar front is a wavy line and it is at one of these waves that a depression will start to form.

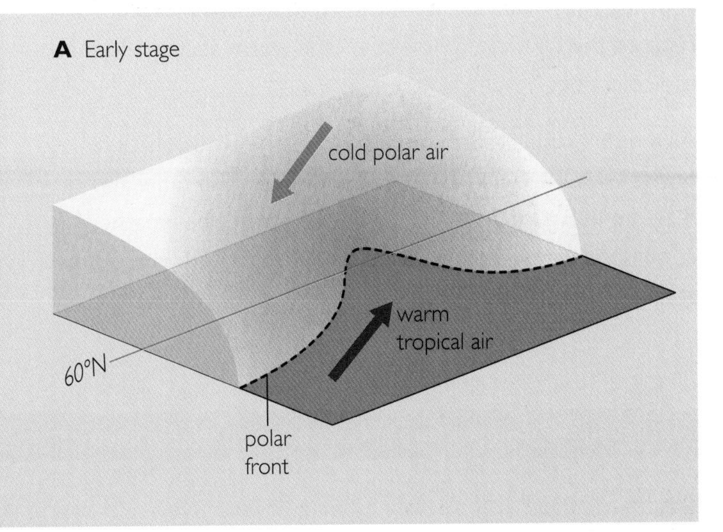

B Open stage

Once a 'kink' has developed on the polar front, winds begin to curve and flow into the centre of the newly developing depression. As air rises at the centre it starts to draw in yet more air from the ground surface. This establishes a spiral-like motion of airflow into the centre of the depression. There are now distinct boundaries between the warm and the cold air. A **warm front** marks the front edge of the warmer air. A **cold front** forms the boundary between the warm air and the colder air behind it. In-between the two fronts is a zone called the **warm sector**.

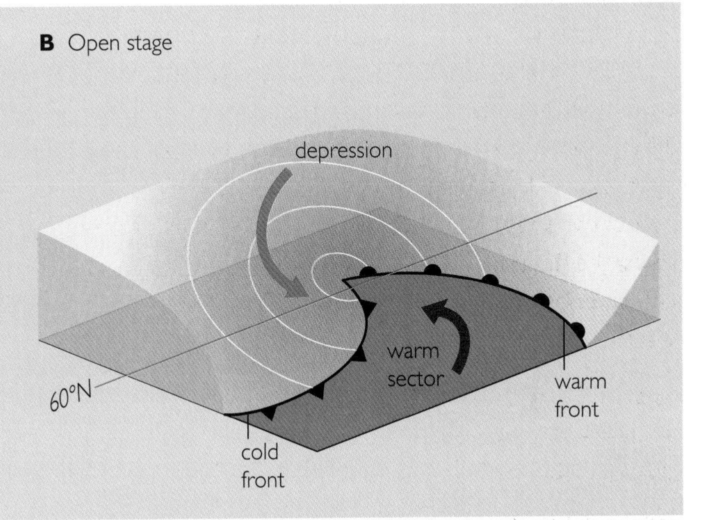

C Occluded stage

As the entire frontal system moves from west to east, the cold front gradually catches up with the warm front. This squeezes the warm sector and forces it upwards where it cools and loses its identity. Notice that where the two fronts meet a single front called an **occluded front** is formed. As the fronts continue to occlude (rather like zipping up two pieces of fabric) the warm sector decreases in size until it disappears altogether. Eventually, the frontal system will lose its identity. Pressure will start to rise (the depression is said to 'fill') and the entire depression will gradually fade away.

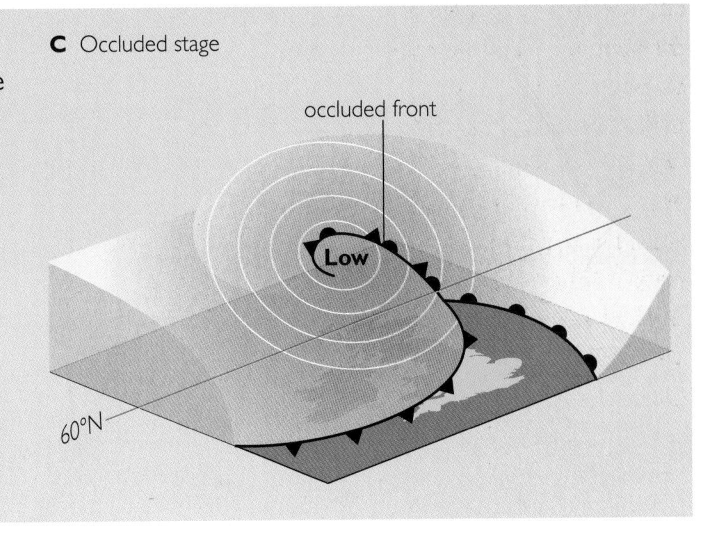

Figure 2.25 *The life-cycle of a depression. The changes take place as the depression moves from west to east towards the UK.* ▲

Weather associated with depressions

Most depressions that reach the UK come from the west or southwest, driven across the Atlantic by the prevailing southwesterly winds. This means that the warm front reaches the UK first, followed by the warm sector and then the cold front.

As a depression system moves across the British Isles it brings with it specific weather conditions. The table and diagram in Figure 2.26 explain and illustrate the weather conditions (temperature, wind, etc.) associated with each stage of a depression.

Did you know?

Depressions can bring severe weather conditions to the UK. On 18 January 2007, a deep depression passing over Scotland brought damaging hurricane force winds of over 75 mph to England. Thirteen people were killed by the storm. In March 2008, a powerful storm battered the south and west of England, causing flooding and leaving thousands without power. The storm, which had winds of up to 80 mph, was caused by one of the deepest weather depressions for several years.

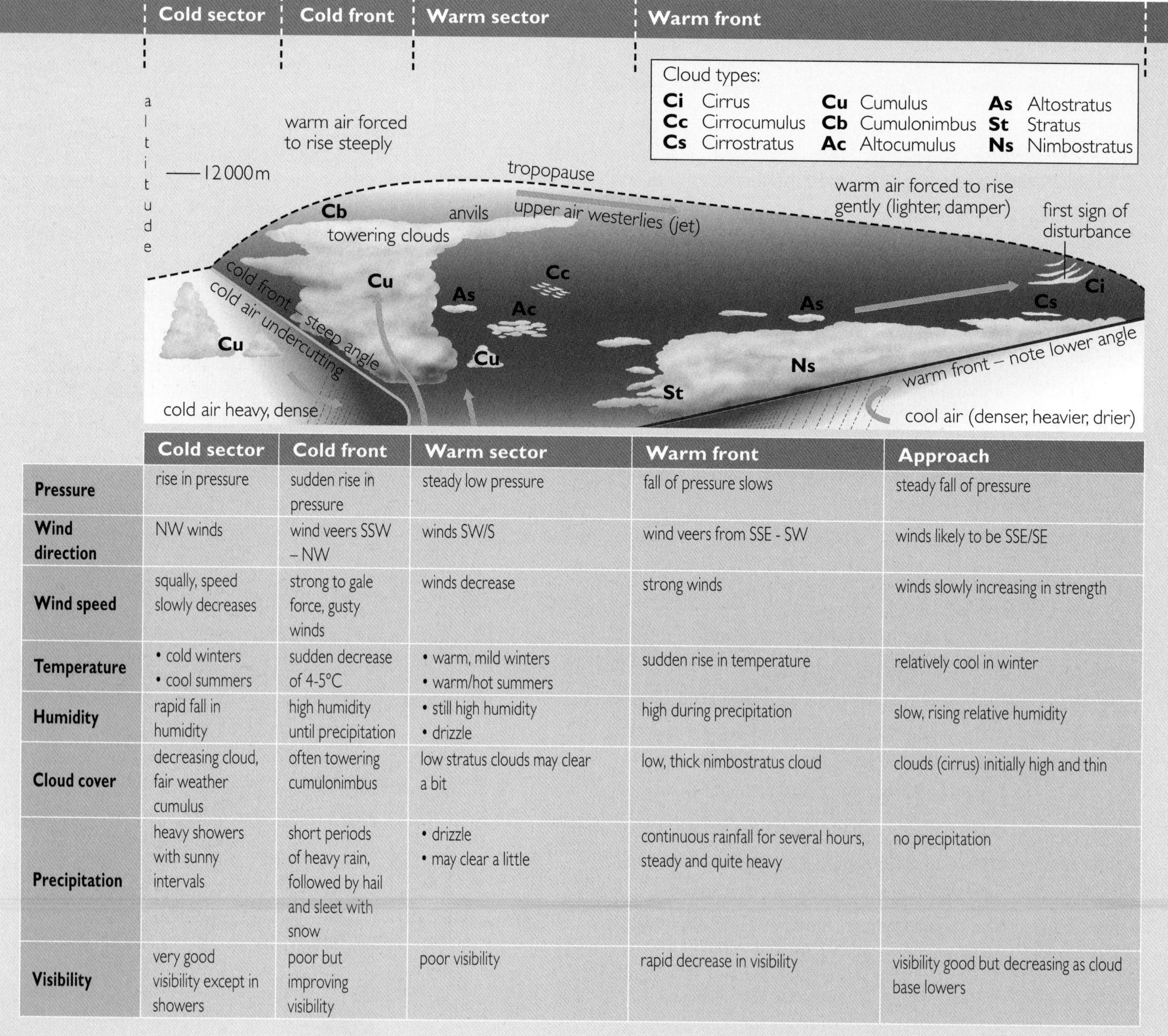

	Cold sector	Cold front	Warm sector	Warm front	Approach
Pressure	rise in pressure	sudden rise in pressure	steady low pressure	fall of pressure slows	steady fall of pressure
Wind direction	NW winds	wind veers SSW – NW	winds SW/S	wind veers from SSE - SW	winds likely to be SSE/SE
Wind speed	squally, speed slowly decreases	strong to gale force, gusty winds	winds decrease	strong winds	winds slowly increasing in strength
Temperature	• cold winters • cool summers	sudden decrease of 4-5°C	• warm, mild winters • warm/hot summers	sudden rise in temperature	relatively cool in winter
Humidity	rapid fall in humidity	high humidity until precipitation	• still high humidity • drizzle	high during precipitation	slow, rising relative humidity
Cloud cover	decreasing cloud, fair weather cumulus	often towering cumulonimbus	low stratus clouds may clear a bit	low, thick nimbostratus cloud	clouds (cirrus) initially high and thin
Precipitation	heavy showers with sunny intervals	short periods of heavy rain, followed by hail and sleet with snow	• drizzle • may clear a little	continuous rainfall for several hours, steady and quite heavy	no precipitation
Visibility	very good visibility except in showers	poor but improving visibility	poor visibility	rapid decrease in visibility	visibility good but decreasing as cloud base lowers

Figure 2.26 *The weather conditions associated with a depression crossing the British Isles* ▲

Depressions on a synoptic chart

Information about the weather can be presented in the form of a specialist map called a **synoptic chart** (Figure 2.27). A synoptic chart contains the following:

- **Isobars** are thin black lines joining points of equal pressure. Notice on Figure 2.27 that they are drawn at 5mb intervals. Remember that the closer the isobars, the stronger the winds.
- **Fronts** are boundary lines that mark distinct temperature differences on either side of the line. On a map, these are shown as thicker black lines with specific symbols to indicate whether it is a warm, cold or occluded front.
- **Weather information** is the recorded details of the weather (cloud cover, temperature, precipitation, etc) shown using symbols which are placed at specific weather stations.

Figure 2.27 shows weather conditions that are typically associated with a depression centred to the northwest of the UK. Notice the following features:

- The isobars are tightly packed around the depression (shown on the map as L). This indicates the presence of very strong winds.
- The air is circulating in the depression in an anticlockwise direction.
- Temperatures on either side of the fronts are markedly different.
- Cloud and rain happen at the fronts because the air is being forced to rise (as Figure 2.26 shows).

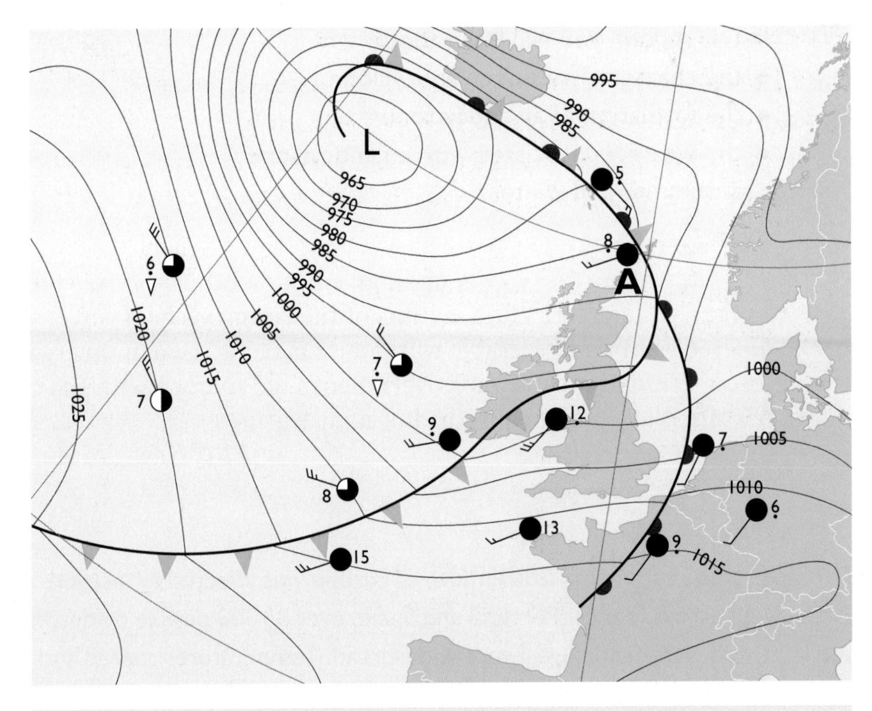

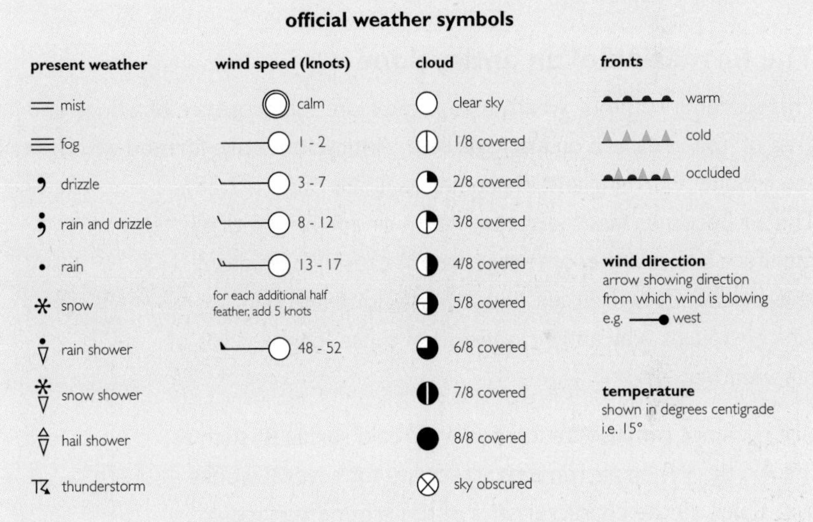

Figure 2.27 *This synoptic chart shows a depression over the UK at noon on 30 November 2007* ▲

ACTIVITIES

1 Study Figure 2.27.

a A weather station close to the centre of the depression (L) recorded a temperature of 4 ° C, total cloud cover, rain and strong south easterly winds at 48-52 knots. Use the symbols in the key to represent this information.

b A student has predicted that, by 12 midnight, the weather recorded at weather station A will be 3°C, showery and with strong winds blowing from the north west. Comment on the accuracy of this prediction.

Anticyclones

In this section you will learn about:

- the characteristics of an anticyclone
- the formation of an anticyclone
- the weather associated with an anticyclone in summer and winter

> France's largest mortuary – set up in an empty food warehouse near Paris – has begun to take in its first victims of the country's heat wave.
> The disused fruit and vegetable warehouse is being used as emergency accommodation as part of the government's bid to cope with the growing number who have perished in the hot temperatures.
>
> *BBC News, 16 August 2003*

For most of the summer of 2003, much of Europe was affected by a severe heat wave. Fires swept across France, Portugal and Spain, over 35 000 people died from the effects of the heat, and water shortages were widespread. Temperatures soared and records were broken – in the UK, the highest-ever temperature of 38.5°C was recorded in Kent.

The formation of an anticyclone

This extreme summer weather resulted from the presence of a huge and largely stationary area of high pressure or **anticyclone**. Anticyclones are formed when air sinks towards the ground, warming and expanding as it falls (Figure 2.28). The air becomes less saturated (warm air can hold more moisture before it becomes saturated) causing humidity levels to fall. This drier air leads to less cloud and rain, which explains why anticyclones often cause long periods of dry weather.

Anticyclones can also be formed over cold surfaces, such as the Arctic. When air remains stationary for several weeks it can pick up the characteristics of the ground surface. Once cooled, the lower layers of the atmosphere will have a tendency to sink, which causes an increase in the air pressure, forming an anticyclone.

Anticyclones can be very long lasting and stubborn to move. Sometimes, they block or divert the more freely moving depressions. This often happens in the winter in Europe when a 'blocking anticyclone' over Scandinavia can cause depressions to get 'stuck' and sit over the North Atlantic.

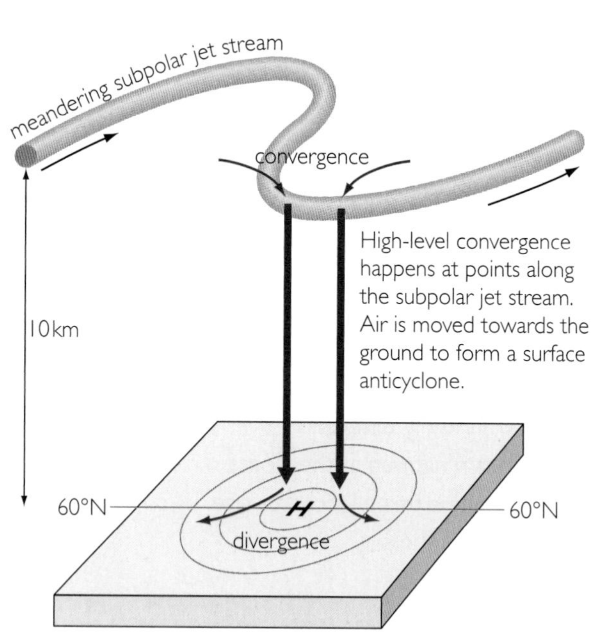

Figure 2.28 *The formation of an anticyclone* ▲

Weather associated with anticyclones

While in general anticyclones bring dry and settled conditions to the UK, there are differences depending on the season (Figures 2.29 and 2.30 opposite). In summer, cloudless skies associated with an anticyclone mean that more of the Sun's energy reaches the Earth, leading to hot days. Evenings are often cool because of the lack of cloud cover to trap the day's heat. Cloudless skies also mean no rain – so periods of drought may occur. However, on very hot days, the hot air may rise quickly and then cool, creating large, black clouds that can bring thunderstorms. An anticyclone that remains stationary for a period of time can lead to heat waves, such as the one Europe faced in 2003.

In winter, anticyclones bring very different weather. The cloudless skies allow the heat to escape into the atmosphere. At night, the ground cools quickly which in turn cools the air above it. The water vapour in the air condenses and freezes, forming frost or fog. The fog can last into the day until it is evaporated by the heat of the Sun.

***Figure 2.29** A winter anticyclone over the British Isles, 28 January 1998 at noon* ▲

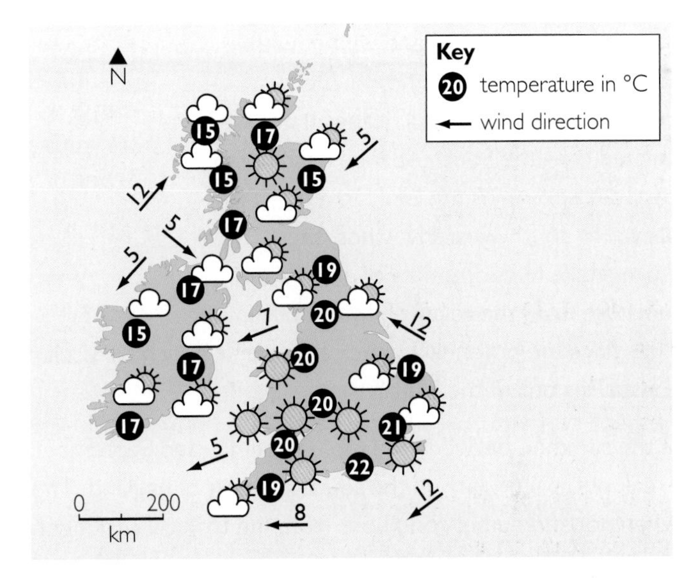

***Figure 2.30** A summer anticyclone over the British Isles, 13 May 1998 at noon* ▲

ACTIVITIES

1 Study Figures 2.29 and 2.30.
   a Describe the weather experienced over southern England on 28 January 1998.
   b Describe and try to account for the different weather conditions experienced over Scotland.
   c What evidence is there in Figure 2.30 that a large anticyclone has settled over the UK?
2 Study Figure 2.31. For this activity you will need a blank outline map of Europe.
   a Make a copy of the isobars from Figure 2.31 on to your outline map. Write the values of the isobars.
   b Draw the long wavy front that is positioned over the Atlantic.
   c Use an atlas to locate and name the locations listed in the table. For each location carefully draw the weather symbol and write the temperature alongside.
   d Write a paragraph giving a reasoned description of the weather associated with this anticyclone.

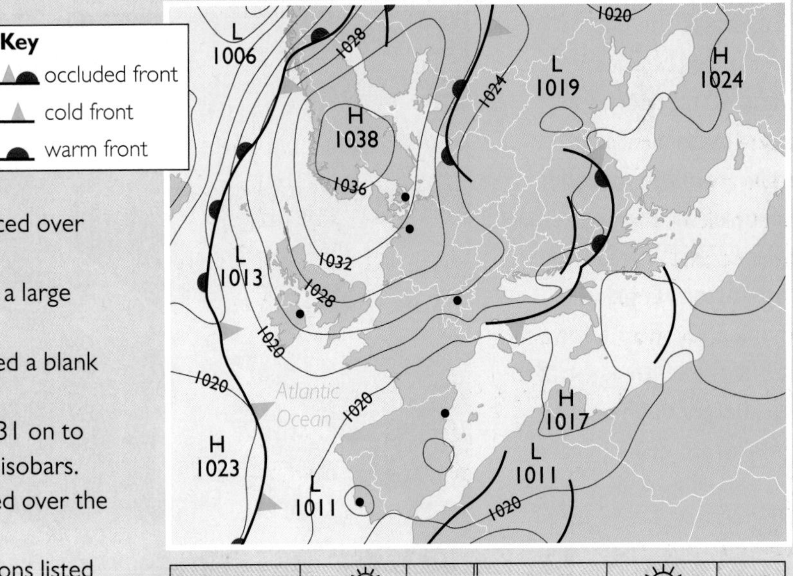

Barcelona		22°C	Zurich		18°C
Copenhagen		19°C	Hamburg		17°C
Dublin		19°C	Lisbon		28°C

***Figure 2.31** A summer anticyclone and its related weather, 30 May 2009* ▲

Storms in the British Isles

In this section you will learn about:
- the different types of storms that affect the British Isles
- the 'Great Storm' of 1987

Storms in the British Isles

Storms are extreme weather events. Stormy weather in the British Isles generally happens in the autumn and winter, with October being a prime month. Deep depressions and even hurricanes that have formed over the South Atlantic head north-eastwards towards the UK. Driven by the prevailing south-westerly winds, these storms take huge amounts of energy and moisture from the relatively warm ocean. They have the potential to pack a powerful punch particularly to western and northern coastal regions of the British Isles.

"We were the first lifeboat on the scene and were greeted by a 10-15 foot wall of water coming down the town, out of the harbour and pushing 30, maybe 50 cars in front of it. There were cars all around us at sea. There was debris everywhere, the air was thick with the stench of fuel. Then another storm came in as we arrived. Lightning was hitting all around us in a big thunderstorm. The rain was so heavy you couldn't see in front of you."

Rescuer Andrew Cameron on the Boscastle floods in 2004

In the summer, between the months of May and September, severe thunderstorms can break out, particularly in the south and east of England. These storms often develop when hot and humid conditions from the tropical continental air mass drift north from the Mediterranean.

Lightning strikes, hail and torrential rain can cause serious hazards to people and property. In June 1996, hailstones up to 50 mm fell across parts of southern Britain, damaging greenhouses and vehicles.

On average, the UK experiences between 35 and 40 tornadoes a year, mostly in central Britain from the midlands to central-southern England, East Anglia and southeastern England (Figure 2.32). Tornadoes in the UK are often associated with sharply defined cold fronts which quickly travel over relatively warm inland areas.

Figure 2.32 *An exception to the rule – while most tornadoes in the UK cause relatively little damage, the 1998 tornado that hit the Sussex town of Selsey caused over £10 million of damage* ▶

CASE STUDY

The 'Great Storm' of 1987

One of the most severe storms of recent times occurred during the early hours of 16 October 1987. A small but intense depression formed over the Bay of Biscay and moved rapidly northwards during 15 October. The storm intensified further as it crossed the English Channel and moved across southern England and the Midlands before reaching the Humber estuary (Figure 2.33). Warnings for heavy rain and strong winds had been issued during the evening, and at 1.35am the Met Office sent a warning to the Ministry of Defence fearing a major disaster that could require military assistance.

Winds gusted to over 100 mph as the storm tore across southeast England, killing eighteen people (Figure 2.34). 15 million trees were blown down, falling on houses, cars, railways and trains. Blocked roads stopped thousands of people from getting to work or school. Power lines were brought down, leaving thousands of homes without electricity. The scale of the damage was tremendous.

Short-term responses were largely focused on a massive clear up operation. Trees had to be cleared from roads, damaged houses had to be repaired and power lines re-connected. It took several days, and in some cases weeks before schools, businesses and the country's infrastructure were operating normally.

The Met Office carried out a review following some criticism that it had not predicted the sheer scale of the storm. Following the review, the Met Office improved the quantity and quality of its weather observations from ships, aircraft, buoys and satellites, particularly over the sea to the south and west of the UK. Computer models were refined to improve forecasting.

The Great Storm of 1987 was an exceptional event, with an occurrence interval of about once every 200 years – prior to 1987, the last recorded storm of such intensity was in 1703.

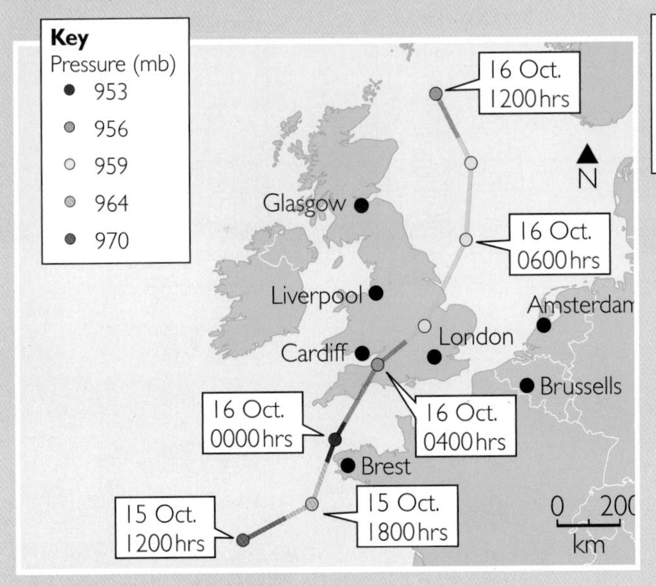

Figure 2.33 *The path of the Great Storm of 1987 as it moved across England* ▲

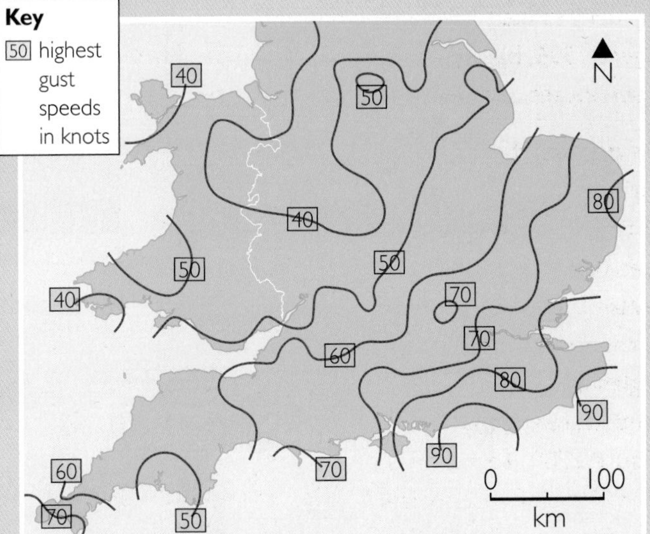

Figure 2.34 *The winds during the Great Storm of 1987 reached over 90 miles per hour in some places* ▲

ACTIVITIES

1 Study Figures 2.33 and 2.34 above.

a In relation to the storm's central track, where were the strongest winds? Support your answer with evidence from Figure 2.34.

b If the most damaging winds had been on the other side of the storm's central track, do you think the impact (and media attention!) would have been as severe? Explain your answer.

c How does Figure 2.34 illustrate the extreme difficulties in attempting to provide accurate forecasts of strong wind gusts?

2.7 Tropical climates

In this section you will learn about:

- the patterns of world climates
- the three low latitude climate zones (equatorial, tropical continental and tropical monsoon)
- the atmospheric circulation and major surface winds and pressure systems in the low latitude climate zones
- the tropical continental climate in West Africa

For many of us in the UK, the 'tropics' is a region that sounds very appealing, particularly during the dark cold days of winter! Images of tropical beaches with white sands and palm trees come to mind, or maybe dense colourful and noisy jungles. Hot sunny days and warm evenings typify our image of these low latitudes. How correct are we in these preconceptions?

Patterns of world climates

Figure 2.35 below shows the world's major climates. Notice that the British Isles lies within the temperate climate zone. Locate the Equator and the tropics of Cancer and Capricorn. Notice that there are three main climate zones between the tropics:

- equatorial – bordering the Equator, roughly 5 °N – 5 °S
- tropical continental – either side of the equatorial climate zone, most obviously in Africa
- tropical monsoon – some coastal areas particularly in South-East Asia

Notice that because of their high altitude, mountain ranges within the tropics, such as the Andes in South America, have their own climate zone.

Figure 2.35 *The world's major climate zones* ▼

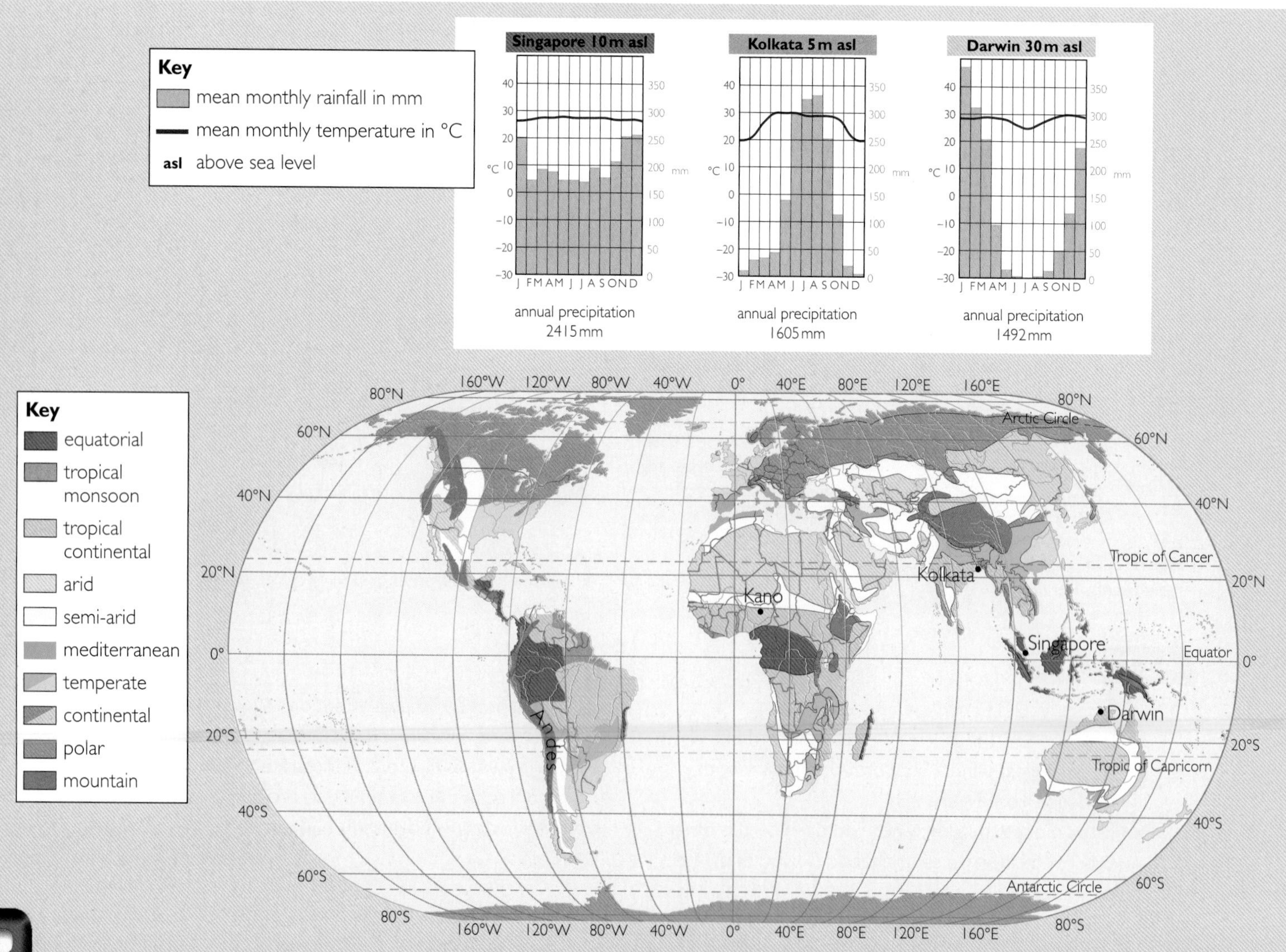

Tropical climates

As Figure 2.35 shows, the three tropical climate zones form a rather complex pattern. This is mainly because of the differences in the distribution of land and sea and the presence of warm and cold ocean currents.

The main driving force deciding the climate in the tropics is the position of the two Hadley cells which form part of the global atmospheric circulation system (Figure 2.15 page 63). In Figure 2.36 notice that a zone of rapidly rising air called the **inter-tropical convergence zone** (ITCZ) marks the boundary between the two Hadley cells.

The position of this boundary is largely determined by the position of the overhead sun. During the course of the year this moves between the tropics – the Tropic of Cancer in our summer and the Tropic of Capricorn in our winter. As the Hadley cells shift north and south so does the ITCZ. This explains why there is rain throughout much of the year in the equatorial zone but a distinct wet and dry season in the tropical continental zone. You can see this for yourself by comparing the climate graphs for Singapore and Darwin in Figure 2.35.

When the ITCZ is in the opposite hemisphere, the sub-tropical high-pressure zone associated with trade winds stretches towards the Equator. Within this broad zone, distinct **sub-tropical anticyclones** form, bringing dry and warm conditions and accounting for the 'dry season' in tropical continental regions.

Monsoon climates are broadly similar to tropical continental climates in that they have a wet and a dry season (see the climate graph for Kolkata in Figure 2.35). The wet season is sometimes referred to as the 'monsoon' and it can bring huge amounts of rain. This is caused by the prevailing winds which blow off the warm oceans towards the land, causing torrential rainfall for several days in countries such as India and Bangladesh (Figure 2.37).

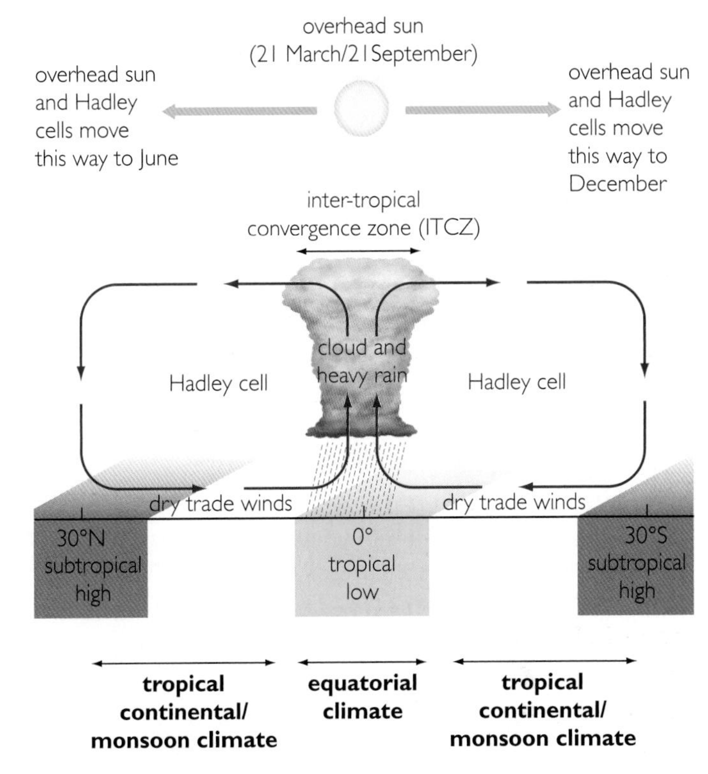

Figure 2.36 *Weather conditions associated with the Hadley cell* ▲

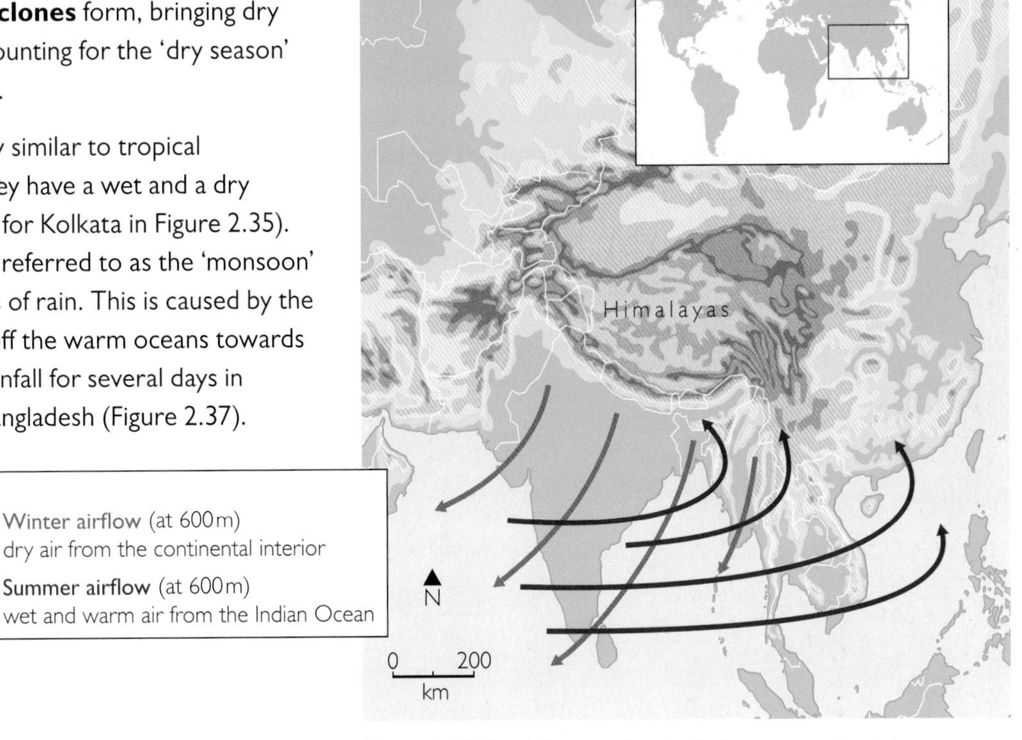

Figure 2.37 *The airflow associated with the monsoon climate in South-East Asia* ▲

CASE STUDY

Tropical climate case study: tropical continental climate in West Africa

The AQA geography specification requires that you study one of the three tropical climates in detail. We will look at the tropical continental climate, which is both varied (interesting!) and relatively straightforward to understand. You can, of course, choose to study one of the other tropical climates if you wish.

Look at Figure 2.38 which shows the range of the equatorial and tropical continental climates in Africa. Figure 2.39 is a climate graph for the Nigerian city of Kano, which has a tropical continental climate. Notice the following features on the graph:

- There is a distinct wet and dry season. The wet season (May to September) is a result of the northward movement of the ITCZ as the overhead sun moves towards the Tropic of Cancer (see Figure 2.40). In September, when the overhead sun migrates south, the ITCZ rain belt moves with it leaving Kano to experience a long dry season from October to April.
- Temperatures are generally high and fluctuate slightly. The sun is high in the sky for much of the year, leading to generally high temperatures. However, when the sun is overhead in the southern hemisphere (December), temperatures do dip slightly in Kano because the sun is at a relatively low angle. Cloud cover during the wet season also cuts down incoming radiation, leading to slightly lower temperatures.
- The range between maximum and minimum temperatures fluctuates during the year. During the dry season, the lack of cloud means that heat readily escapes during the night. This explains the relatively low minimum temperatures in December and January. During the cloudier wet season, there is a smaller range between the maximum and minimum temperatures.

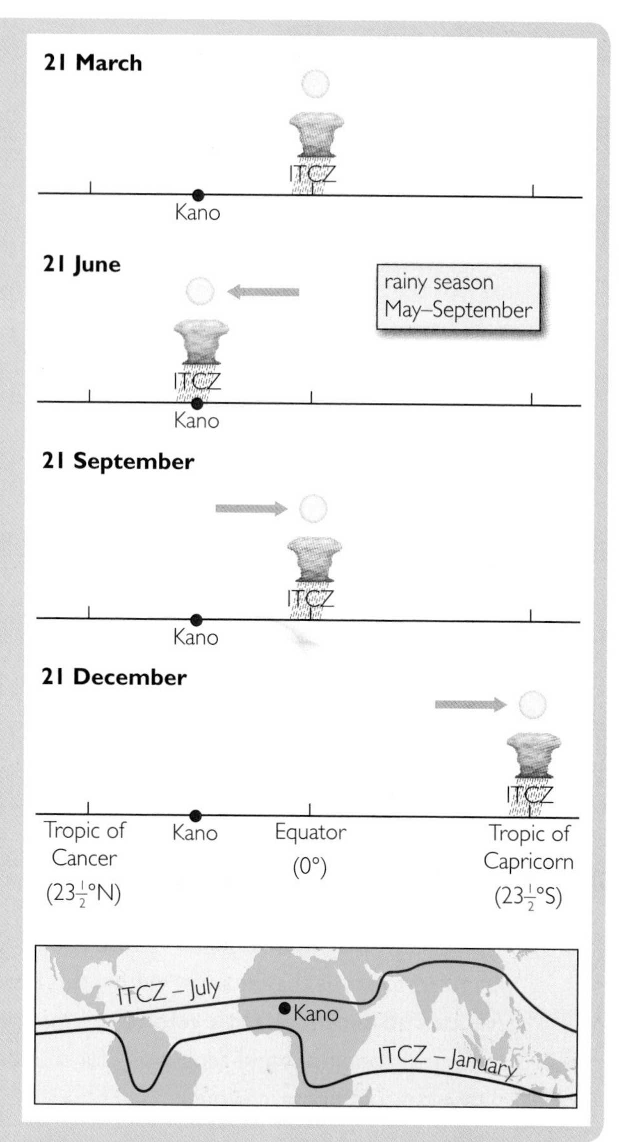

Figure 2.40 *The illustrations show the inter-tropical convergence zone (ITCZ) migrating over Africa. The map shows the average position of the ITCZ in January and July* ▲

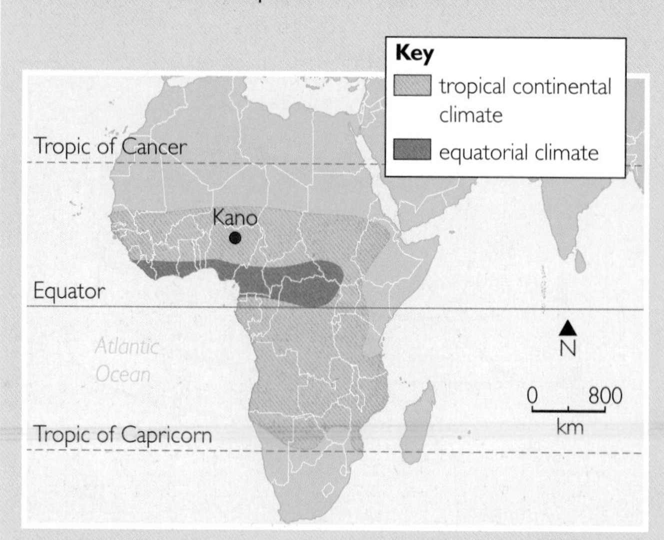

Figure 2.38 *The climate zones in Africa* ▲

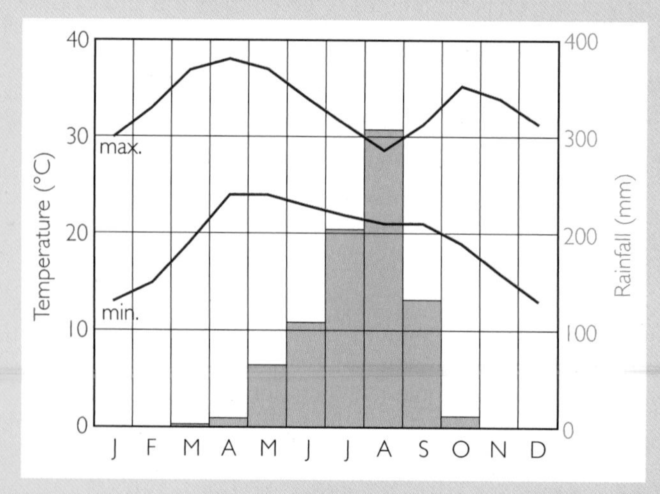

Figure 2.39 *This climate graph shows the rainfall and average temperatures (high and low) of Kano, Nigeria* ▲

During the wet season rain usually appears as heavy showers, often with thunder. The high temperatures over the land create unstable rising air which combines with the low pressure of the ITCZ to trigger torrential downpours. Flooding is a real threat at this time.

During the dry season a strong, dry wind often blows from the north. This is the West African trade wind, otherwise known as the Harmattan. Occasionally the Harmattan brings dusty conditions to northern Nigeria, having come from over the desert regions to the north (Figure 2.41).

Figure 2.41 *Passengers stand outside of their taxi waiting for a sandstorm caused by the Harmattan wind to pass* ▼

ACTIVITIES

1. Using Figure 2.35, describe the distribution of the three low latitude climate zones.
2. Use Figures 2.35 and 2.36 to explain the latitudinal position of the equatorial climate.
3. What are the characteristics of the tropical monsoon climate?
4. Use Figure 2.37 to help you explain why parts of South-East Asia experience very heavy rainfall during the summer.
5. Complete a detailed study of the tropical continental climate experienced by Kano, Nigeria. Use the text and figures in this section together with photos and other information from the Internet. Present your study in a format of your choice – for example, it could be an information poster, an electronic report, or a PowerPoint presentation. Your study should include:
   - the location of Kano in relation to the tropical continental climate zone and the ITCZ
   - a description of the climate in Kano (wet and dry season, Harmattan winds, etc)
   - the cause of the tropical climate (migration of the ITCZ)
   - the impact of the climate on people and human activities, for example agriculture. Consider how life differs between the dry and wet seasons.

Tropical cyclones

In this section you will learn about:

- the characteristics, formation and development of tropical cyclones
- the hazards associated with tropical cyclones when they make landfall
- the approaches adopted to reduce these hazards
- the contrasting cases of two tropical cyclones: Hurricane Katrina (2005) and Cyclone Nargis (2008)

The giant swirl of cloud in Figure 2.42 is instantly recognisable as a hurricane or tropical cyclone. But this is not just any hurricane. It is Hurricane Katrina, the storm that caused more economic damage in the USA than any previous natural disaster and killed over 1800 people.

Despite the awesome power of a hurricane and the destruction that it can cause, it is a perfectly natural weather event and has to be expected in certain latitudes at certain times of the year. Indeed, hurricanes play an important role in the redistribution of heat across the world, transferring warm conditions from the tropics towards the poles. Without hurricanes, the weather conditions and climate of the world would be very different.

Figure 2.42 *A satellite image of Hurricane Katrina* ▲

What are the characteristics of tropical cyclones?

A **tropical cyclone** is a huge and extremely violent tropical storm that can extend to about 500 km in diameter. Also known as **hurricanes** (in the North Atlantic) and **typhoons** (in South-East Asia), tropical cyclones can cause extensive damage and loss of life to coastal regions in many parts of the tropics. By definition, a tropical cyclone must have average wind speeds in excess of 120 kph (75 mph) – hurricane force!

Look at Figure 2.43. It shows a cross-section of a tropical cyclone. Notice that there is a degree of symmetry around its central point or **'eye'**. The most powerful and damaging part of a tropical cyclone is the menacing bank of cloud that rings the central eye. This is called the **eye wall** and people who have experienced a cyclone say that it 'hits like a train'! As you can see from Figure 2.43, cloud and rain extend in a series of waves well beyond the eye wall. Within a tropical cyclone, **tornadoes** are commonly formed. Their highly localised nature makes them difficult to predict and their impact can be highly destructive.

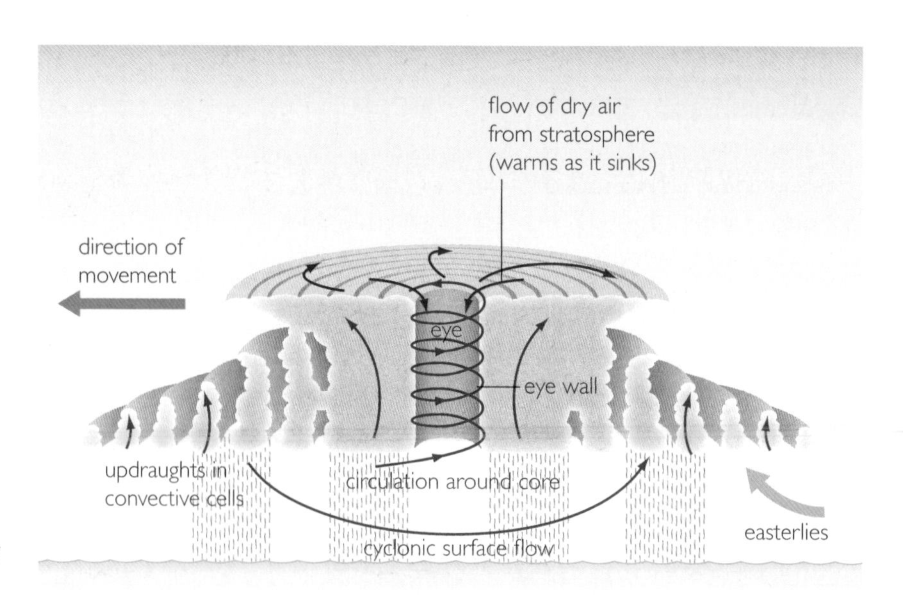

Figure 2.43 *A cross-section of a typical hurricane over the North Atlantic* ▶

Where are tropical cyclones formed?

The vast majority of tropical cyclones are formed in the tropics (see Figure 2.44). This is because they will only form under certain conditions:

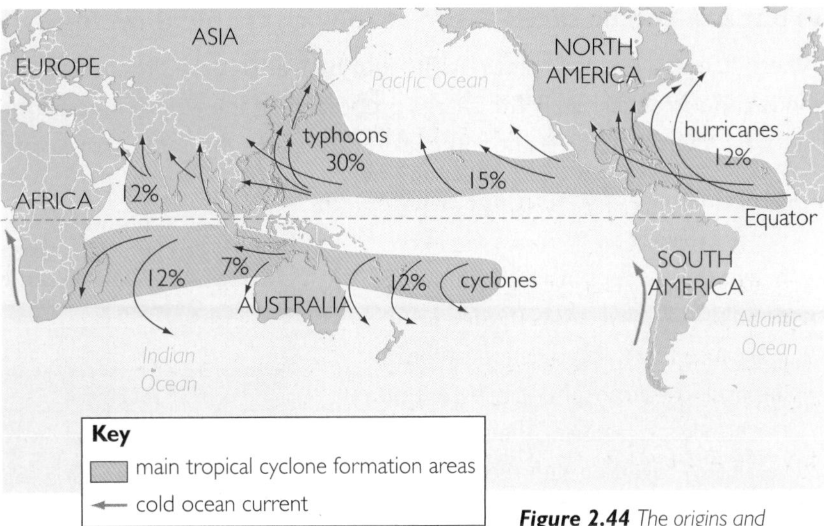

Figure 2.44 *The origins and common paths of tropical cyclones. The numbers represent the annual percentage of cyclones in each region.* ▲

- **Oceans** – tropical cyclones get their moisture and energy from the oceans. They form and then continue to develop over ocean areas and peter out when they reach land.
- **High temperatures** – a sea surface temperature in excess of 26°C is needed for tropical cyclones to form. This explains why they are formed in low latitudes during the summer, when temperatures are at their highest.
- **Atmospheric instability** – tropical cyclones are most likely to form in regions of intense atmospheric instability, where warm air is being forced to rise. The ITCZ, where two limbs of the Hadley cell converge, is a perfect spawning ground for tropical cyclones.

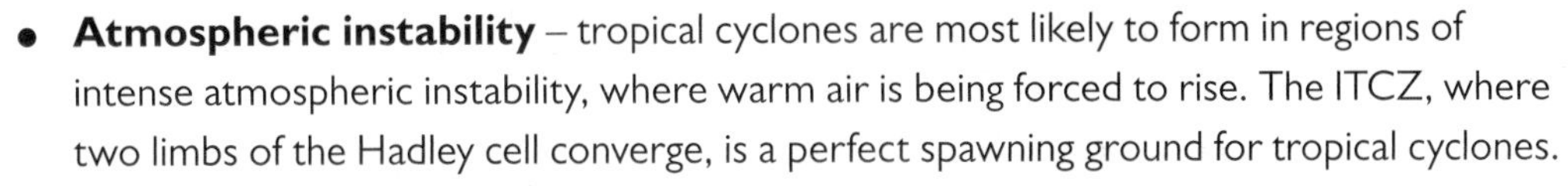

- **Rotation of the Earth** – a certain amount of 'spin' is needed to initiate the characteristic rotating motion of a tropical cyclone. As we saw in earlier sections, the influence of the Earth's rotation on surface phenomena is called the Coriolis effect. This increases with distance away from the Equator and explains why hurricanes do not usually form in the region 5 °N – 5 °S of the Equator.
- **Uniform wind direction at all levels** – winds from different directions at altitude prevent a tropical cyclone from gaining height and intensity. The vertical development is effectively 'sheared off' by the multi-directional winds.

Once a tropical cyclone has started to form it will soon develop into a distinct and clearly defined spinning storm. Warm moist air rises rapidly in its centre replaced by air drawn in at the surface (Figure 2.45). A central vortex develops as more and more air is drawn in and rises. The very centre of the storm is often characterised by a column of dry sinking air. This creates the familiar 'eye' of the storm.

Figure 2.45 *The early stage in the formation of a tropical cyclone* ▼

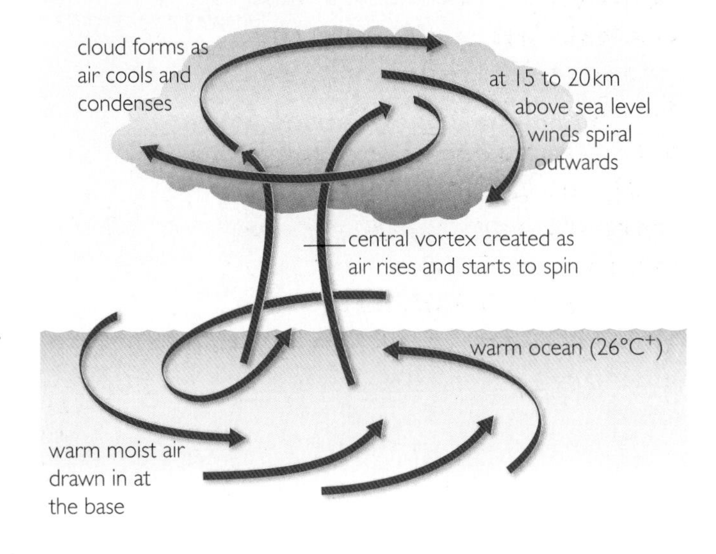

As the air rises, it cools rapidly – this leads to condensation and the formation of towering cumulonimbus thunderstorm clouds. Sometimes, a number of isolated thunderstorms will join together to make a single giant cyclone. The latent heat that is released during condensation effectively powers the storm.

Whilst moving over the oceans, driven by the prevailing winds, a tropical cyclone will continue to grow and develop. Only when it reaches land and the supply of energy and moisture is cut off will the storm start to decay. Should it move back out over the ocean, it will become re-invigorated.

What are the hazards associated with tropical cyclones?

When a tropical cyclone makes landfall it brings with it a deadly cocktail of high seas, strong winds and torrential rain. With a large proportion of the world's population living close to the coast, there is a significant potential for loss of life and damage to property.

- **Storm surge**. This is a surge of high water typically up to about 3 m in height that sweeps inland from the sea, flooding low-lying areas (Figure 2.46). It is caused by a combination of the intense, low atmospheric pressure of the hurricane (allowing the sea to rise vertically) together with the powerful driving surface winds. Most of the people who die during tropical cyclones do so as a result of storm surges.
- **Strong winds**. By definition, a tropical cyclone packs a powerful punch with average wind speeds in excess of 120 kph (75 mph). Wind gusts of over 250 kph have been recorded at the eye wall. The strong winds are capable of causing significant damage and disruption, for example, by tearing off roofs, breaking windows and damaging communication and transport networks.
- **Torrential rain and flooding**. The warm humid air associated with a tropical cyclone produces very large amounts of rainfall, often in excess of 200 mm in just a few hours. This can cause short-term flash flooding, especially in urban areas, as well as somewhat slower river flooding as the storm moves inland.

Saffir-Simpson Hurricane Scale				
Category	**Wind speed**		**Storm surge**	
	mph	(km/h)	ft	(m)
Five	≥156	(≥250)	>18	(>5.5)
Four	131–155	(210–249)	13–18	(4.0–5.5)
Three	111–130	(178–209)	9–12	(2.7–3.7)
Two	96–110	(154–177)	6–8	(1.8–2.4)
One	74–95	(119–153)	4–5	(1.2–1.5)
Additional classifications				
Tropical storm	39–73	(63–117)	0–3	(0–0.9)
Tropical depression	0–38	(0–62)	0	(0)

The Saffir-Simpson scale allows the intensity of different storms to be compared. It is of limited value in assessing the impact of a particular hurricane because it does not take into account the area affected by a storm. It also does not take into account rainfall. So, a low category hurricane that hits a densely populated urban area can be far more damaging than a high category storm that makes landfall in a more remote region.

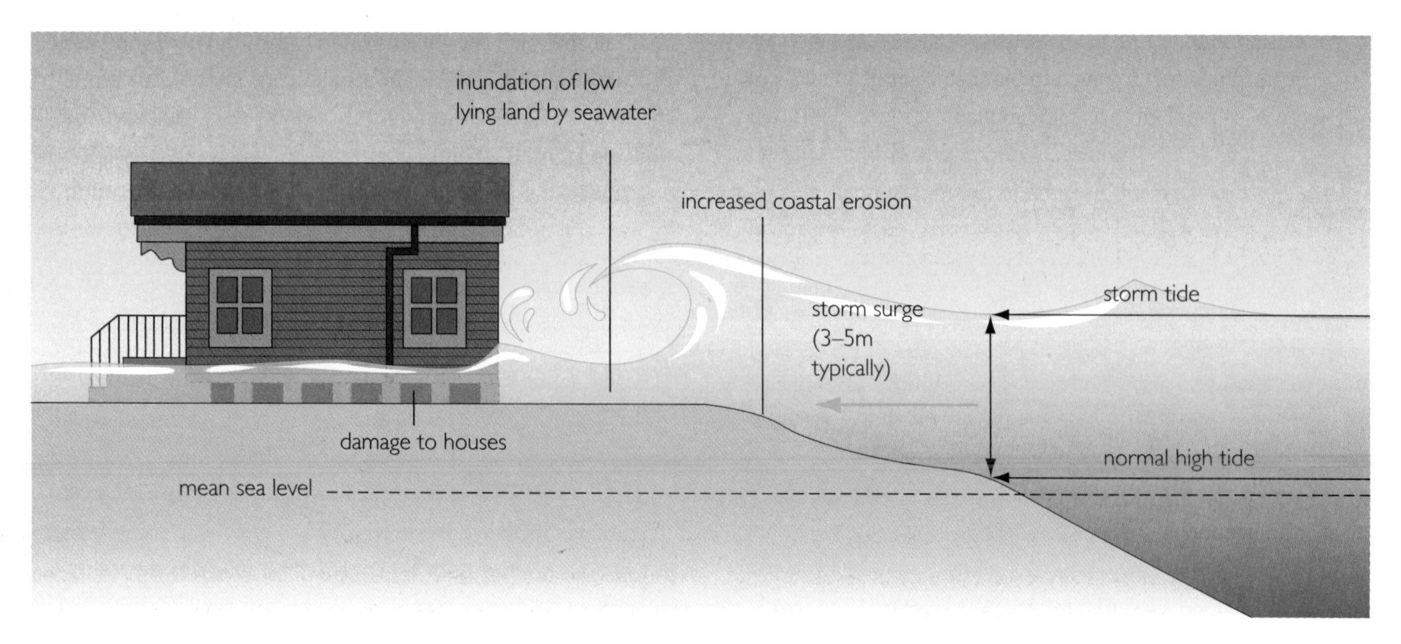

Figure 2.46 *In addition to loss of life, storm surges swamp agricultural land with saltwater and debris, pollute freshwater supplies and destroy housing and infrastructure. The increase in coastal erosion also can lead to the undermining of buildings and highways.* ▼

Reducing the hazards

There are several approaches that can be adopted to reduce the hazards associated with tropical cyclones. These range from forecasting, prediction and evacuation to constructing barriers, shelters and defences.

Monitoring and forecasting

Today we use satellites and other technology, such as radar, to identify and track tropical cyclones. Computer models based on historical data allow scientists to predict the likely course or 'track' of an individual storm. In the USA and the Caribbean, a 'hurricane watch' is issued for areas where hurricane force winds are a serious possibility within 36 hours. This is upgraded to a 'hurricane warning' when the hurricane is expected to reach land in the next 24 hours or less.

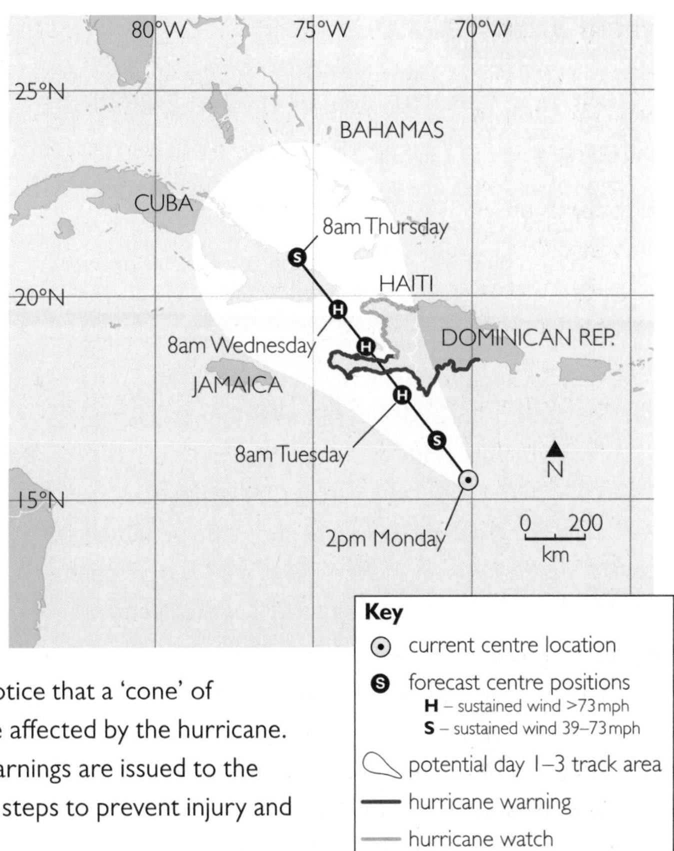

Figure 2.47 *The track of Hurricane Gustav in 2008* ▲

Look at Figure 2.47. It shows the early track of Hurricane Gustav as it passed through the Caribbean in August 2008. Notice that a 'cone' of prediction has been plotted to show the area most likely to be affected by the hurricane. Notice also that warnings have been issued for Haiti. These warnings are issued to the general public and to government authorities so they can take steps to prevent injury and loss of life.

Preparedness

Education and preparedness are an effective way to reduce the impact of a tropical cyclone. If the people at risk know what the dangers are they can take steps to avoid them. This can involve making minor structural improvements to buildings (e.g. strong doors and windows), preparing emergency supplies and planning evacuation routes. In Florida, evacuation routes and cyclone shelters are clearly signposted.

Land use planning

In some coastal regions planning regulations have been used to reduce the vulnerability of people and property to the effects of tropical cyclones. Often this involves allowing only low-value land uses (e.g. recreation) to occupy the coastal strip. In parts of northeastern Florida, coastal properties are raised above the ground on stilts and have non-residential functions on the ground floor, such as a garage or storage area.

Structural approaches

The sheer scale of a tropical cyclone means that there is little that can be done to physically reduce its impact without committing a huge amount of money. It is, however, possible to offer some protection to coastal areas from storm surges by soft engineering schemes (planting trees and building up beaches) or hard engineering, such as constructing sea walls (Figure 2.48).

Figure 2.48 *The sea wall flood defence at Corpus Christi, Texas* ▲

CASE STUDY

Reducing the impacts of tropical cyclones in Bangladesh

Tropical cyclones are a real threat in Bangladesh. In 1970, the Bhola Cyclone (so named after the island that was devastated by the storm) killed up to 500 000 people. In 1991, some 138 000 people were killed by another powerful cyclone. In response to these deadly events, a number of initiatives were introduced:

- tropical cyclones are monitored, with warnings issued over the television, radio and Internet
- in remote village communities, designated wardens help to spread the warnings and guide people to safety
- concrete cyclone shelters (Figure 2.49) have been constructed to provide a safe refuge for people threatened by flooding. These shelters, often raised up on stilts, serve as schools or community centres at other times.

In 2003, Bangladesh was hit by another powerful cyclone.

Figure 2.49 *A cyclone shelter in Bangladesh* ▲

Despite killing 3000 people, Cyclone Sidr had a greatly reduced impact due to the initiatives adopted by the government.

CASE STUDY

Hurricane Katrina and Cyclone Nargis — more alike than different?

In 2005, Hurricane Katrina (Category 4) struck a densely populated urban area in the USA, a relatively wealthy country. In contrast, in 2008 Cyclone Nargis (Category 3) struck a largely rural region in Myanmar, a relatively poor country in South-East Asia. Many of the impacts and responses to the cyclones clearly reflected the differences in these two areas.

Impacts

Hurricane Katrina is the costliest hurricane ever recorded in the USA and the sixth most powerful to hit the country. It brought tremendous devastation to New Orleans, displacing over a million people from their homes and flooding 80% of the city. Many coastal communities were devastated as the storm surge swept almost 20 km inland from the coast. In total, over 1800 people were killed and many thousands were injured. Economic damage totalled some $90bn.

Three years later, Cyclone Nargis brought widespread destruction to the low-lying Irrawaddy delta region

Figure 2.50 *Flooding of the Irrawaddy delta region in Myanmar following Cyclone Nargis* ▲

of Myanmar where 95% of all homes were destroyed (Figure 2.50). An estimated 2 to 3 million people were displaced and 140 000 people killed. Economic losses estimated to be about $10bn were far lower than those in the USA.

Responses

The short-term responses to the two events were broadly similar. The first priority was search and rescue carried out by individuals within their local communities. Those who survived needed access to medical care, fresh water, food and shelter. In New Orleans, despite the fact that 80% of the population had been evacuated, some 60 000 people had remained behind – many keen to secure their homes.

In Myanmar, help took a long time to arrive. For several days aid organisations and foreign governments were not even allowed into the country to provide assistance. Many remote communities were completely cut off by the floodwaters and had to fend for themselves in appalling conditions.

Figure 2.51 *The construction of new flood gates in New Orleans* ▲

In the long-term, there was considerable government support for the people of New Orleans. Damaged properties were repaired or rebuilt, services such as electricity and water were restored and quite quickly things got back to a sense of normality. Repairs and improvements were made to the levees to prevent future breaching (Figure 2.51). However, four years later, several thousand people were still being accommodated in temporary trailer parks.

In Myanmar, improvements were very slow and piecemeal. Many communities had experienced almost total destruction of homes, services and transport networks. Fields of crops were ruined by salt-water and sewage pollution, affecting peoples' livelihoods and food supply. The government was simply overwhelmed by the scale of the task and it was often down to the local communities to do what they could to help themselves (Figure 2.52).

Figure 2.52 *Residents rebuild their homes after Cyclone Nargis* ▲

ACTIVITIES

1 Conduct your own enquiry to make a comparison between Hurricane Katrina and Cyclone Nargis. Use the information in this section together with additional material from articles, books or the Internet. Your research should:
   - Use development statistics to compare the USA and Myanmar.
   - Compare and contrast the short-term and longer-term responses. Consider the different scales of response – individuals and local communities, government and international responses (i.e., foreign countries, non-governmental agencies, and aid agencies).

2 Having completed your study of the two cyclones, answer the following essay question: How do variations in an area's income level affect the social, economic and environmental impact of a cyclone?

Internet research

Miami is one of America's wealthiest and fastest-growing cities, yet it is in the firing line for hurricanes. Using the Internet, do your own research on hurricanes in Miami. Be sure to look at why Miami is at risk from hurricanes and what is being done to reduce the impact of future hurricanes.

Local urban climates

In this section you will learn about:

- the concept of urban climates as being distinctly unique
- the urban 'heat island' effect
- the impact of urban areas on precipitation
- air quality issues and pollution in urban areas
- air movement in urban areas

Beijing Olympics: The Battle Against Pollution

Key facts on China's fight to deliver clear blue skies for the Olympics

£8.6 billion: amount spent on improving Beijing's air quality and environment
28 million: trees planted in and around Beijing in the last two years
1.3 million: cars taken off Beijing's roads since 20 July
60000: workers laid off by a steel factory, Beijing's biggest polluter
200: number of the worst polluting factories in Beijing that have been re-located ahead of the Olympics
13: Beijing's position on the World Bank's list of the most polluted cities on the planet

Adapted from an online article in *The Telegraph*, 2 August 2008

Figure 2.53 *Prior to the 2008 Olympic Games, citizens in Beijing wore face masks to protect themselves against the high levels of air pollution* ▲

In the run-up to the Beijing Olympics of 2008, the Chinese government was extremely concerned about the levels of pollution in Beijing. People wearing face masks was hardly the image of a modern and developing country they wanted to portray to the world (Figure 2.53). So the government embarked on an ambitious and aggressive campaign to clean up pollution in the city and improve the air quality.

Up to 300000 heavy-emission vehicles, mostly lorries, were banned from the city centre and up to 2 million vehicles were prevented from entering the city. Vehicle emission standards were imposed, construction projects were stopped and coal was replaced by natural gas as the main source of generating power. The campaign was effective and the athletes and spectators were, for the most part, able to enjoy a largely smog-free Games.

High levels of pollution, and the dirty smog associated with it, is just one trait that characterises large cities around the world, giving them a distinctive urban climate.

What is an urban climate?

Weather conditions and, over a long period of time, climate can vary a great deal between a large urban area and the surrounding countryside. Look at Figure 2.54 to see some of these differences. You can now see why it is extremely unlikely that you will win money by betting on snow falling in central London on Christmas Day! As Figure 2.54 shows, urban areas have their own distinctive climate.

Weather feature	Urban areas compared to rural areas
Length of sunshine	5-15% less
Annual mean temperature	0.5-1.0°C higher
Winter maximum temperatures	1-2°C higher
Occurrence of frosts	2-3 weeks fewer
Relative humidity in winter	2% lower
Relative humidity in summer	8-10% lower
Total precipitation	5-10% more
Number of days with rain	10% more
Number of days with snow	14% fewer
Cloud cover	5-10% more
Occurrence of fog in winter	100% more
Amount of condensation nuclei	X10 more

Figure 2.54 *Some of the differences in the climates of urban and rural areas* ▲

There are several reasons for these differences:

- Urban areas have extensive dark surfaces (roads, roofs, etc) that absorb heat during the day and release it slowly at night.
- People and human activities such as transport, industry and power stations generate their own heat.
- The lack of vegetation in urban areas reduces the amount of moisture in the air. This is intensified by drains and sewers which remove surface water quickly.
- The presence of tall buildings and narrow streets and the patterns of their arrangement interferes with airflow, causing turbulence and gustiness.
- Pollution from vehicles and industry, together with human dust, increases the density of condensation nuclei, increasing the likelihood of cloud and fog.

Hot in the city: the urban heat island

Research has shown that noticeable differences in temperature exist between urban and rural areas. For example, the temperature difference between London and the surrounding area can be as high as 6°C or more (Figure 2.56). It is this 'dome' of warm air that develops over urban areas that has led to the term urban heat island. Figure 2.57 shows the urban heat island for the city of Chester in northwest England.

The higher temperatures in urban areas also account for other weather and climatic characteristics. For example, Kew in central London has an average of 72 more frost-free days than rural Wisley just 32 km away. In Berlin, there is a 20% chance that snow in the countryside will occur as rain or sleet in the city centre because of its higher temperatures.

The differences in temperature and related weather conditions are at their most extreme under stable anticyclonic conditions. Under clear skies, heat loss from the ground will be at its most intense. The darker surfaces in cities hold onto heat and release it slowly compared to natural surfaces in the countryside. In calm conditions, there will be little mixing of air, heightening contrasts between urban and rural environments.

Figure 2.55 *Artificial surfaces and glass-fronted buildings in Canary Wharf, London contribute to higher urban temperatures* ▲

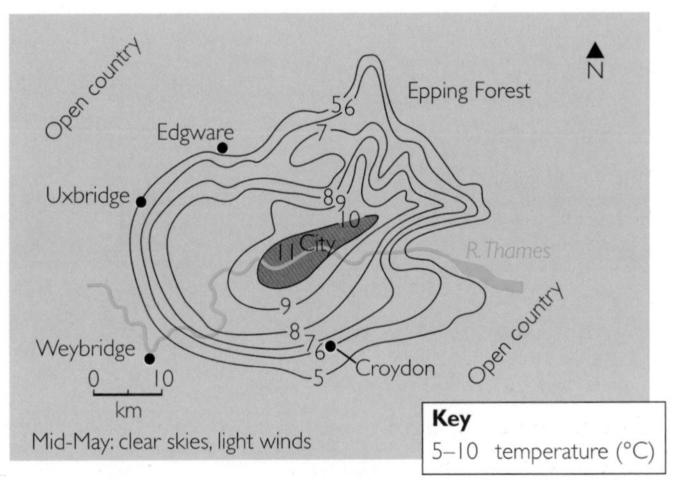

Figure 2.56 *London's 'heat island'* ▲

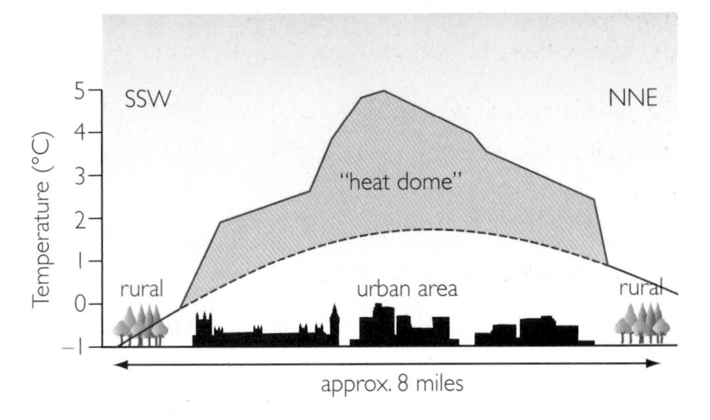

Figure 2.57 *A heat island in Chester, Cheshire* ▲

Variations in precipitation

Look back at Figure 2.54. Notice that urban areas have more precipitation and more days of rain than rural areas. Notice also that there is more cloud and, consequently, less sunshine in urban areas.

The main reason for these trends is that urban areas tend to make more **condensation nuclei**, the tiny particles that are essential for water droplets to form. Despite the fact that humidity levels are lower in urban areas (due to less vegetation and more drains and sewers), the greater concentration of condensation nuclei means that there is more cloud and rain as well as fog and mist (Figure 2.58).

There are several sources of urban condensation nuclei, exhaust from cars and lorries, industrial pollution and construction, for example. Dust created by human activity also adds to the general mix. In the past, the burning of coal in homes and industry created a huge amount of coal dust that was pumped into the atmosphere over urban areas.

The heat island effect, especially in the summer, can lead to convective activity (showers, thunderstorms, hail, etc.) over large cities. During unstable atmospheric conditions rapidly rising air may trigger periods of heavy rainfall and even thunderstorms. Cities with tall buildings can also have a slight orographic effect (similar to hills), causing air to rise. Nurnberg, Germany has recorded 14% more thunderstorms than the surrounding countryside, and in St Louis, in the United States, there are 20-30% more thunderstorms downwind of its industrial zone than in the surrounding rural area.

As Mumbai, India expanded rapidly in the 1960s, the city's industrial region had 15% more rain than the surrounding countryside. It might be reasonable to assume that similar trends may well exist today in cities that are developing rapidly, for example, in China and India, where levels of atmospheric pollution are high. However, in many other cities where pollution levels have been declining, there may well be a reduced effect.

Figure 2.58 *Early morning mist over London* ▲

Urban air pollution and health

In the past London was renowned for its 'pea soupers', the name given to the city's winter fogs. Dense cold air associated with anticyclones trapped pollutants, making a pollution dome (Figure 2.59). This situation often reflected an atmospheric condition called a **temperature inversion**. This is when instead of the usual situation of air becoming colder with altitude, the situation is reversed – the temperature is 'inverted' – so a layer of warm air sits above the cooler air at the ground.

- **Particulates** are very small particles, such as of dust or soot, which are given off when oil, gasoline, and other fuels are burned. They can remain in the atmosphere for long periods, where they are a major part of air pollution and smog.

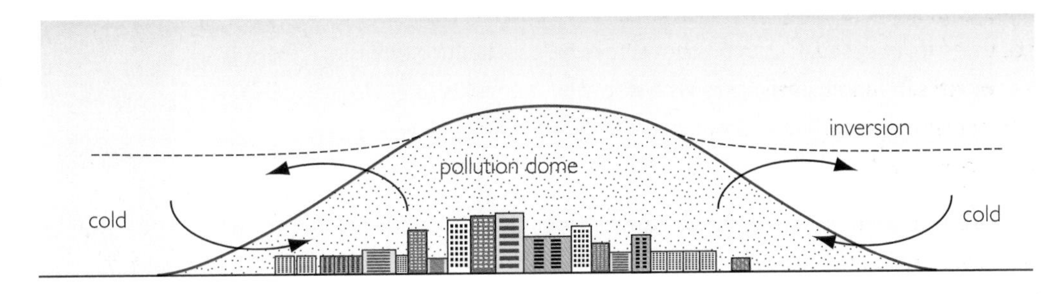

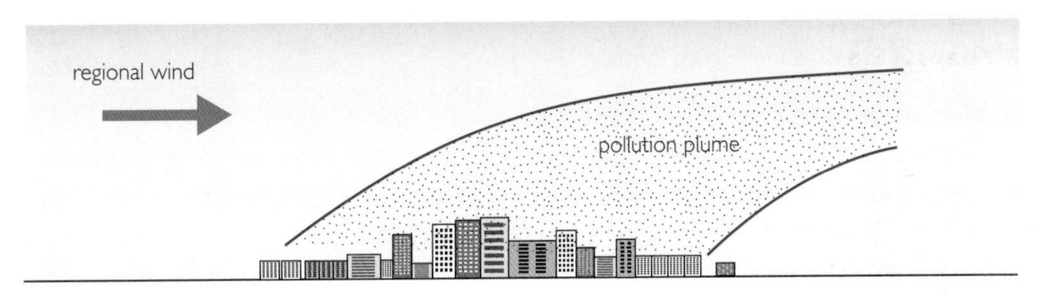

Figure 2.59 *The formation of a pollution dome* ▲

In December 1952, London experienced a dense fog mixed with particulates (primarily coal dust), creating what is known as **smog**. With visibility down to less than 10 metres, day turned to night. The effect of the highly polluted fog on health was profound. During the next three months, 12 000 more deaths were recorded than during the same period in the previous year. Most of these people died from chest infections, almost certainly linked to the December smog.

In addition to particulates, gases also add to pollution levels in towns and cities. Coal-fired power stations produce huge amounts of gas, including sulphur dioxide, that can lead to acid rain. Some 38% of Chinese cities are affected by acid rain associated with the burning of coal.

The burning of petrol and diesel in cars and lorries release gases including nitrous oxides and carbon monoxide. Not only are these primary gases harmful to human health, but they can also react with sunlight to form a nasty cocktail of secondary gases. This forms a type of pollution called **photochemical smog**. It is known to make some medical conditions, such as asthma and bronchitis worse – some of the pollutants have also been linked to cancer.

Throughout the world many cities are reported to have seriously high, and dangerous, levels of air pollution. The Bangladesh capital Dhaka has some of the worst incidents of air pollution in the world. This is due to the large number of poorly maintained cars along with a largely unregulated industrial zone close to the city centre. Levels of sulphur dioxide are extremely high, causing acidic precipitation in the form of rain and smog, particularly in the winter.

Did you know?

The Chumash tribe of Native American's referred to the area that forms present-day Los Angeles as the 'valley of smoke'. This was because smoke from open fires had a tendency to hang over the valley rather than spreading out. It should come as no surprise, then, that modern day Los Angeles, with its 'smoke' from vehicles and industry, has high levels of air pollution.

Reducing air pollution in cities

Reducing air pollution in cities is a major challenge for the future, particularly in countries experiencing rapid industrial and economic growth. Strategies usually involve a mixture of technological innovations aimed at reducing emissions, land use planning, vehicle restrictions and legislation.

Technological innovations – Fitting devices to exhausts and industrial chimneys can reduce gas and particulate emissions. One such device is an electrostatic precipitator, which uses an electrical charge to remove particles from gases. Catalytic convertors help to reduce emissions from car exhausts. Increasingly, cars with non-polluting fuels (e.g. electric cars) are being developed as alternatives to those requiring hydrocarbons.

Land use planning

- Urban planning can help to reduce levels of urban pollution. In Curitiba, Brazil, zones for heavy industry have been established outside the city centre and away from populated areas. To reduce the amount of travel – and therefore cut vehicle exhaust pollution – there has been a deliberate policy of mixing low-polluting land uses such as housing, light industry and retailing (Figure 2.60).

Figure 2.60 *A mixture of low-polluting land uses in Curitiba, Brazil* ▲

Vehicle restrictions

- In order to reduce the number of vehicles in a city's centre, a number of cities operate road pricing or, as in London, congestion charging (a daily charge for driving into parts of a city centre).
- In Singapore, vehicle ownership is heavily taxed and motorists are charged electronically using an in-car smart card for driving into the city centre.
- In some cities, registration number plates are used to restrict cars entering city centres, for example, even numbers on some days and odd numbers on others.

Legislation

- Many countries have adopted air quality control legislation, setting minimum levels of emissions and air quality.
- In the UK, air quality standards are set by the Department for Environment, Food and Rural Affairs (DEFRA). Monitoring is widespread and air quality warnings are issued when pollution levels are expected to be dangerously high. This is most likely during the summer when photochemical smog may occur.
- In the USA, since the 1960s there have been a series of Clean Air Acts. Studies conducted between 1970 and 2006 show that they have been successful in reducing the levels of a number of pollutants. For example, particulate emissions fell by 80% and sulphur dioxide emissions fell by 50%. Despite these successes, the Environment Protection Agency (EPA) recognises the need to reduce levels of photochemical smog in many of the country's large cities.

Internet research

Barcelona's Heat Island

The aim of this research activity is to produce a short case study of Barcelona's heat island. Your study should include maps and diagrams.

1. Access a map of Barcelona's heat island at http://geographyfieldwork.com/BarcelonaHeatIsland.htm . This map is dated and timed at 5 February 1986 (22h 35m). To identify specific areas, use Google Maps to find a map centred on Barcelona. The terrain map is probably the best.
   - **a** Describe the pattern of temperatures (isotherms). Be sure to name locations in your description.
   - **b** What evidence is there of a heat island over Barcelona?
   - **c** Using evidence from the map, suggest reasons for the temperature patterns you have described. Consider variations in relief and river valleys.
2. Using the information you have gathered, write a summary paragraph or two on Barcelona's heat island.

Urban winds

Tall buildings can greatly interfere with airflow – they disturb the lower layers of the atmosphere, slowing the overall speed of the wind over an urban area (Figure 2.61). On the ground the wind can become gusty. You have probably experienced this for yourself when walking through a large town or city. You may even have seen small vortexes developing like mini-tornadoes whisking litter up into the air.

Building shape, spacing and orientation are important controls on wind speed and direction. (Figure 2.62). Narrow streets with tall buildings either side can produce an 'urban canyon' with winds being funnelled along the street. Approaching the centre of a CBD (central business district) where the streets become progressively narrower, wind speeds may increase as airflow is effectively 'squeezed' – this is called the venturi effect.

Figure 2.61 *In urban areas, airflow is interrupted by tall buildings* ▼

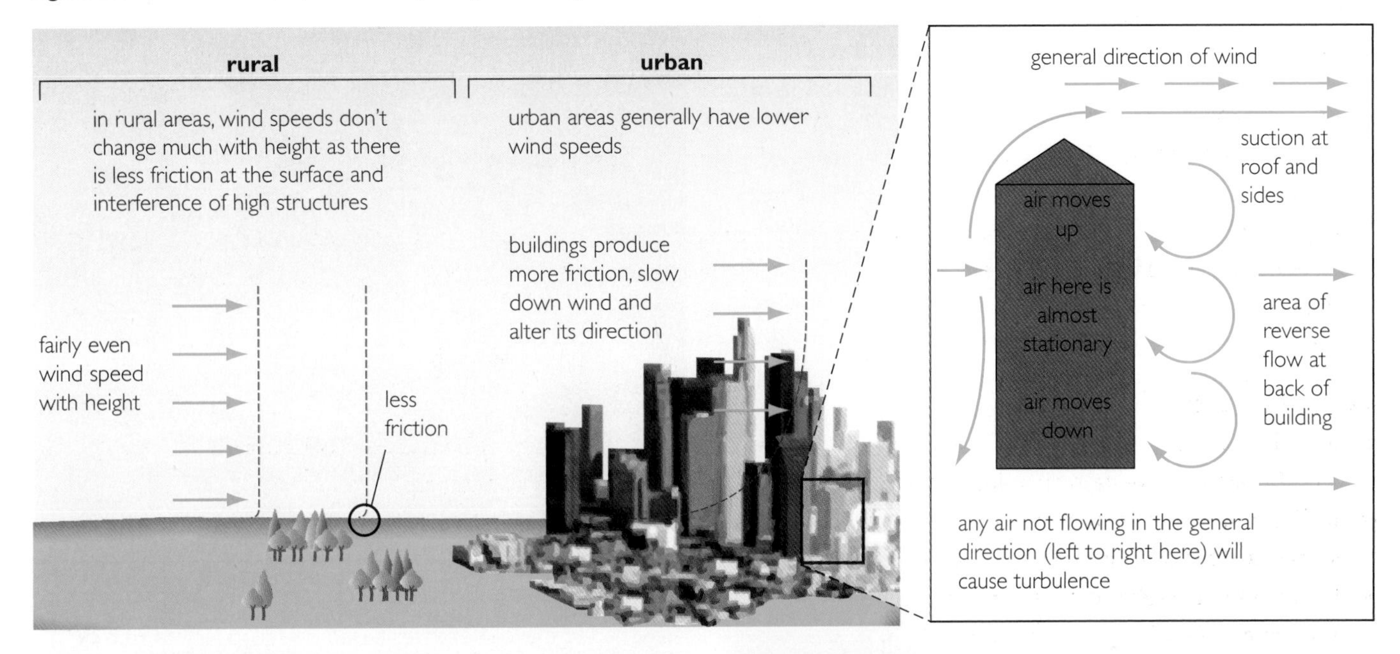

Figure 2.62 *The size, shape and position of the buildings in downtown Jacksonville, Florida disrupt the flow of air* ▲

ACTIVITIES

1 Study Figure 2.54.
   a Why do urban areas tend to experience less sunshine than rural areas?
   b How and why are winter conditions different in urban and rural areas?
2 Study Figures 2.56 and 2.57.
   a Under what atmospheric conditions would you expect an urban heat island to be most pronounced and why?
   b What impact does the urban heat island have on people living in the middle of a large city?
3 Study Figure 2.59.
   a What is a temperature inversion and how can it lead to the creation of a pollution dome?
   b What are the main sources and types of urban pollution?
   c What is the difference between 'smog' and 'photochemical smog'?
4 Study Figure 2.62. How might the people of Jacksonville be affected by the variations in airflow that have been described?

2.10 Climate change: evidence and causes

In this section you will learn about:

- climate change in the last 18000 years
- evidence for recent climate change
- evidence for global warming
- possible causes of climate change and global warming, including the enhanced greenhouse effect

For centuries, local people in Estonia were puzzled by the huge boulders strewn about their coastline. How did they get there? Where did they come from? A legend developed to explain their presence – the boulders were thrown there by the giant king, Kaleviopoeg, the dents in some of the three metre high stones, his fingerprints.

We now know that these boulders are in fact called glacial erratics, brought by the vast ice sheets that spread south to cover much of northern Europe about 18000 years ago (Figure 2.63). As the ice sheets melted and drew back, they simply dumped the boulders on the ground. Unlike the finer sediments that were washed away by meltwater, these heavy boulders remained where we find them today.

Figure 2.63 *Glacial erratics in Lahemaa National Park, Estonia* ▶

Evidence for climate change

In studying climate change it is important to consider the sources used to measure temperature. Temperature has only been measured using thermometers since about the 1850s. In order to study temperatures before 1850 scientists have had to make inferences based on secondary or **proxy data**. Inevitably this can lead to different interpretations of the data.

Secondary (proxy) evidence sources

- **Dendroclimatology** – This is the study of the links between tree growth and climate. Generally, it involves looking at annual tree rings – although a tree makes a new ring each year, the width of the ring is not always the same. In wet, warm years trees grow well so the ring will be wide. But in cold years, the tree will not grow as much, so the ring will be narrower.
- **Ice cores** – Ice cores have been taken from holes over 2000m in length from glaciers and ice sheets in Greenland and Antarctica. Ice cores allow scientists to infer temperature trends going back some 400000 years.
- **Pollen analysis** – Collecting pollen helps scientists reconstruct vegetation from the past, which is then used to estimate past climates and in particular temperature. It is difficult to infer absolute temperatures from pollen analysis but scientists can suggest changes, such as the climate becoming wetter or drier. Large-scale temperature changes are marked by clear changes in the types of pollen that existed.
- **Ocean sediments** – When organisms, such as plankton, that live on the ocean surface die, they drift down and collect on the ocean floor. By analysing these organisms we can learn about the surface of the oceans at the time, for example the amount of salt in the water.
- **Historical records** – There are many types of historical records, such as ancient writings or government reports, that describe characteristics of the weather and climate. Phenomena related to climate, such as when in the year birds began to migrate, can also be learned from these records.

While much of the information from these sources is based on casual observation and not scientific study, it does provide useful supporting evidence for trends that have been identified using other methods.

Evidence for long-term climate change

Erratics are just one sign that in the recent geological past our climate was quite different than it is today. Throughout much of northern Europe and North America there are many landscape features formed in glacial or periglacial (the edge of a glacier) conditions, when average temperatures were much lower than they are today. In the UK, upland areas such as the Lake District, Snowdonia and the Cairngorms bear witness to the immense landscape-forming power of ice.

Elsewhere in the world, there are other signs that the climate has changed. In the Middle East there are vast 'fossil' aquifers in countries such as Egypt and Jordan. These underground reservoirs were formed some 8000 years ago when the rainfall in this dry region was much higher than it is today.

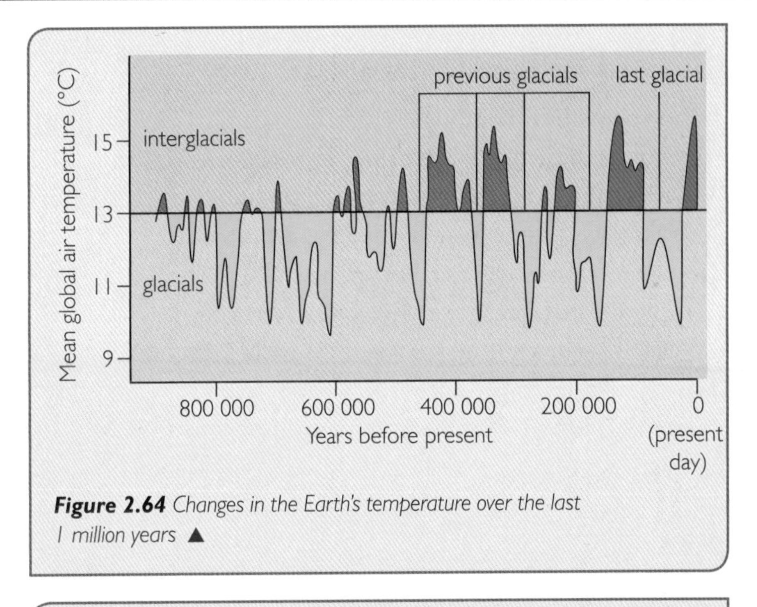

Figure 2.64 *Changes in the Earth's temperature over the last 1 million years* ▲

Climate change over the last 18 000 years

Look at Figure 2.64. It shows the pattern of temperatures for the last one million years. Notice that temperatures fluctuated greatly and that there were cold periods (glacials) when ice advanced and warmer periods (interglacials) when conditions were much as they are today. This period of alternating glacials and interglacials ranged from about 2 million years ago until about 12 000 to 10 000 years ago when we entered our current warm period.

Now look at Figures 2.65 and 2.66. Notice the following temperature trends.

- A rapid rise in temperature about 10 000 years ago marking the end of the last ice advance.
- From about 5000 – 3000BC, during the period known as the Holocene Maximum, temperatures were several degrees higher than they are today.
- Colder and wetter conditions followed in Europe between 900 – 500BC.
- Temperatures warmed again during the Medieval Warm Period (approximately 1000 – 1300AD) during which time the Vikings colonised a warmer and more hospitable Greenland.
- Temperatures then dipped from 1500 – 1850 to form the Little Ice Age. This colder period killed off the Viking communities in Greenland.
- Since about the 1950s, and particularly since the 1980s, the trend has shown a rapid warming – this is the current issue known as global warming.

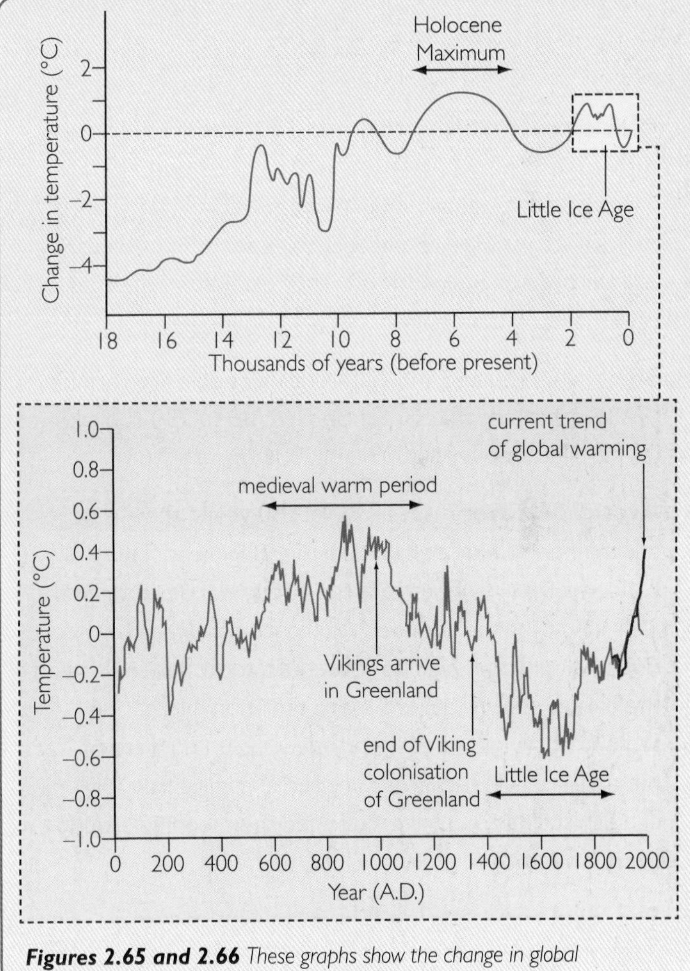

Figures 2.65 and 2.66 *These graphs show the change in global temperature during the last 18 000 years* ▲

Did you know?

The longest running temperature record in the world is for Central England and began in 1659.

Evidence for recent global warming

- **Instrumental readings** – Measurements recorded directly using thermometers go back to the 1850s. Whilst this suggests the availability of reliable data from that date, this is not necessarily the case. Recording stations are by no means widespread across the world or reliably calibrated and operated. As such, slight variations in temperatures could easily result from a range of instrumental and operational inaccuracies.

  The Intergovernmental Panel on Climate Change (IPCC) has concluded that average global temperatures have risen by 0.74°C during the last 100 years and by 0.5°C since 1980 (Figure 2.67).

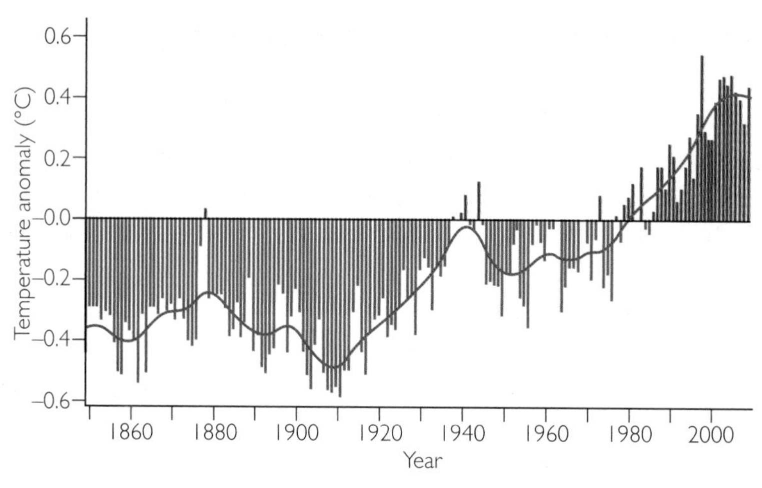

Figure 2.67 *Recent trends in global warming show that despite the significant monthly variations shown by the zig-zag pattern, the overall trend shown by the purple line is upwards* ▲

- **Glacier retreat and shrinking ice sheets** – There is strong scientific evidence that many of the world's glaciers have been retreating over the last 50-100 years. The World Glacier Monitoring Service reported a significant retreat of glaciers since 1980. They estimate that up to 25% of global mountain glacier ice could disappear by 2050.

  Both the Greenland and Antarctic ice sheets are shrinking. NASA estimates that Greenland lost up to 250 cubic km of ice per year between 2002 and 2006 (Figure 2.68).

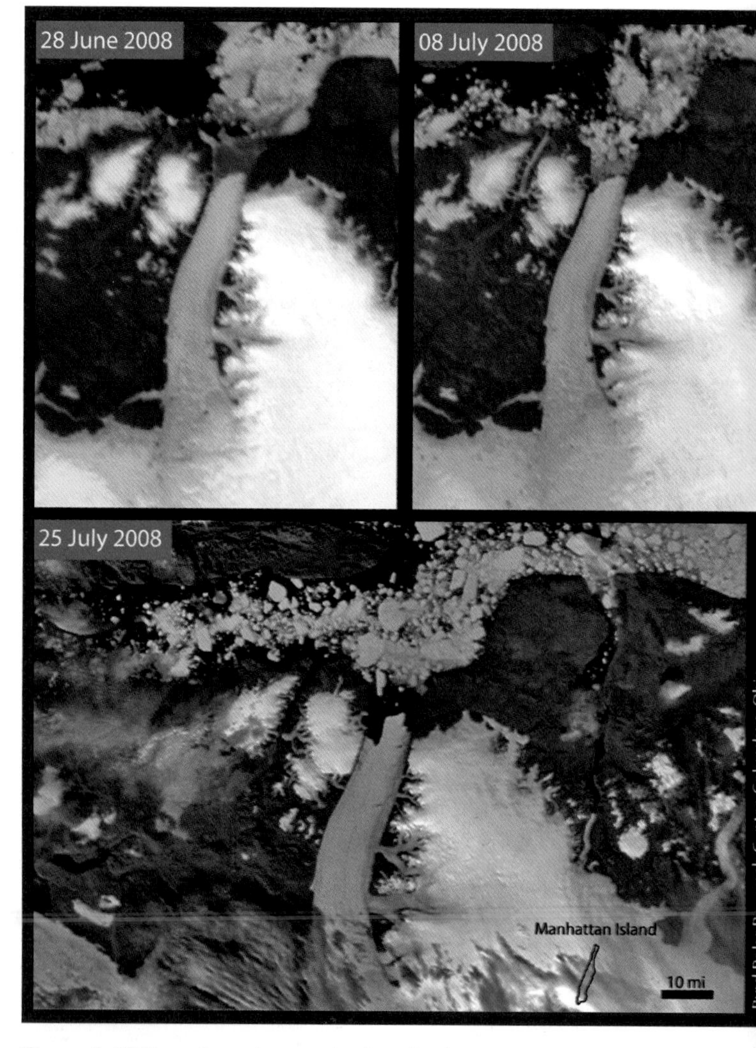

Figure 2.68 *These three photographs show the dramatic breakup of the Petermann Glacier in Greenland* ▲

- **Arctic ice cover** – Over the last 30 years, the Arctic ice has thinned to almost half its earlier thickness. There are concerns that in a few years the Arctic will be completely ice free during the summer. As the ice continues to thin less solar radiation will be reflected back to space. Instead, the darker sea will absorb more radiation, increasing temperatures still further and speeding up the rate of melting. Ports in the Arctic are remaining ice free for longer and fish such as cod have extended their feeding grounds further north.

- **Sea level rise** – Sea levels have risen by an average of 1.8 mm per year over the last century, and recent satellite data suggests an increasing rate to over 3 mm per year in the last ten years.

- **Record summer temperatures** – Worldwide, the 1990s was the warmest decade of the millennium, and 2006 was the warmest year on record in the UK. Some scientists have suggested that the 2003 European heat wave is evidence of global warming.

Possible causes of climate change and global warming

- **Plate tectonics** – over millions of years the continents have moved across different climate zones. Although this certainly accounts for long-term geological climate change, it does not explain shorter-term changes.
- **Solar cycles** – short-term cycles in solar output have been identified, and some scientists have suggested that this factor may be responsible for the Little Ice Age.
- **Variations in orbit** – slight variations in the Earth's orbit and tilt affect the distribution of radiation on the Earth.
- **Volcanic eruptions** – large volcanic eruptions throw huge amounts of particulate matter into the atmosphere, which reduces the amount of sunlight that reaches Earth, causing temperatures to cool.
- **Ocean circulation** – short-term changes in ocean circulation (El Niño and La Niña) are known to impact the climate every few years. Longer-term variations may have an influence on climate change, as the oceans are a massive store of energy.

The enhanced greenhouse effect

The main difference between past episodes of climate change and the recent trend of global warming is the impact of people and human activities. Many scientists believe that people have been at least partly responsible for the recent warming trend through a process called the **enhanced greenhouse effect**.

The enhanced greenhouse effect occurs when human activities increase the ability of the atmosphere to absorb the Earth's radiation, which then increases temperatures (Figure 2.69). This happens when human activities release more greenhouse gases, such as carbon dioxide, methane and nitrous oxides, which are most effective at absorbing long-wave radiation from the Earth.

One of the most compelling pieces of evidence to support this theory is the measured increase in carbon dioxide in the atmosphere, recorded by direct monitoring on Hawaii and corroborated by ice core data (Figure 2.70). While there is still considerable debate about the causes of global warming, the IPCC have stated that global climate change is 'very likely' to have a human cause.

Greenhouse Gas	Sources
Carbon dioxide – accounts for an estimated 60% of the 'enhanced' greenhouse effect. Global concentration of carbon dioxide has increased by about 30% since 1850.	• Burning fossil fuels (e.g. oil, gas, coal) in industry and power stations to produce electricity • Car exhausts • Deforestation and the burning of wood
Methane – very effective in absorbing heat. Accounts for 20% of the 'enhanced' greenhouse effect.	• Decaying organic matter in landfill sites and compost tips • Padi rice farming • Farm livestock • Burning biomass for energy
Nitrous oxides – very small concentrations in the atmosphere by up to 300x more effective in capturing heat than carbon dioxide	• Car exhausts • Power stations • Agricultural fertilisers • Sewage treatment

Figure 2.69 *The greenhouse effect* ▲

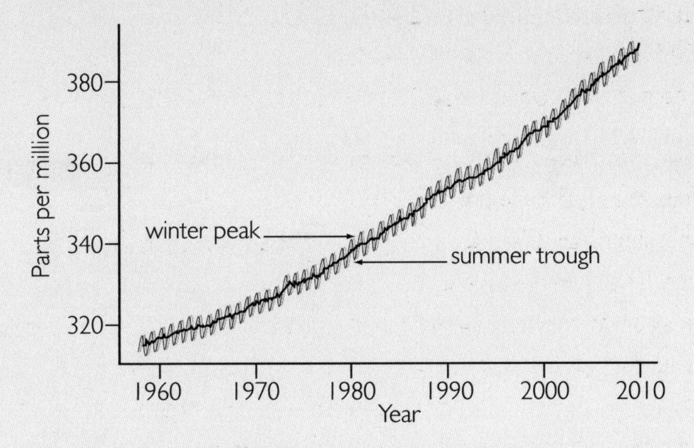

Figure 2.70 *This graph shows rise in carbon dioxide levels since 1960. The zigzag line is caused by seasonal changes – during the winter plants die back and decay, putting carbon dioxide back into the atmosphere. In summer months, plants grow, using up carbon dioxide – this is shown by the troughs in the graph.* ▲

ACTIVITIES

1. Study Figure 2.65.
   - **a** Describe the changes in the pattern of global temperatures from 18 000 to 2000 years ago.
2. Now focus on the temperature changes since 1860 (Figure 2.67).
   - **a** Do you think these temperature values are likely to be more reliable? Explain your answer.
   - **b** Do you think the graph offers powerful evidence to support recent global warming? Explain your answer.
3. Understanding the time scale is critical when studying trends in the climate. Why is it misleading, perhaps even irresponsible, for the media to respond to a short heat wave by suggesting that it is evidence of global warming?

2.11 Global warming: impacts and responses

In this section you will learn about:

- the positive and negative impacts of global warming
- the possible impacts of global warming on the world, including the UK and Africa
- responses to the threat of global warming at individual, national and international levels

"I stand before you as a representative of an endangered people. We are told that as a result of global warming and sea-level rise, my country, the Maldives, may sometime during the next century, disappear from the face of the Earth."

President Maumoon Abdul Gayoom of the Maldives

Many people consider global warming to be the greatest threat of our time. Others consider it to be hyped-up by the media and environmental groups. These sceptics are quick to try to discredit or find loopholes in the scientific data. While few people disagree that the world is becoming warmer – the evidence is beyond doubt – there is still considerable debate regarding the role of people and human activities. Either way, it seems certain that the climatic changes that are already underway will have a significant impact on people and natural environments in the future.

Global impacts of global warming

With average global temperatures predicted to rise by 0.28-0.4°C by 2010, there will be significant, but largely unpredictable, changes to the world's climates. Look at Figure 2.71. It is a map produced by the Met Office in 2009 showing the likely pattern of temperatures around the world if the average global temperature increased by 2°C. Notice that there are significant variations – some places are expected to warm by the predicted amount, but many others are either more or less severely affected.

One reason for the differences is the complexity of the relationship between atmospheric and oceanic circulations. The knock-on effects of the temperature changes around the world will be complicated, and almost impossible to predict with a high degree of accuracy. While it is reasonable to predict some of the possible effects of global warming, it is unreasonable to think that these predictions will all prove to be right.

Figure 2.71 *The regional variations of an average 2°C rise in global temperature (Met Office)* ▶

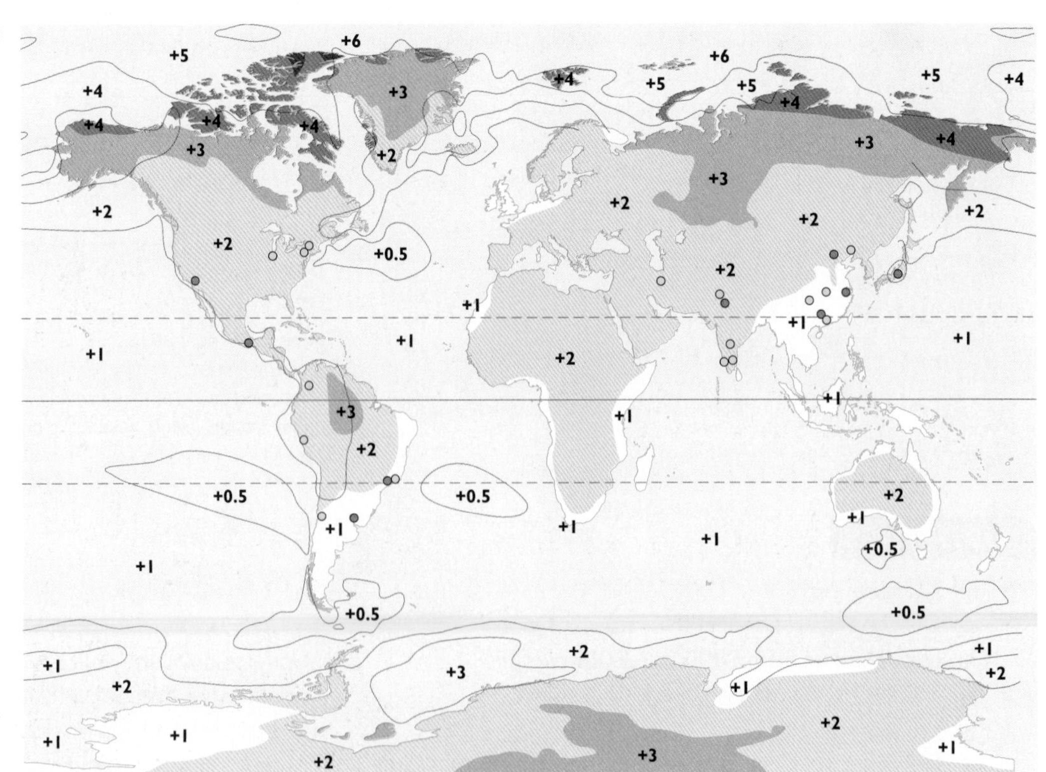

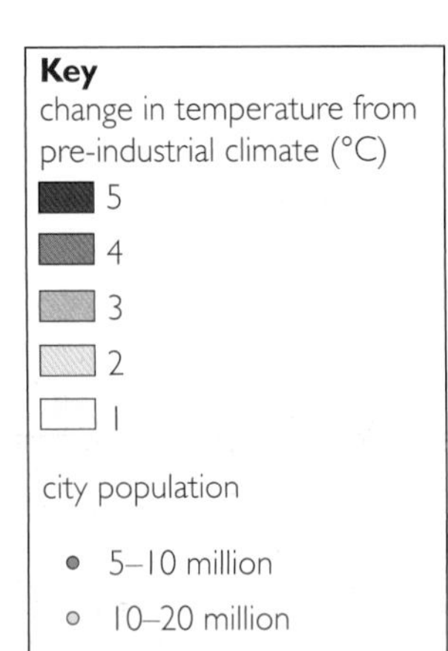

If the world does warm as the scientists predict, there will probably be a number of major global impacts. Not all of them are negative!

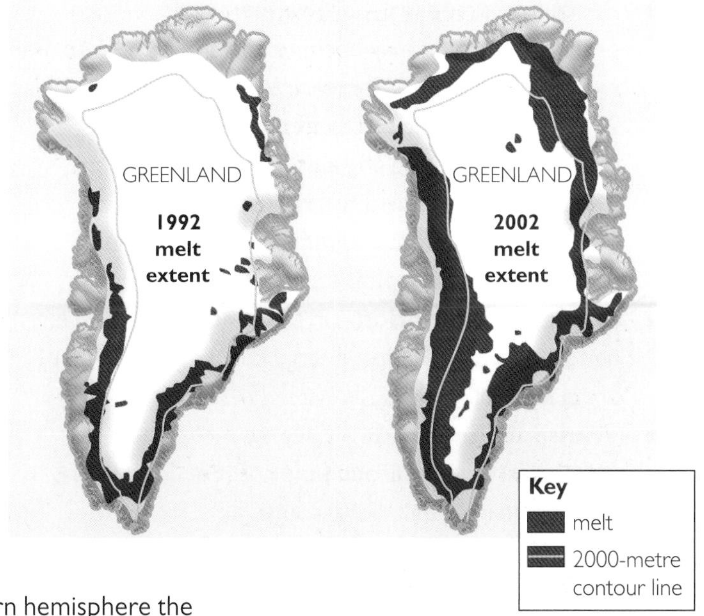

Figure 2.72 These two maps show the shrinking of the ice sheets in Greenland in just 10 years ▲

- **Sea level rise** – the Intergovernmental Panel on Climate Change (IPCC) predicts that by 2010 sea levels may rise by 28-43 cm. Some scientists suggest that this is a conservative estimate, and that sea levels might rise in some places by up to 1 m.

  The world's deltas, such as the Nile and the Mississippi, will also be under threat. The impact of storm surges associated with tropical cyclones will be more extensive and will penetrate further inland.

- **Vegetation belts** – a warming world will have an impact on the latitudinal and altitudinal distribution of the world's vegetation belts. For example, in the northern hemisphere the Mediterranean biome will stretch much further north, perhaps even into southern Britain. Other vegetation belts, such as deciduous woodland and coniferous forest, will also move further north. As the vegetation belts move, agriculture will change as will the distribution of birds, animals and insects. Whole ecosystems will shift as the climate changes. How fast this will happen and how successfully species will be able to adapt is unknown.

- **Melting of ice** – it is reasonable to assume that in a warmer world more surface ice will melt. The Arctic ice may melt completely as might a large number of the world's glaciers, threatening water supplies in countries such as Nepal and China. Melting of the great Greenland and Antarctic ice sheets will probably increase too (Figure 2.72). This will lead to higher sea levels, and will threaten many species that live in these areas. Impacts on the ocean circulation systems could also be significant.

- **Melting of permafrost** – warmer temperatures threaten to melt areas of permafrost in Siberia, Canada and Alaska (Figure 2.73). In Siberia, melting permafrost is leading to the release of stored methane gas, worsening the enhanced greenhouse effect.

- **Agriculture and food supply** – agricultural productivity is likely to increase in some parts of the world, but may well decrease in other areas, such as Africa, where drought may be more of a problem. As climate changes, so the pattern of farming will change. Mediterranean crops (tomatoes, grapes, etc) will be grown further north than they are today and the cereals belt (wheat, barley, etc) will also shift north into northern Europe. Parts of Siberia and Canada, currently unsuitable for farming, may become productive regions.

Figure 2.73 Buildings sink into one another in Canada's Yukon Territory as the permafrost underneath them melts ▼

- **Pests and diseases** – in a warmer world, certain pests and diseases may become more of a problem. For example, malaria may increase to affect an additional 280 m people as the warmer conditions create widespread breeding grounds for mosquitoes.
- **Tropical storms and hurricanes** – these may become more severe and more frequent, posing a greater threat to tropical coastlines.
- **Energy consumption** – in a warmer world with less need for central heating, energy consumption in many areas may fall.
- **Transport and trade** – if sea ice melts, the Northwest Passage around the coast of Canada may become an ice-free trading route.

Impacts of global warming in the UK

Figure 2.74 describes some of the possible impacts of global warming on the UK. Notice that there are both advantages and disadvantages. Agriculture seems likely to become more productive, with intensive farming shifting north into Scotland. In the south, recently introduced crops, such as olives and grapes, could become more extensive. New crops typical of Mediterranean climates, such as oranges and peaches, may well become established in the far south.

Perhaps the most significant negative impact relates to the increased risk of flooding associated with sea level rise. Coastal habitats will be threatened and rates of cliff erosion will increase. The Thames Barrier will no longer be effective and will need to be replaced.

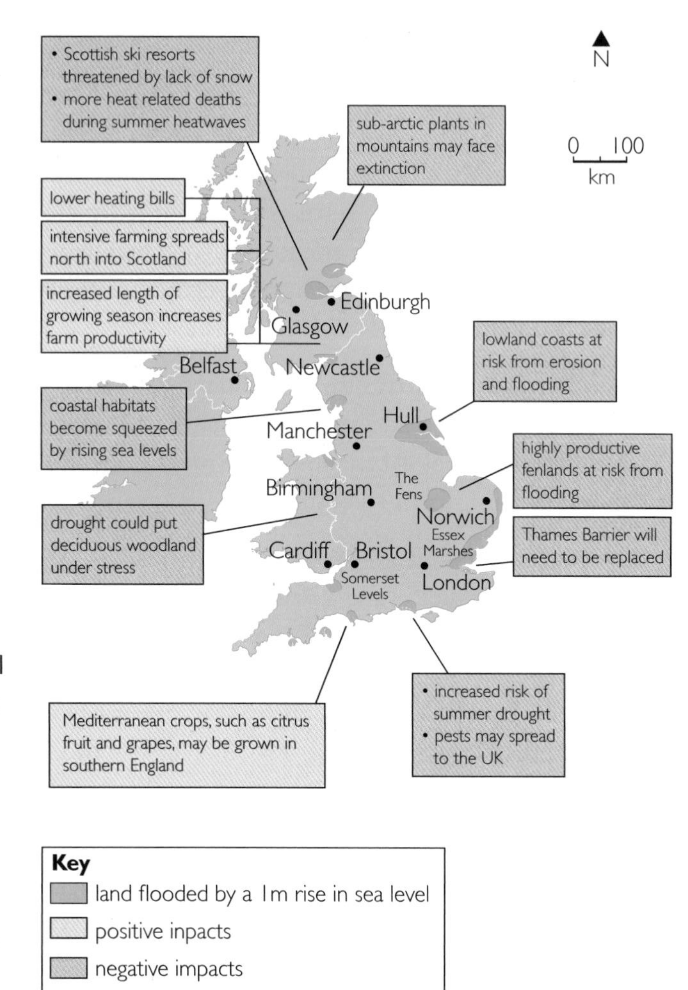

Figure 2.74 *Some of the possible positive and negative effects of global warming on the United Kingdom* ▲

"Africa is one of the continents least responsible for climate change and is also least able to afford the costs of adaptation. Africa will remain vulnerable even if, globally, emissions peak and decline in the next 10 to 15 years."

South African Environment Minister
Marthinus Van Schalkwyk, 2008

Impacts in tropical continental environments

In Section 2.7 we studied the tropical continental climate, focusing on Kano in Nigeria. How might this environment be affected in the future by global warming?

Much of this climatic region is at risk from global warming. Indeed, the United Nations has stated that Africa is probably at greatest risk from food shortages and desertification. The reasons for this include:

- Temperatures are expected to rise and rainfall patterns may change, leading to more extreme events such as droughts, floods, fires and famines.
- Water supplies will be threatened and the food supply will probably be reduced.
- Changes in the amount and distribution of rainfall could spell disaster for millions of people and affect animal habitats.
- Migration routes for animals and birds within Africa and between Africa and other continents could be affected.
- In some areas, the wet season may become more intensive, unleashing devastating floods.

Responses to global warming

It is possible to consider responses to global warming at three different scales: individual, national and international.

Individual

Whilst individual actions may seem to be of limited use given the sheer scale of the problem, the collective action of many individuals, particularly in influencing national and international decision-makers, can be significant. Individuals can become involved in campaigns and can use their initiative to drive forward new innovations. Individuals can take a stand for what they believe to be right.

Some individual actions may include:

- Walking or cycling to work or school, or making use of public transport (Figure 2.75).
- Conserving energy in the home and at work by using energy saving lightbulbs, switching off electrical items rather than putting them on 'standby' or choosing to buy low energy appliances.
- Putting into practice the three Rs – reduce, re-use and recycle.
- Insulating homes and using double-glazing.
- Paying a carbon offset fee when making bookings for airline flights.

National: the UK response

The UK government has set a target of reducing carbon emissions by 60% by 2050. In addition, it has committed itself to generating 10% of electricity using renewable energy by 2010. In order to meet these targets, the government has introduced a number of initiatives:

- Introducing congestion charging in London and Edinburgh to reduce the number of private cars driving into the city centres and encourage people to make use of public transport.
- Promoting the use of alternative energy sources, particularly wind power. In 2010 the government granted nine offshore wind farm licences that could generate a quarter of the UK's electricity needs.
- Encouraging local authorities to establish park-and-ride schemes in towns and cities.
- Encouraging people to recycle materials and discouraging them from dumping waste in landfill sites. Industry is being encouraged to cut waste by reducing packaging.
- Promoting the use of energy conservation.
- Promoting the use of biofuels to power vehicles thereby reducing the use of petrol and diesel.
- Creating the Office of Climate Change (OCC) in 2006 to work across government departments and coordinate responses to the threat of global warming.

Did you know?

The United States is considering the use of giant mirrors in space to reduce global warming. Scientists have suggested that if 1% of solar energy is reflected back to space this would be the equivalent to the warming generated by all greenhouse gases emitted since the Industrial Revolution.

◀ **Figure 2.75** *The 'Walking Bus' project in Kent gives parents a safe and easy way to get their children to school without using a car*

International

The Kyoto Protocol (1997)

In 1988, various international organisations met to assess the scientific data on climate change, the outcome of which was an international agreement called the **Kyoto Protocol**.

The Protocol set out a number of commitments:

- By 2012, high-income countries would reduce their collective emissions by 5.2% from 1990 levels, with each country being committed to a particular figure.
- A range of emissions would be included, not only carbon dioxide.
- These commitments would be calculated on a net basis, considering 'sinks' as well as 'sources'. Each country must credibly measure its contribution and meet its commitment.
- Countries may work with one another to meet their commitments, such as through regional agreements.

Over 170 countries have signed the protocol accounting for over 60% of the carbon emissions. The USA's refusal to sign was largely based on the government's belief that having to cut emissions would hurt the country's economy.

- **Carbon sinks** absorb more carbon than they release.
- **Carbon sources** release more carbon than they absorb

CASE STUDY

The Maldives: Coping with rising sea levels

The Maldives are a group of tiny islands in the Indian Ocean some 500 km southwest of India (Figure 2.76). Most of the 200 or so inhabited islands are at or just above sea level (Figure 2.77). The highest point on the Maldives is only 2.4 m high! Climate models suggest that most of the islands will be uninhabitable by 2030, and by 2070 will be submerged as global warming causes sea levels to rise.

The 380 000 inhabitants of the Maldives face a very uncertain future, apart from the one inevitable outcome. Various strategies have been put forward by the government to cope with the impending rise in sea level:

- Restoration of coastal mangrove forests. The tangled roots of these trees will gather sediment and will, at least in the short term, provide a buffer to rising sea levels and storm waves.
- Construction of sea defences. A 3 m sea wall is being built around the capital city of Male.
- Building more houses on the very highest ground and relocating people to these islands. Some houses can be built on stilts.
- Artifical islands could be constructed up to 3 m high. A floating golf course is being planned to enable tourism to continue in the future!
- Relocate the population of the Maldives to a new territory, such as in Sri Lanka or India.

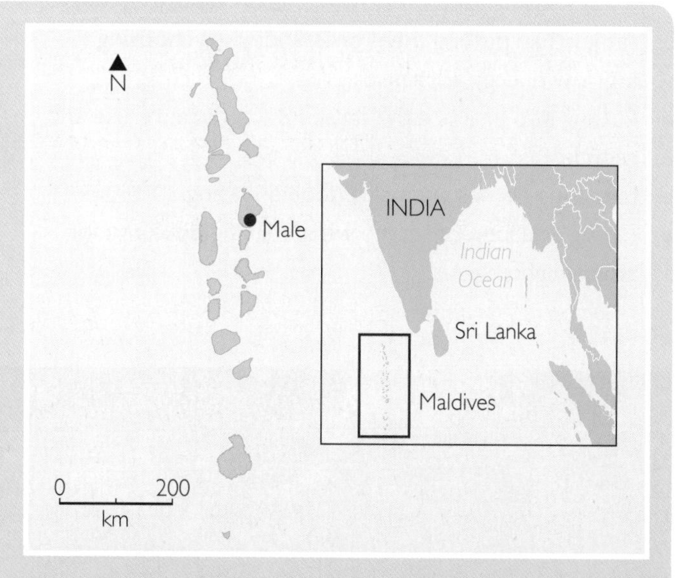

Figure 2.76 *The Maldives* ▲

Figure 2.77 *It's easy to see how rising sea levels threaten these low-lying islands* ▲

Emissions (carbon) trading

One of the outcomes of the Kyoto Protocol has been **emissions trading**. In order to meet their overall target, individual governments have to set emissions targets for organisations within their country. If an organisation not only meets but even beats its target, then in effect it has some 'spare' emissions that can be traded. Other organisations can then buy these **carbon credits** as an alternative to having to install expensive equipment or undertake other actions in order to meet their own targets.

This same principle can be applied internationally, with countries buying carbon credits from other countries.

As carbon 'sinks' can be used to help meet targets, schemes such as planting trees to remove carbon dioxide from the atmosphere can also be used to 'buy' **carbon credits**.

As this chapter has shown, global warming is an international issue. Carbon emissions and the enhanced greenhouse effect are not restricted to individual countries. The only long-term strategy for addressing the issue, then, is to establish and enforce international agreements.

Figure 2.78 *In Brazil, fire is used to clear land in a rainforest for ranching. An interesting recent approach to reducing carbon emissions has involved the international community paying governments not to carry out deforestation. This initiative, called* ***Climate Protection Payments****, establishes a value for an area of forest and pays governments not to carry out deforestation. In this way emissions can be reduced and important natural habitats protected.* ▲

ACTIVITIES

1 Study Figure 2.71. You may also need to use an atlas.
   - **a** Describe the overall pattern of temperature increase. What seem to be the main trends?
   - **b** Much of central and northern Africa are predicted to have a rise of 2°C. Given the very fragile and marginal nature of this already arid region, how do you think this temperature increase will impact on the environment and on the lives of the people living there?

2 Using examples, describe some of the advantages of global warming. Do the advantages outweigh the disadvantages?

3 Work in pairs to discuss the following questions concerning responses to global warming:
   - **a** Can individuals really make a difference?
   - **b** Should the UK government do more and, if so, what?
   - **c** Do you think emissions trading is a good idea, or is it just an excuse for wealthy countries and organisations to do nothing?
   - **d** How can the United States be convinced of the need to sign up to international agreements?

Internet research

Use the Internet to find out how countries other than the UK are responding at a national level to the threat of global warming. Consider countries such as the USA, Australia and Germany. The Maldives make an excellent example for further study, given their precarious position.

air masses Large bodies of air with distinct characteristics of temperature and humidity. They develop over stable areas where they adopt the characteristics of the ground surface, such as cold and dry (Arctic) or warm and wet (Tropics)

albedo Term used to describe the reflectivity of a surface, whereby white surfaces such as ice and the tops of clouds are highly reflective (high albedo) and darker surfaces such as forests and roads are less reflective (low albedo)

atmosphere The air around us, the atmosphere is a layer comprising gases (e.g. nitrogen and oxygen), liquids (e.g. water) and soils (e.g. volcanic dust)

atmospheric heat budget A model that describes the overall balance between incoming and outgoing heat (radiation) from the Earth-atmosphere system

climate The long-term average weather, usually calculated over a period of 30 years

cold front On a synoptic weather chart, a line drawn to mark the forward edge of colder air approaching and undercutting warmer air

condensation nuclei Microscopic particles such as dust and volcanic ash that form the nucleus of raindrops and are essential for the process of condensation

coriolis effect The effect of the Earth's rotation on, for example, the movement of air (wind) or the direction of 'spin' and tracks of tropical cyclones (hurricanes)

depressions Areas of low atmospheric pressure usually associated with cloudy and wet conditions resulting from the rising of air

enhanced greenhouse effect The increase in the natural greenhouse effect resulting from the emission of additional greenhouse gases from human activities

global warming The term used to describe the recent rise in global average temperatures attributed largely to human activities, such as burning fossil fuels and deforestation

greenhouse effect The absorption by gases and liquids of long-wave (terrestrial) radiation emitted from the Earth forming a 'blanket' of warmth in the lower atmosphere

hurricanes Intense tropical storms (tropical cyclones) formed over warm oceanic areas in the summer

insolation Term used to describe the amount of short-wave radiation (from the sun) that reaches the ground surface having passed through the atmosphere

inter-tropical convergence zone (ITCZ) The zone of convergence and subsequent rising air at the surface boundary of the two Hadley Cells in equatorial regions. This zone is associated with the formation of tropical cyclones (hurricanes)

isobars On a synoptic weather map, lines joining points of equal pressure

latent heat In the atmosphere, heat released or absorbed when a change of state takes place in water. Latent heat is released when water condenses and is absorbed when evaporation occurs

occluded front On a synoptic chart, a type of front (boundary between air masses) formed when a cold front catches up and merges with a warm front

particulates Solid particles held in the atmosphere, often emitted from factories, power stations, and vehicle exhausts, often considered to be a form of pollution

photochemical smog A type of air pollution involving the formation of harmful chemical substances in the air by the effects of sunlight. For example, nitrous oxides from vehicle exhausts can react with sunlight to form low-level ozone

polar front The boundary between the Polar Cell and the Ferrel Cell at approximately 60° N/S associated with uplift and instability (cloud and rain)

pressure gradient On a synoptic weather map, the difference in pressure between two points, indicated by the closeness of the isobars. A steep pressure gradient (isobars close together) will result in strong winds, and a gentle pressure gradient (isobars far apart) low wind speeds

smog A type of air pollution associated with calm conditions in winter, involving a mix of particulates (e.g. soot) and fog

solar radiation Incoming short-wave energy from the sun

storm surge A rise in the level of the sea and subsequent inundation of low-lying coastal land, caused by the passage of an area of intense low pressure, such as a hurricane

sub-tropical anticyclone A broad zone of high pressure resulting from sinking air at the boundary between the Hadley Cell and Ferrel Cell, found at about 30° N/S

synoptic chart A map or chart that makes use of standard symbols to show features of the weather

temperature inversion A reversal in the normal trend for temperature to decrease with increasing altitude. Temperature inversions commonly occur during clear, calm nights when dense cold air sinks beneath warmer less dense air in valley bottoms, often forming frost and fog

terrestrial radiation Heat emitted from the Earth's surface in the form of long-wave radiation

tornadoes Intense localised storms associated with rapidly rising air currents and violent thunderstorms

urban heat island The warmer conditions experienced by a city compared with the neighbouring countryside

warm front On a synoptic weather map, a line drawn to mark the forward edge of warmer air approaching and over-riding colder (denser) air

weather The day-to-day state of the atmosphere, such as temperature, precipitation, etc.

Exam-style questions

EXAMINER'S TIPS

Section A

1 (a) Study Figure 2.9 (page 60). Assess the role of ocean currents in the world's weather and climate. *(7 marks)*

> **(a)** Use the map to identify particular ocean currents that have an impact on weather and climate.

(b) Explain the processes responsible for the formation of the depressions that affect the UK. *(8 marks)*

> **(b)** Focus on the processes responsible for the formation of depressions. Consider using diagrams to support your answer. Make sure you use geographical terminology.

(c) With reference to one example, discuss the factors contributing to the impacts of a storm event in the UK. *(10 marks)*

> **(c)** Make sure you refer to your chosen case study in detail. Consider the relative significance of a selection of factors (physical and human) that determined the impacts of the storm.

2 (a) Outline the characteristics and formation of an urban heat island. *(7 marks)*

> **(a)** Describe the characteristics of a heat island, perhaps using a diagram, and explain how it is formed. Try to use the appropriate geographical terminology.

(b) Study Figure 2.54 (page 88). Explain the reasons for the contrasts in two local climatic conditions between urban and rural areas shown in the table. *(8 marks)*

> **(b)** Look at the table and select only two contrasting local climatic characteristics that you feel confident in writing about. In your explanation, use the correct terminology and refer to processes.

(c) With reference to examples, outline the causes of pollution in urban areas and discuss alternative strategies to reduce the problem. *(10 marks)*

> **(c)** Make sure your answer is balanced between identifying the causes and discussing the strategies. Use geographical terminology.

Section C

1 To what extent is the impact of a tropical revolving storm (hurricane) determined by human rather than physical factors? *(40 marks)*

> **(1)** Focus on the effects (impacts) of hurricanes and use case studies (you should have studied at least two examples) to weigh up and discuss the human and physical factors. Make sure that you answer the question – 'to what extent?'

2 Assess the evidence for global climate change and consider the likely impacts of local, national, and international responses in tackling the issue. *(40 marks)*

> **(2)** The first part is straightforward, but don't spend too long on it. Make sure you address each of local, national, and international responses and focus on the likely impacts (success) of each level of response.

Ecosystems: change and challenge

An orang-utan

Where in the world could this photo have been taken and why?
What changes might this orang-utan face?
How might the removal of trees affect the natural forest ecosystem?
Should countries be allowed to cut down their hardwood trees?
How might deforestation increase the rate of global warming?

Introduction

The living world: varied, fascinating, important. Natural ecosystems provide us with food, clothing, shelter, and industrial raw materials – so we depend on them for our very survival.

People have an important part to play in managing the natural world. In some places, poor management – such as clear felling of forests – has led to long-term damage and even the extinction of species. Increasingly, we are concerned about conservation and stewardship as the way forward in a more sustainable world.

In this chapter you will learn about the mechanisms and interactions associated with natural ecosystems in the UK and elsewhere in the world. You will learn about the threats and opportunities faced by the natural world. And you will consider sustainable options to ensure the future well-being of our planet.

Books, music, and films

Books to read

Understanding Environmental Issues by Susan Buckingham & Mike Turner
Climate Wars by Gwynne Dyer
The Burning Season by Andrew Revkin
The Global Casino by Nick Middleton

Music to listen to

'Big Yellow Taxi' originally recorded by the song's writer Joni Mitchell, other versions by Counting Crows and Amy Grant
'Earth Song' by Michael Jackson
'The Price of Oil' by Billy Bragg
'The Trees' by Pulp

Films to see

Home
Fern Gully – the Last Rainforest
State of the Planet (a tv series presented by David Attenborough, 2001, BBC)

About the specification

'Ecosystems: change and challenge' is one of three Physical Geography options in Unit 3 Contemporary Geographical Issues – you have to study at least one.

This is what you have to study:

Nature of ecosystems

- The structure of ecosystems, energy flows, trophic levels, food chains, and food webs.

Ecosystems in the British Isles over time

- Succession and climatic climax, illustrated by one of: lithosere, psammosere, hydrosere, or halosere.
- The characteristics of the climatic climax: the temperate deciduous woodland biome.
- The effects of human activity on succession, illustrated by one plagioclimax, such as a heather moorland.

The biome of one tropical region

(Savanna grassland, tropical monsoon forest, or tropical equatorial rainforest.)

- The main characteristics of the biome.
- Ecological responses to the climate and soil moisture budget – adaptations by vegetation and animals.
- Human activity and its impact on the biome.
- Development issues in the biome, including aspects of biodiversity and the potential for sustainability.

Ecosystem issues on a local scale: impact of human activity

- Changes in ecosystems resulting from urbanisation.
- Urban niches and the colonisation of wasteland: the development of distinctive ecologies along routeways.
- The planned and unplanned introduction of new species, and the impact of this on ecosystems.
- Changes in the rural-urban fringe.
- Ecological conservation areas.
- One case study of an ecological conservation area.

Ecosystem issues on a global scale

- The relationships between human activity, biodiversity, and sustainability.
- The management of fragile environments (conservation versus exploitation)
- Two contrasting case studies of recent management schemes (from within the last 30 years) in fragile environments.

Characteristics of ecosystems

In this section you will learn about:

- the components of an ecosystem
- energy flows and nutrient cycling in ecosystems

The significance of ecosystems

Ecosystems are central to the study of **biogeography**. They are dynamic, ordered and highly integrated communities of plants and animals together with the environment that influences them (Figure 3.1). Ecosystems occur at all scales, from the smallest patch of lichen on a rock to global **biomes** such as tropical equatorial rainforest.

- **Biogeography** is a vital link between human and physical geography studying ecosystems, soils and vegetation.

Figure 3.1 A deciduous woodland ecosystem ▼

The whole of life on Earth itself is, in effect, an ecosystem. The significance of ecosystems for humans cannot be underestimated – they provide us with our basic food resources, without which we would not survive. But human activities are changing and threatening ecosystems. If we want to ensure a sustainable future for life on Earth, it is essential that we understand and successfully manage the world's ecosystems.

The structure of ecosystems

Every ecosystem is in balance (equilibrium) – the so-called 'balance of nature'. This balance is between living organisms in a biotic environment and inorganic non-living substances in an abiotic environment.

The biotic living environment is made up of both plants and animals:

- plants include all living vegetation, but also dead plant matter which is decomposing
- animals include all fish, birds, insects, micro-organisms (such as bacteria) and mammals, including people.

The total weight of all these living organisms is called the **biomass**.

The abiotic, non-living environment includes all inorganic substances and other environmental influences. These are the chemical and physical components of an ecosystem. They include:

- minerals in the soil released by weathering of rock
- water and gases (such as air) in the soil
- relief and drainage of the land
- climatic variables such as wind, light and seasonal patterns of precipitation and temperature.

Ecosystems are systems of inputs, outputs, stores and flows (transfers). They are open systems in that energy and living matter can both enter and leave.

Inputs include:

- seeds blown into the area
- animals that arrive from elsewhere
- water from rivers or precipitation.

But the most important input is solar energy from the sun (**insolation**) driving **photosynthesis** and so allowing plants to grow.

Outputs involve anything moved out of the ecosystem, such as:

- animals that leave the area
- water lost though river runoff, overland flow, soil throughflow, groundwater flow and evapotranspiration
- **leaching** of nutrients from the soil.

Stores include nutrients held in the biomass, litter and soil, while **flows** include all transfers of energy or nutrients (see Figure 3.2).

Energy flows and nutrient cycling

All life depends on the sun because **insolation** is the energy source which drives ecosystems. Photosynthesis and chlorophyll (in green plants) convert carbon dioxide and water into carbohydrate and tissue which increases their biomass. Oxygen, also essential to supporting life on Earth, is released as waste. But nutrients are also required for plant growth and these are recycled from one store to another in a process called **nutrient cycling**.

Gersmehl's simplified diagrams demonstrate this well (Figure 3.2). Proportional circles show the relative importance of nutrient stores contained in the biomass, litter and soil.

- The biomass store contains all living plant and animal matter in the ecosystem.
- The soil store contains minerals from rock in addition to **humus** from decomposed plant and animal remains.
- The litter store sits on top of the soil and contains both dead and decaying plant and animal material.

Proportional arrows then represent the relative importance of the nutrient flows between them – the thicker the arrow, the greater the flow.

Did you know?

The term ecology comes from the Greek word oikos – meaning 'home'.

ACTIVITIES

1. Ecosystems are said to represent 'the balance of nature'.
   - **a** Explain what is meant by this statement.
   - **b** Identify ways in which both natural events and human activities can destabilise this balance.
2. Figure 3.2 shows the transfer of plant nutrients in an ecosystem. Briefly explain the nature of the nutrient transfer:
   - **a** from soil to biomass
   - **b** from biomass to litter
   - **c** from litter to soil.
3. Figure 3.2 also shows the nutrient cycles in both temperate deciduous woodland and the tropical equatorial rainforest biomes. Explain the differences between the two biomes in terms of the amount of nutrients in each store.

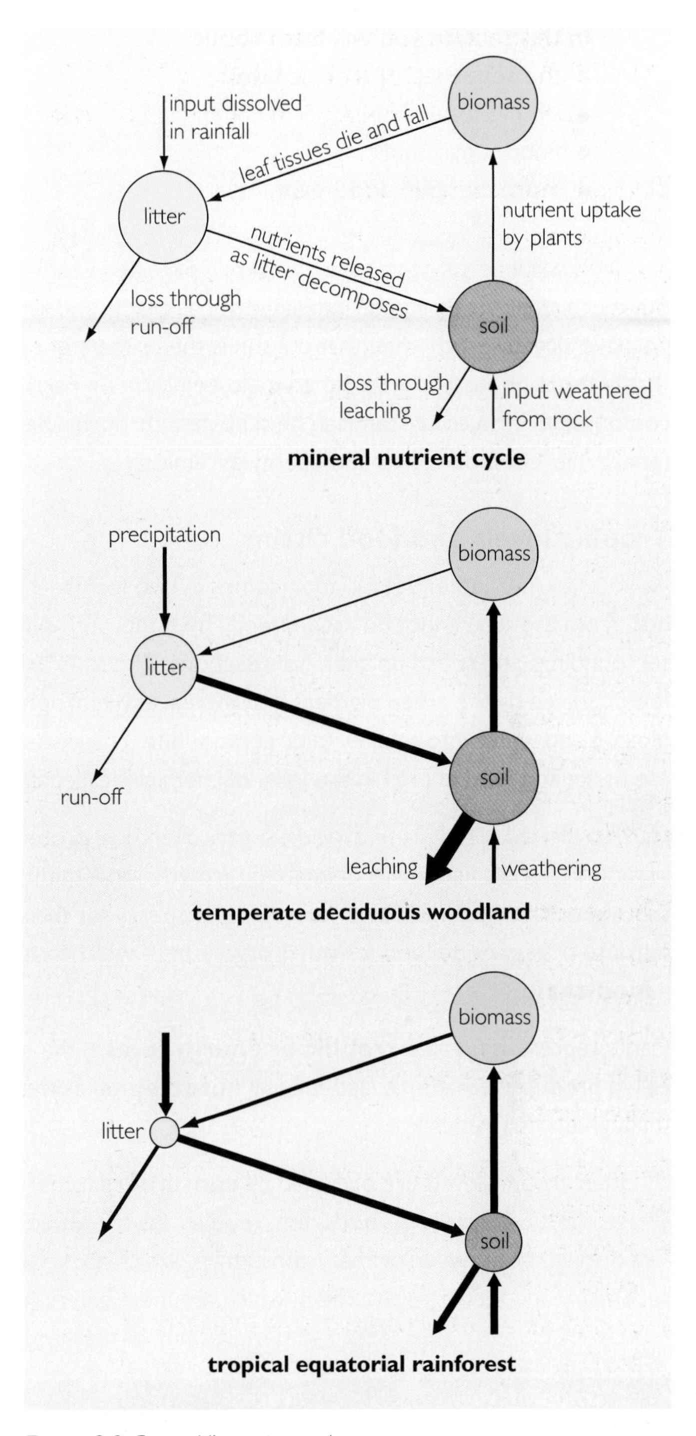

Figure 3.2 *Gersmehl's nutrient cycles* ▲

Internet research

A key concept to grasp early in any study of ecosystems is that of nutrient cycling. Using an Internet search engine, explore the term with specific reference to tropical equatorial rainforests.

Trophic levels, food chains and food webs

In this section you will learn about:

- the sequence of trophic levels
- simple food chains
- energy pyramids
- more complex food webs

The links within ecosystems are complex – particularly at larger scales such as in biomes. But much of **ecology** can be explained in basic facts and simple structures. For example, we have already established that the sun is the primary energy source for ecosystems and that without photosynthesis there would be no life on Earth. Likewise, the links between components in an ecosystem can be simplified through a basic understanding of **trophic** (energy) levels, food chains and energy pyramids.

- **Ecology** is the scientific study of organisms and their environment.

Trophic levels and food chains

As you saw in Chapter 2, insolation comes in two forms – heat and light energy. This heat from the sun cannot be used directly by plants and animals – but it does warm their surroundings and speed up chemical reactions. The sun's light energy is far more significant. It is captured by the green pigment in plant leaves (chlorophyll) which converts carbon dioxide and water into their organic compounds. These so-called 'building blocks' of plants are tissue and food energy in the form of chemicals called carbohydrates.

Did you know?

The transfer of light energy from insolation to food energy in an ecosystem is only 1% efficient!

Carbohydrates contain amino acids, sugars, starches, proteins, fats and vitamins – all the organic materials needed by animals for growth, movement and reproduction. Plants, therefore, form the basis of all nutrition and energy for the whole ecosystem. This is because they provide food for other organisms – which in turn feed others in what we call a **food chain**.

Plants represent the first **trophic or energy level** in the chain (Figure 3.3). They are called producer organisms (self-feeding **autotrophs**) as they produce their own food through photosynthesis.

All other trophic levels are occupied by **consumers** (other-feeder **heterotrophs**). These include all animals – birds, fish, reptiles, insects and mammals, including humans. We distinguish between primary consumers, which are vegetarian (**herbivores**), and secondary and tertiary consumers, which are meat eaters (**carnivores**).

Figure 3.3 *Trophic levels and simple food chains* ▼

Trophic level	Process	Examples		
Level 1 – producers	self-feeding **autotrophs** produce energy by photosynthesis	grass	oak leaf	phytoplankton
Level 2 – primary consumers	vegetarian **herbivores** eat plants	earthworm	caterpillar	zooplankton
Level 3 – secondary consumers	meat eating **carnivores** eat animals	house sparrow	blue tit	fish
Level 4 – tertiary consumers	top predators (**omnivores**) eat smaller animals	tawny owl	hawk	great white shark

The second trophic level is where herbivores eat producers directly. The third trophic level sees small carnivores feeding on the herbivores. Finally, the fourth trophic level is occupied by the larger carnivores. These top predators, known as **omnivores**, eat both plants and animals, and so have two sources of food.

All food chains also include **reducer organisms**. These operate at all trophic levels. They complete the flow of energy through the chain by returning any remaining nutrients to the soil to support new plant growth. Reducer organisms fall into two groups:

- **Detritivores** are animals which eat dead and decaying organisms. Examples include lice, earthworms and vultures.
- **Decomposers** are organisms which cause the decay and breakdown of dead plants, animals and excrement (Figure 3.4). Examples include bacteria and fungi.

Figure 3.4 *Decomposers at work* ▲

Energy pyramids

In any study of an ecosystem, nutrients should be thought of as being recycled, whilst energy is lost. A staggering 90% of energy is lost at each trophic level, mostly in animal respiration, movement and excretion. Consequently, insolation is progressively lost along the food chain and at each higher trophic level fewer organisms can be supported. These energy losses are illustrated using **energy pyramids**. For example, thousands of leaves at the first trophic level may feed hundreds of caterpillars at the second. But hundreds of caterpillars feed only one blue tit. So only a fraction of the energy taken in by one organism reaches the next.

Energy pyramids are also used to great effect in demonstrating the extravagant food and energy waste of protein-rich diets in high income countries (HICs) compared with the cereal-based diets of low income countries (LICs) (Figure 3.5).

Figure 3.5 *In a LIC 2 million kg of grass feeds 2000 people, whereas in a HIC the same amount of grass feeds only one person* ▼

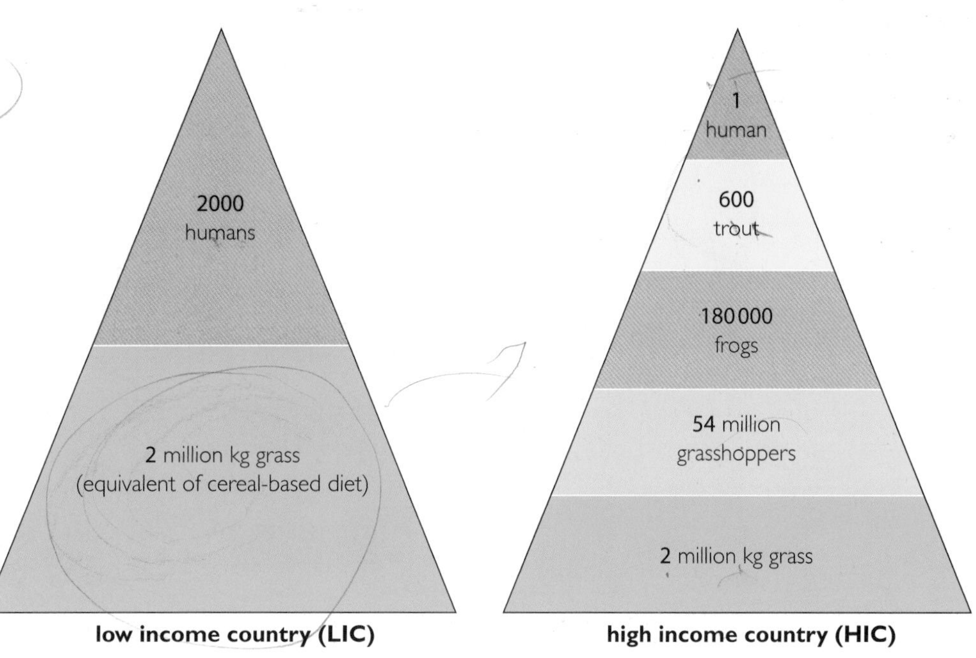

Food webs

Simple food chains are useful in explaining the basic principles behind ecosystems. But, most food chains show only one species at a particular trophic level (see below). If this were truly the case, then as this species was consumed all of the organisms in succeeding levels would be threatened by this loss of a food source. In reality this will rarely be the case, not least because most animals have multiple sources of food. And, any one species of plant or animal will be eaten by a variety of different consumers. In most ecosystems, therefore, such is the variety of plants and animals at each trophic level that a large number of food chains will be operating all at the same time and interconnecting with each other. The resulting complex network of linked food chains is called a **food web**.

> **Did you know?**
>
> *The term 'ecosystems' originates from the phrase 'ecological systems'.*

Marine ecosystems

Marine (aquatic) ecosystems illustrate food webs well (Figure 3.6). When the basic requirements for life (water, nutrients, heat, light) are considered, the sea provides more favourable conditions for organic production than land. In the sea, there is no water shortage, with abundant oxygen and carbon dioxide also readily available.

Temperature variations in the sea are less marked than on land and the transparency of the sea allows a thicker photosynthetic zone for phytoplankton, algae and seaweed at the first trophic level. The second trophic level consists of herbivores, such as zooplankton and the common periwinkle, which then eat the primary producers. The third trophic level includes fish, crabs and seals (carnivorous and omnivorous), with the fourth trophic level consisting of omnivores, such as polar bears and humans.

Figure 3.6 *A simplified marine food web* ▼

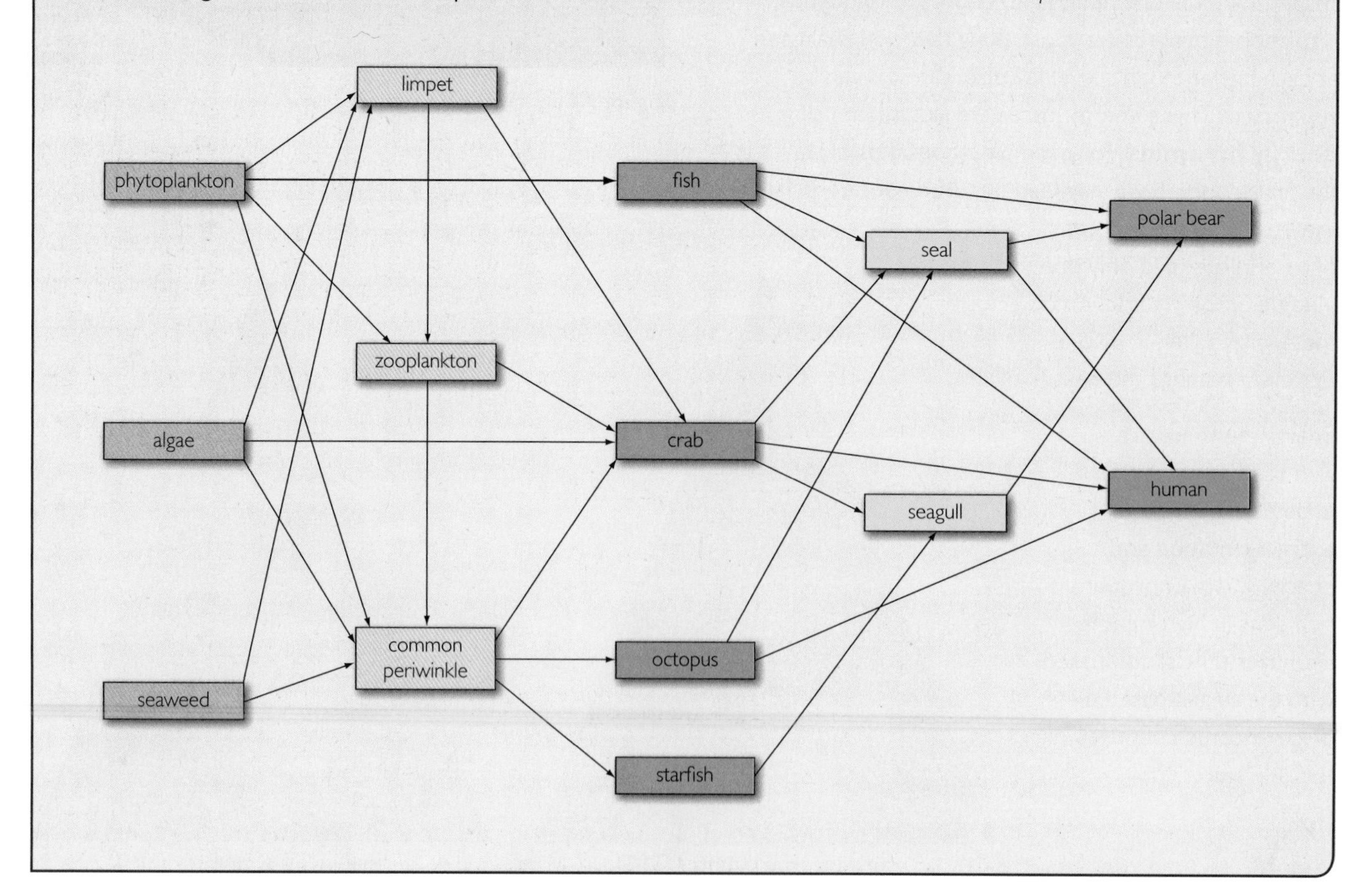

ACTIVITIES

1 Draw and label a simple food chain showing the four trophic levels, ending with humans.
2 Figure 3.5 shows contrasting energy pyramids.
   a Explain why there must always be a greater biomass of plants than of animals in an ecosystem.
   b Compare and contrast the losses of energy at each trophic level.
   c Explain the effect of this energy loss on humans.
3 Study Figure 3.6 and the diagram to the right which shows a selection of organisms in a typical English deciduous woodland ecosystem. Draw a four-column table listing all of the woodland organisms from the diagram, shown at their correct trophic levels. Complete the table to create a food web (similar to Figure 3.6) by drawing arrows linking each feeding interaction. An example has been started below.

Level 1 producers	Level 2 primary consumers	Level 3 secondary consumers	Level 4 tertiary consumers
oak tree leaves →	caterpillar →	wood mouse	
acorns → (caterpillar)	(caterpillar) →	robin	

Vegetation succession and climatic climax

In this section you will learn about:
- primary and secondary vegetation successions
- lithoseres and hydroseres

Ecosystems, especially on land, are complex. This is because so many elements of their environment are interacting and changing over time. The vegetation within any ecosystem reflects the interconnecting characteristics of its **habitat** – characterisics such as climate, soils, relief and human activities. As a result, specific plants will evolve with characteristics appropriate to their specific habitats – for example **hydrophytes** (such as water hyacinth) in water and **xerophytes** (such as cacti) in deserts (Figure 3.7).

Primary and secondary vegetation successions

Plant communities not only vary from area to area, but evolve and become more complex over time. This development is called a **vegetation succession** (or seral progression).

- Each stage in a vegetation succession is called a **sere**.
- If it is allowed to continue without being disturbed, a vegetation succession will end up in a state of perfect adaptation to, and equilibrium with, the environment at the time. This state of equilibrium is called the vegetation climax.
- The vegetation climax is sometimes called the **climatic climax vegetation** because climate is the main controlling factor.

There are two types of vegetation succession:
- Primary successions (priseres) happen on surfaces that have never had any vegetation growing on them. For example, on bare rock that has been exposed either by retreating glaciers or lava flows after volcanic eruptions, and on the sand dunes made by onshore winds at the coast.
- Secondary successions (subseres) happen on surfaces that have already been covered by vegetation (colonised), but have since been changed or destroyed. Fire from lightning strikes, landslides or human activities such as **deforestation** can all lead to a secondary succession.

Figure 3.7 *The fleshy stem of this Saguro cactus helps the plant survive in the hot and dry Arizona desert* ▲

- **Plant communities:** the various plants growing together in a particular habitat.

- **Climatic climax vegetation** is the final stage of a natural vegetation succession where all the plants are in equilibrium with the climatic and environmental conditions of the area. This plant community will remain so long as its habitat does not change.

Development of a vegetation succession

All vegetation successions pass through a sequence of stages or seres, starting with a **pioneer community** (**colony**) (Figure 3.8). This is a **ground layer** of hardy plants such as lichens and mosses that can grow without soil. As weathering breaks down the rock and dead plant remains are decomposed into humus by bacteria, a **field layer** of herbs and grasses begin to grow in the immature soil. These are taller than the ground layer, so become dominant until fast-growing shrubs take over. Taller plants will always dominate smaller ones by capturing light for photosynthesis that would otherwise reach plants lower down. They will also provide shelter to allow other plants to become established. Consequently, each new seral stage will show both an increase in the height of plants and in the number of species. Over time, the **shrub layer** will ultimately be taken over by a **tree layer** of taller, slower-growing trees. Providing that the environmental factors do not change, the vegetation should eventually reach a state of equilibrium at this point – the climatic climax vegetation.

This whole sequence can take anything from a few decades to thousands of years depending on the environmental circumstances. But in reality, human activities often change the natural vegetation succession, whether by clearing the climax vegetation or preventing the succession getting to this stage. As a result **plagioclimax communities** are more likely to occur (see later in this chapter).

Figure 3.8 *Natural vegetation succession along a raised beach – a lithosere* ▼

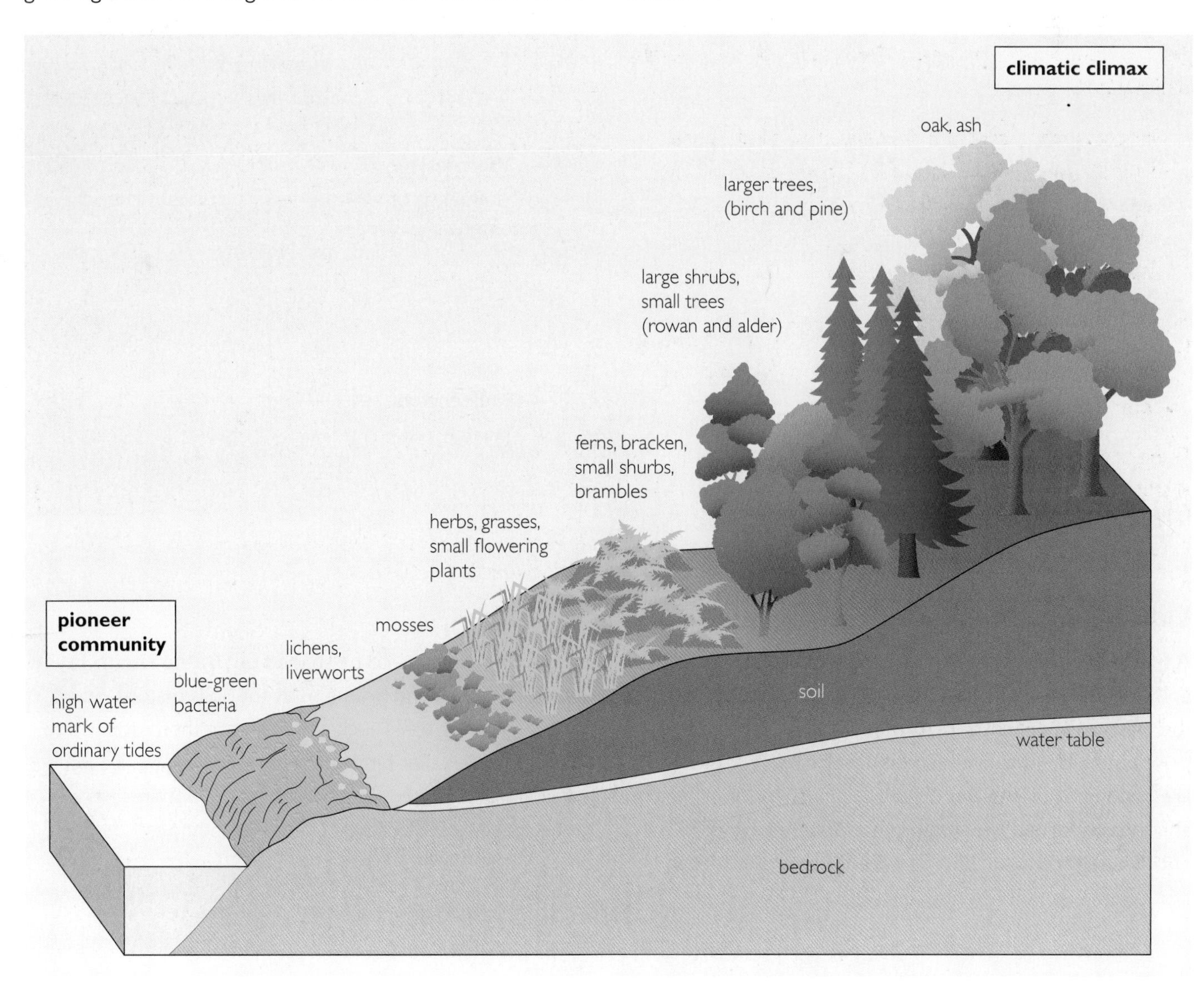

Throughout much of Britain there are four generally accepted environments where primary vegetation successions at a local scale could progress to the climatic climax vegetation of temperate, deciduous oak-ash woodland (Figure 3.9). Xeroseres (**lithoseres** and **psammoseres**) are found on dry land, and, hydroseres (**haloseres** and **hydroseres**) are formed in water.

Did you know?

Remnants of some of the last remaining natural climatic climax vegetation in northern Britain – Caledonian (Scots) pinewood – is preserved in the Royal Balmoral Estate, Upper Deeside, Aberdeenshire.

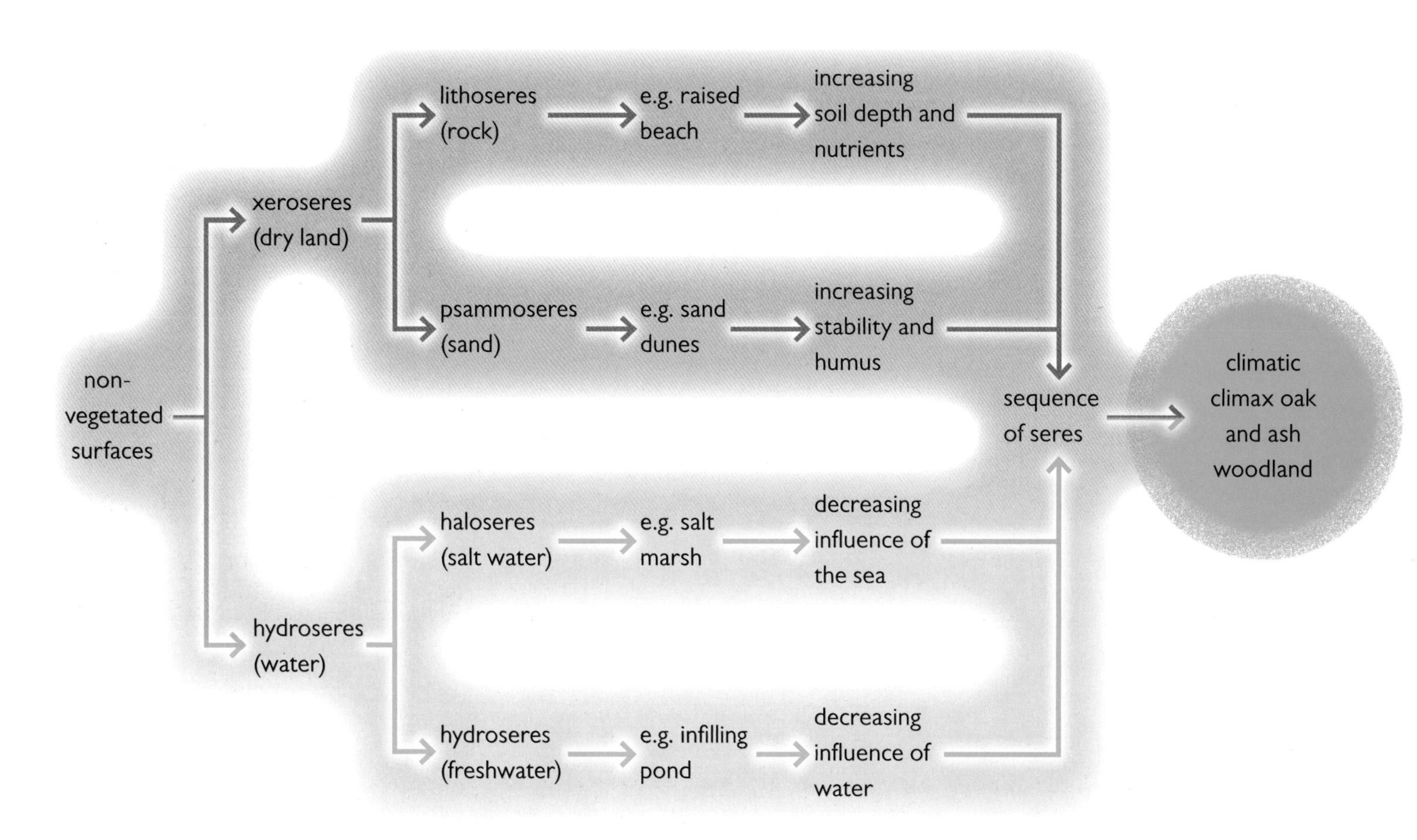

Figure 3.9 *Primary vegetation successions leading to climatic climax in Britain* ▲

Lithoseres

A **lithosere** occurs when bare rock is progressively colonised by plants as soils develop. Raised beaches on the west coast of Scotland, formed since the last Ice Age from both falling sea levels and **isostatic** uplift of the land, are good examples of this (Figure 3.8). Lichens, liverworts and mosses form a **ground layer** pioneer community encouraging soil formation. The **field layer** of herbs and grasses then follow – to be shaded out by the **shrub layer** of ferns, bracken and brambles as the soils mature. Small but fast-growing trees such as rowan can then establish to be followed by taller birch and pine trees. Finally, slower-growing oak and ash trees emerge as the dominant **tree layer** species of the climatic climax vegetation.

Hydroseres

Hydroseres develop where freshwater environments such as ponds and lakes silt-up over time. The spores of algae and mosses, blown on to the water surface, create a pioneer community of rafts of floating vegetation. Floating and submerged water weeds, including lilies, can then develop – trapping sediment at the water's edge to allow marsh plants such as reeds and rushes to establish.

Continuing sedimentation of both silt and plant debris will slowly build up to eventually rise above the water level to produce a fen (**carr**) of small shrubs and trees, including willow and alder. Finally, the climatic climax of oak and ash trees can emerge – taking over the whole site once the pond silts up entirely (Figure 3.10).

Figure 3.10 *Primary vegetation succession in a hydrosere* ▼

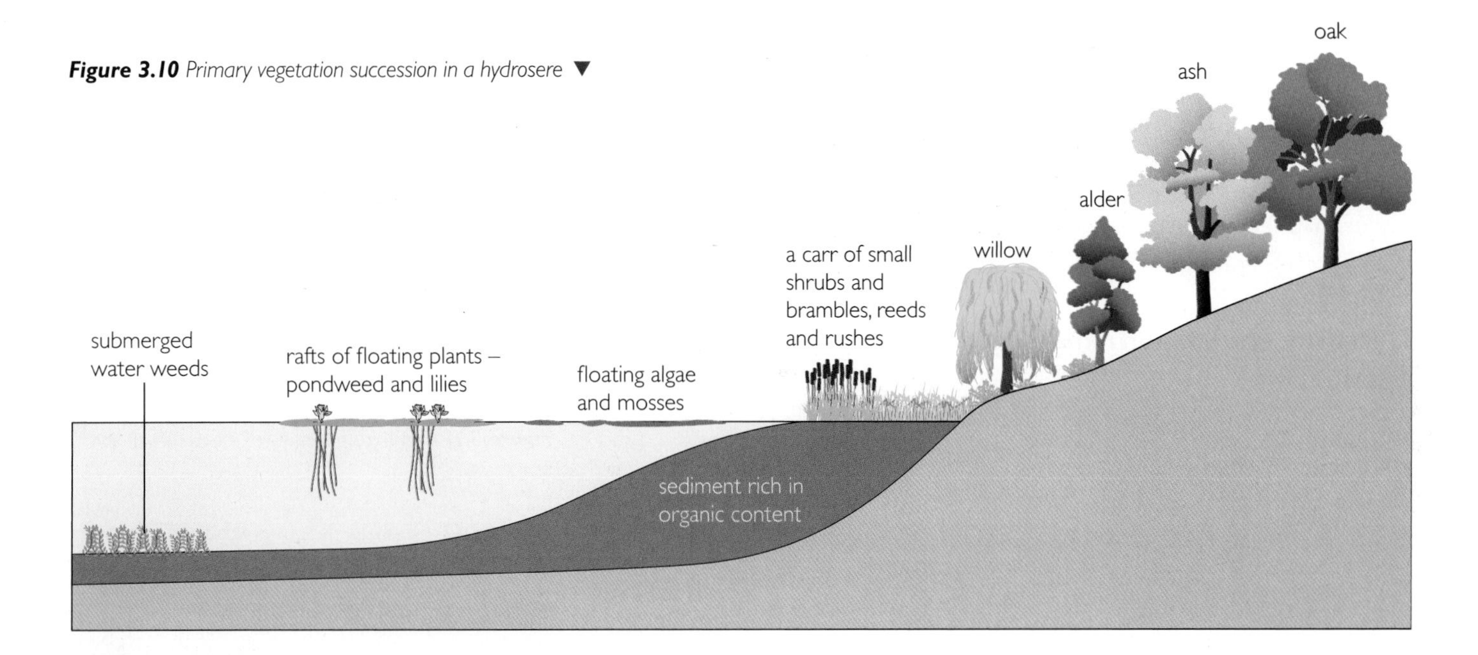

ACTIVITIES

1 Outline the main difference between primary and secondary vegetation successions. Use the Internet to find some photographs to illustrate your account. For example, investigate the impacts of the eruption of Mt St Helens (1980) on the surrounding landscape or the effects of a wildfire (e.g. Australia 2009, Russia 2010).

2 The illustration to the right shows a small British freshwater lake and its margins.

a Describe the climatic climax vegetation.

b Why are there no green plants rooted in the middle of the lake?

c What is the function of decomposer organisms found in the water and mud?

d Describe and explain the seral stages in the secondary vegetation succession which would occur if grazing was abandoned in the grassland area.

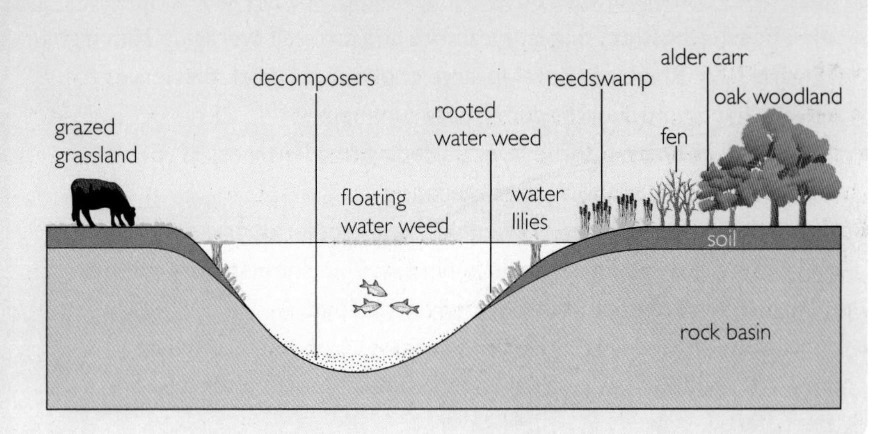

3.4 The temperate deciduous woodland biome

In this section you will learn about:

- the dynamic equilibrium and organic productivity of biomes
- stratification of vegetation in a temperate deciduous woodland biome
- brown earth soil

As we saw earlier in this chapter, biomes are large-scale ecosystems, such as deserts or rainforests. Biomes are mainly determined by climate (climate decides what type of vegetation will grow) but soil types play a role too. As a result, maps showing world climate zones, climax vegetations or soil types have very similar patterns.

But these maps are inevitably generalised – they show overall patterns, especially of vegetation and soil zones, rather than the wide range of local variations that factors such as relief, drainage and economic development cause.

Energy	Biome	NPP (g/m²/year)
high	tropical equatorial rainforest	2200
	temperate deciduous woodland	1200
average	tropical savanna grassland	900
	temperate coniferous forest	800
	Mediterranean woodland	700
	temperate grassland	600
low	tundra	140
	hot deserts	90

Net primary productivity (NPP) is the amount of energy absorbed or fixed by plants minus that which is lost through plant respiration

Figure 3.11 *Organic productivity of the world's major biomes* ▲

Dynamic equilibrium and organic productivity

Each biome, like any ecosystem, is in a state of balance with its environment. This balance is called a **dynamic equilibrium** because any change in the climate or soil changes the characteristics of the vegetation and animal life. Biomes are best compared by their organic productivity (the amount of organic matter they produce over a specific period of time), hence descriptions such as high-, average- and low-energy biomes are commonly used (Figure 3.11).

The temperate deciduous woodland biome

Temperate deciduous woodland – the climatic climax vegetation of much of Britain – is a high-energy biome, second only to the tropical equatorial rainforest in productivity (see page 122). Vegetation in a temperate deciduous woodland usually forms separate layers called **stratification**.

- Oak trees dominate, with other tall deciduous species such as lime, elm, beech, chestnut, maple, sycamore and ash – all averaging 20m high (Figure 3.12). The trees develop large **crowns** of broad, thin leaves to absorb maximum sunlight during cool summers.
- Below the **canopy**, a shrub layer (including smaller trees) of rowan, holly, hazel and hawthorn compete for light.
- Beneath this, a field layer of brambles, bracken, ferns, grass and flowering plants, such as bluebells, flourish in spring and early summer before the leaf canopy above has fully developed.
- Finally, the ground layer is limited to mosses and lichens growing amongst the thick layer of leaf litter which is important for ensuring deep, fertile soils.

Figure 3.12 *Temperate deciduous woodland, Hayley Wood in Cambridgeshire* ▼

Brown earth soil

The soil that develops in a temperate deciduous woodland is called "brown earth" soil. It is usually deep, well drained, fertile and supports a large amount of **fauna** (Figure 3.13).

Decomposers break down the thick layer of leaf litter to create rich, dark brown humus, which earthworms and rodents then mix into the layers (horizons) below – this mixing also helps to **aerate** the soil. Unless precipitation is particularly heavy, there is little leaching of nutrients out of the soil, which makes it slightly acidic and ideal for plant growth. Nutrient cycling is also efficient – deciduous trees need large amounts of nutrients, and the annual leaf fall ensures that nutrients are returned to the soil.

These ideal soil-forming conditions make it easy for the deep roots of deciduous trees to find moisture and nutrients, with little danger of becoming water-logged. This makes the agricultural potential of these soils enormous.

Most temperate deciduous woodland in the UK has been cleared for farming, urban development or for use as fuel and building materials. Only small pockets remain, such as Hayley Wood in Cambridgeshire, shown in Figure 3.12, which is now managed as a nature reserve.

- **Deciduous** plants shed their leaves in the autumn. This allows them to reduce **transpiration** when less moisture is available (much of the water in the soil is frozen) and helps protect the branches from snow and ice damage.

Figure 3.13 *Rich, crumbly brown earth soil* ▲

Did you know?

*In Britain, temperate deciduous woodland vegetation gives way to more **coniferous** species the further north you go. As such, the climatic climax vegetation across much of Scotland is temperate coniferous forest.*

ACTIVITIES

1 Using an atlas, compare and contrast world maps that show climate zones, natural vegetation and soil regions.
   - **a** What connections do you see between these three factors?
   - **b** Find the area where you live. What type of vegetation is there?
   - **c** Why might this type of vegetation not in fact be widespread in your local area?

2 Draw and annotate a sketch diagram to show vegetation stratification in a temperate deciduous woodland biome. Make sure that you clearly identify the ground, field, shrub and tree layers. The diagrams on pages 115 and 117 will help you.

The effects of human activity on a heather moorland

In this section you will learn about:

- arresting factors
- plagioclimax communities
- heather moorland management

As we saw on page 114, the final stage of a natural vegetation succession, where the plants are all in equilibrium with the climatic and environmental conditions of the area, is called the climatic climax vegetation. But in reality, human activities often affect plant succession – whether by clearing the climax vegetation or preventing the succession ever getting to that stage.

Human activities that interfere with the natural vegetation succession are called 'arresting factors'. These include deforestation, **afforestation**, ploughing, clearance by burning, animal grazing and trampling. The climax vegetation that results from these activities is called a **plagioclimax**. Put simply, a plagioclimax is a climax vegetation created by the actions of people.

Heather moorland

Heather moorland is found in many upland areas of Britain, such as the North Yorkshire Moors, and is an excellent example of a plagioclimax community (Figure 3.14). Despite being widespread and appearing to be a completely natural form of vegetation, in fact, it is heavily managed by individuals and organisations.

Some three thousand years ago, these areas were covered in climatic climax deciduous woodland on fertile brown earth soils. But once the woodland was cleared for farming (for both growing crops and grazing) the soils deteriorated. This left the naturally efficient nutrient cycling of deciduous woodland broken, exposing the ground to the heavy upland rainfall. Soils were both eroded and leached of their nutrients – leaving thin, acidic, and less fertile peaty **podsols**.

As a result, the upland areas became colonised by hardier plants, such as grass, bracken and heather, that could cope with the new conditions. These plants, called mixed moorland vegetation, could be maintained by sheep grazing. But the North Yorkshire Moors and other upland areas, such as the Eastern Scottish Highlands, are now being deliberately managed, mainly with controlled burning, to help keep a plagioclimax community of heather moorland.

Figure 3.14 *Upland heather moorland* ▼

Heather moorland management by burning

Heather is a valuable **evergreen** forage plant and is a major food source for hardy hill sheep. It is also the staple diet of red grouse – a bird that is important to the highly profitable shooting industry.

Heather is a woody shrub that can live in a wide range of climates and soils. As shown in Figure 3.15, its natural growth cycle follows four phases:

1. The **pioneer phase** sees heather seedlings establish themselves, with the root systems growing more quickly than the shoots.
2. The **building phase** is the most productive and valuable phase. The plants grow quickly, with plenty of flowers forming. This increases the biomass and helps the plants form continuous ground cover.
3. The **mature phase** follows as small gaps in the ground cover start to occur and the plants become woodier.
4. The **degenerate phase** sees the growth slowing down. The heather's maximum height is reached and the oldest branches die off. At this stage, there are large gaps in the ground cover and the value of the heather is limited.

Controlled burning keeps as much of the moorland as possible in the most productive phase – the building phase. This helps to keep the amount of edible green shoots at their highest.

The surface heather is burnt off in a 10 to 15 year rotation in small areas of about one hectare, which allows the burning to be carefully controlled. It also provides feeding areas for breeding grouse, with unburnt nesting cover nearby. Normally, within each square kilometre there are about six burning patches, resulting in a patchwork quilt pattern of heather at various stages of re-growth.

If the burning was stopped, the moorland would be invaded again by mixed moorland vegetation. A natural secondary vegetation succession would follow of scrub, birch, and in the long-term, a climax of deciduous woodland.

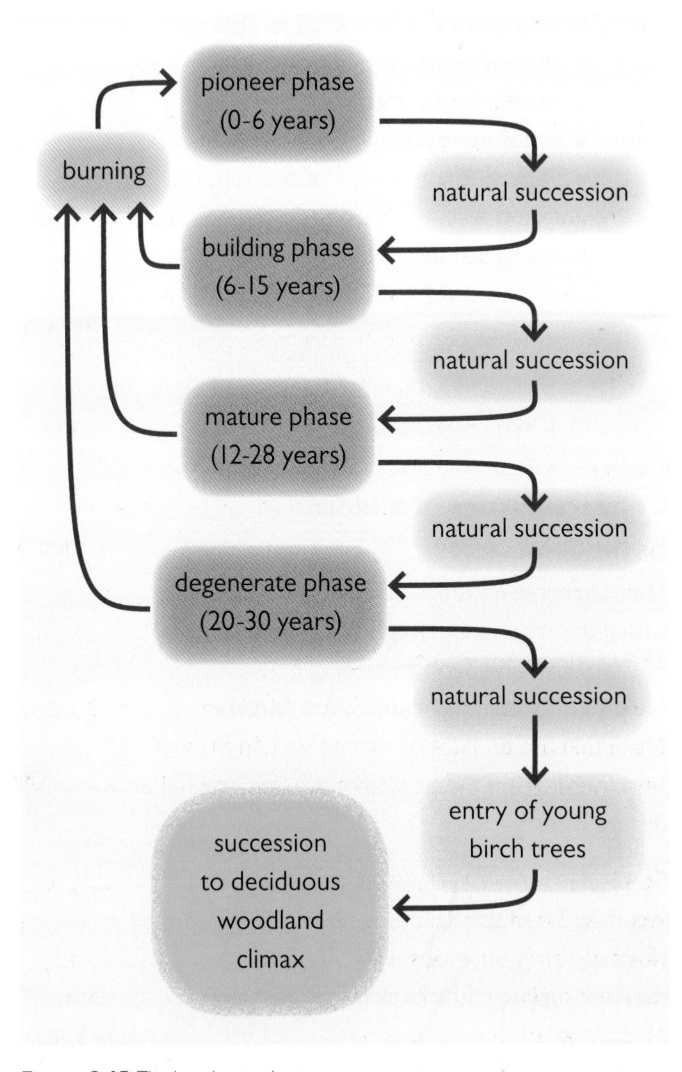

Figure 3.15 *The heather cycle* ▲

Did you know?

The 'Glorious Twelfth' in Britain refers to one of the busiest days in the shooting season. This is because August 12th sees the start of the shooting season for red grouse.

ACTIVITIES

1. Just like fox hunting, grouse shooting is a controversial topic. But its significance to the local economy, in often remote rural communities, cannot be underestimated. Write a paragraph examining the arguments ***for*** and ***against*** grouse shooting.
2. With the aid of simple sketches and/or photos from the Internet, describe the management practices that keep heather as a plagioclimax vegetation community in moorland areas.

3.6 Characteristics of tropical equatorial rainforest

In this section you will learn about:

- the distribution of the tropical equatorial rainforest biome
- the equatorial climate
- vegetation and soils of the tropical equatorial rainforest
- biological interdependency

"A tree is a wonderful living organism which gives shelter, food, warmth and protection to all living things. It even gives shade to those who wield an axe to cut it down." **Buddha**

Characteristics of a biome

Distribution

The rainforest biome is mostly located within the tropics, between the Tropic of Capricorn and Tropic of Cancer. The **tropical equatorial rainforest biome** lies within the equatorial climate belt, an area five degrees either side of the Equator (Figure 3.16).

Tropical equatorial rainforests now cover less than 5% of the Earth's surface – 200 years ago they covered twice as much. Yet they still support 50% of all living organisms on Earth.

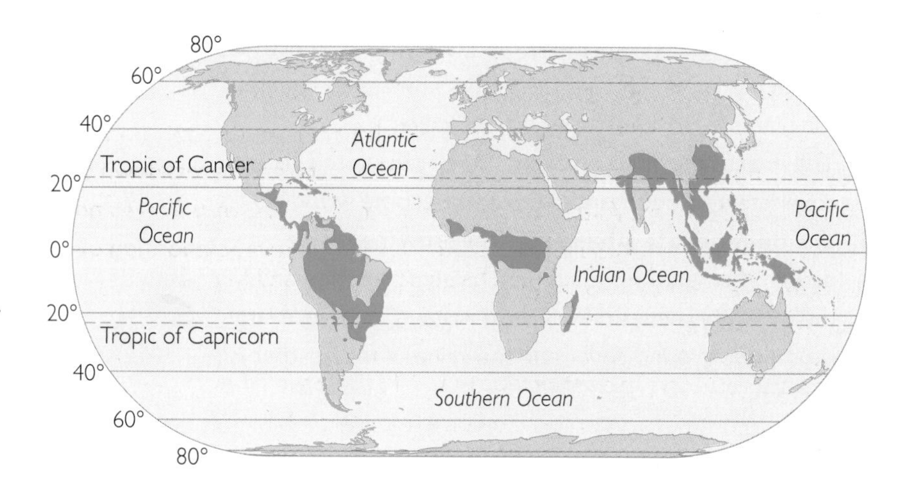

Figure 3.16 *Distribution of the tropical equatorial rainforest biome* ▲

Equatorial climate

The equatorial climate supports the richest **biodiversity** on the planet. The two ingredients necessary to support life in abundance – warmth (sunlight) and moisture – are ever-present. Some characteristics of an equatorial climate include:

- A low **diurnal** temperature range. The average daily temperature is about 28 °C. Temperatures rarely fall below 22 °C at night, while cloud cover restricts most daytime temperatures to 32 °C or lower.
- A low annual temperature range. This can be as low as 2 °C.
- A high annual rainfall. Warm, moist and unstable air is forced to rise at the inter-tropical convergence zone (ITCZ) (see page 64). This results in daily convectional storms that contribute to the 2000 mm of rainfall each year. The violent storms occur in the late afternoon after intense heating during the day.
- A year-round growing season. There are no defined seasons; insolation is evenly distributed throughout the year, with each day having around 12 hours of daylight.
- High humidity. Rapid evapotranspiration from swamps, trees and rivers creates a sticky and oppressive heat.

- **Biodiversity** is short for biological diversity. It is the number and types of organisms in an ecosystem, region or environment.

Did you know?

To help to remember the correct order of the Tropics, note that the Tropic of CaNcer lies North of the Tropic of Capricorn!

Vegetation

The ideal growing conditions provided by the equatorial climate result in the tropical equatorial rainforest being the most productive biome on the planet. It is characterised by an unmatched biodiversity. In a tropical equatorial rainforest there may be an almost incomprehensible 50 million species of animals, and as many as 480 species of tree per hectare (the size of a rugby pitch)!

The permanent rich green landscape of the equatorial tropical rainforest is deceiving – the trees are deciduous, so they lose their leaves. The year-round growing season, however, means that trees can shed their leaves at any time of the year. The constant fight for sunlight, which all plants need for photosynthesis, results in distinct stratification of the vegetation (Figure 3.17).

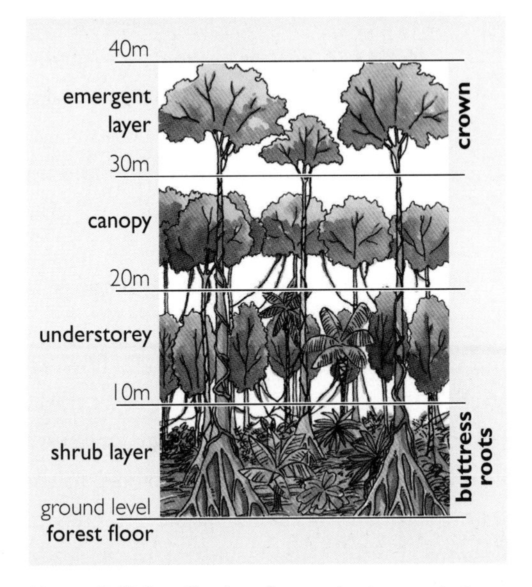

Figure 3.17 *Stratification of vegetation in a tropical equatorial rainforest* ▲

Soils

The hot and humid climate allows for rapid chemical weathering of the bedrock and creates perfect conditions for the rotting and breakdown of the large amount of leaf litter. Nevertheless, the soils are very fragile and depend on continuous leaf fall for nutrients. The rapid decomposition of the litter layer and the work of **biota** (such as ants) result in a thin humus layer (Figure 3.18).

80% of the nutrients in a tropical equatorial rainforest originate from the vegetation. The nutrient cycle of the rainforest is incredibly efficient at recycling these nutrients for sustainable growth. If the vegetation is removed, however, the heavy rainfalls cause nutrients to leach and the fertility of the soil is quickly lost.

Figure 3.18 *Profile of the iron-rich laterite soil commonly found in tropical equatorial rainforest* ▼

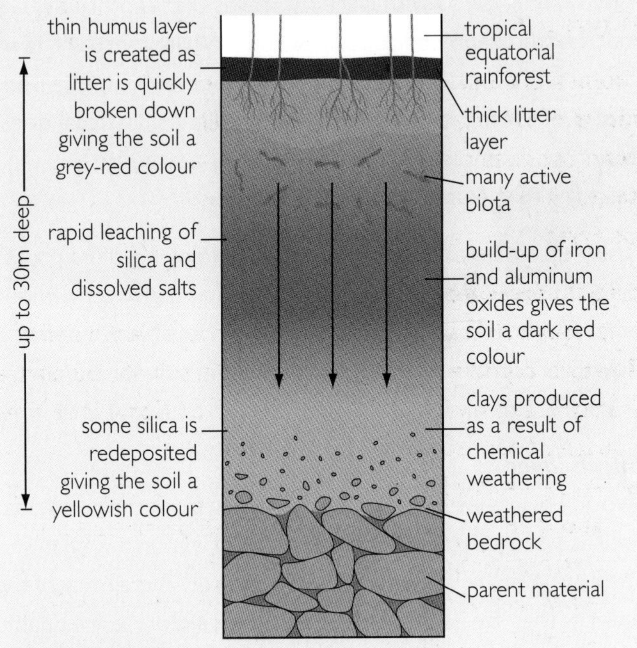

Biological interdependency – Brazil nut plantations

Plantations growing Brazil nut trees have not always been successful. Without the presence of Euglossine orchid bees, the trees cannot be pollinated. In other words, Brazil nut reproduction is dependent on the orchid bee. This is an example of **interdependence**, where one species is dependent on another species for survival. It is an important characteristic of the tropical equatorial rainforest and helps to explain why the effects of deforestation, when **keystone species** may be lost, are so catastrophic.

- **Keystone species**: an organism that links many other species together, much like the keystone of an arch.

ACTIVITIES

1. Why is the tropical equatorial rainforest biome described as having a high level of biodiversity?
2. How does climate account for the nature of the vegetation and soil in a tropical equatorial rainforest?
3. Using an atlas, locate two cities in Brazil: Belem and Manaus. Explain why Belem has a smaller temperature range but higher annual precipitation total than Manaus.

Adaptations to climate and the soil moisture budget

In this section you will learn about:

- adaptations of vegetation and animals to climate
- adaptations of vegetation and animals to a soil moisture budget

A walk into a tropical equatorial rainforest is to enter a magical and unfamiliar place. In 1752, Europeans identified the origin of large and very beautiful flowers attached to a tree's bark as a separate species that they imagined was rooted inside the tree trunk. They were wrong. So called **cauliflory** trees are carefully adapted to the needs of the tropical equatorial rainforest biome. They bear their own flowers on their trunks, which after dark emit a pungent smell. The flowers prove irresistible to bats that feed on the sweet nectar and pollinate the flowers (Figure 3.19).

Figure 3.19 *The beautiful flowers of a cauliflory tree* ▲

The forest floor

As little as 1% of sunlight passes through the vegetation layers to the forest floor. The few plants that live here are mostly small herbs with large flat leaves that capture as much of the dappled light as possible. Whilst there are few competing plants on the forest floor, such is the specialised **niche** of these plants that when more light than usual does enter, for example through the death of a large tree, it quickly kills the undergrowth.

Rising up to 2m above the forest floor, thick buttress roots help to spread the weight of the towering tree over a wide area. Other trees rely on stilt or prop roots, which point downwards from the main trunk into the soil, for support (Figure 3.20).

Shrub layer

Woody plants with many stems growing from their base, as well as younger trees, characterise this layer. In a way, it is the equivalent of a rainforest supermarket! A majority of the animals that live in a tropical equatorial rainforest are unable to climb. As such, herbivores compete for fallen seeds, fruit, nuts and leaves. Omnivores, unable to climb to safety from possible predators, have developed strategies to survive, including speed, stealth or camouflage. For example, the thick skin on the back of the neck of the Brazil tapir and its excellent swimming ability offer some protection from predatory crocodiles or cats, such as the cougar or jaguar.

Figure 3.20 *Stilt roots help to spread the weight of the tree. This is a particularly important adaptation where soil is thin or the ground marshy, such as in a mangrove swamp as shown here* ▶

Understorey

Slender and less substantial trees extend above the shrub layer. Whilst the trees only receive the hazy light that breaks through the canopy, they are far more tolerant of shade than most trees and still manage to thrive. The interlocking spindly branches of the trees form a 'spaghetti junction' of green corridors, along which lightweight animals can travel. One exception is the orang-utan of Borneo and Sumatra (see page 133), which despite its large weight, is able to swing from one bending branch to another.

Figure 3.21 *The striking green layer of the canopy interspersed by the emergent layer* ▲

An almost unnatural gap exists between the top of the understorey trees and those above them in the canopy. Growth of understorey trees is restricted to several metres below the base of the canopy. This allows them to receive a more even spread of sunlight and avoid some or all of the upper part of the tree, namely the crown, remaining in shadow.

Canopy

This is the most productive part of the forest and where photosynthesis is at its greatest. Each tree is carefully adapted to take full advantage of maximum exposure to the sun. Each of the huge mushroom-shaped crowns has an enormous photosynthetic surface – a leaf area of a mature tree may amount to ten times the area of the ground beneath it!

Heavy daily convectional rainfall has the potential to saturate leaves in the canopy, dissolving and washing away nutrients from the leaf surface. Fungi, mosses and algae, or **epiphylls**, may also grow in the puddles that form on the leaves, stealing sunlight from the leaf surface. Around 80% of canopy tree species grow **drip tip** leaves that help them shed water quickly and efficiently.

Emergent layer

These enormous trees tower above the canopy roof of the forest. The trees are particularly hardy to the elements, having to cope with the power of the sun, stormy winds and heavy rain. As any moisture is soon evaporated, one adaption to the climate is the development of thicker, non-drip tip leaves.

- **Symbiotic relationship**: close interaction that benefits both species involved in the relationship. The species often depend on this special relationship in order to survive and prosper.

ACTIVITIES

1. Sketch a copy of one of the vegetation adaptations discussed in this section or one of your own choice. The Internet is an excellent source of information on vegetation adaptation. Annotate your sketch to show how the vegetation is adapted to the equatorial climate and the soil moisture budget.
2. Describe how sunlight is the driving force in the layering of the tropical equatorial rainforest. Use examples to support your answer.
3. "Plants and animals of the tropical equatorial rainforest share a **symbiotic relationship**." To what extent do you agree with this statement? Include named examples of plants and animals in your answer.

3.8 The impact of human activity

In this section you will learn about:

- the causes of tropical equatorial rainforest deforestation
- the economic, environmental and social impacts of tropical equatorial rainforest deforestation
- the causes and impacts of deforestation in Amazonia

Did you know?

- *95% of global deforestation occurs in the tropics.*
- *Tropical equatorial rainforests may hold up to 90% of the world's biodiversity.*
- *On average, a species becomes extinct every half hour in the tropical equatorial rainforest.*
- *More than 25% of the world's population rely on forest resources for their livelihoods.*

What is greed? Is it possible for greed to be more than simply a selfish appetite at whatever cost? Arguably, greed has driven the deforestation of the world's tropical equatorial rainforest (Figure 3.22). Yet, this greed has funded the economic development of many LICs and met our demand for cheap manufactured goods and food.

Cause	Details	Examples
To provide space for farming. Most tropical equatorial rainforests are found in LICs that have rapidly growing populations. Land has to be cleared to farm and for building additional settlements (see below).	Traditional **slash-and-burn** shifting cultivation is very wasteful of space. **Cattle ranching** exposes new pastures to soil erosion and produces only low-quality meat. Transnational companies often run these ranches to produce meat for fast-food burgers sold in HICs. **Plantations** of cash crops, such as oil palm and soya bean, require more and more land to be cleared.	Yanomami Amerindian tribes in western Brazil. Southern margins of Amazonia. Soya grown in Amazonia is used to feed chickens in the UK which then become chicken nuggets for fast-food chains.
Logging to provide timber. The increasing demand for valuable hardwood timbers in HICs provides a reliable and essential source of income.	The continued demand for furniture, building materials and other wooden products has encouraged felling of valuable equatorial hardwoods such as mahogany, teak and rosewood.	Japan alone accounts for a staggering 11 million cubic metres of equatorial hardwood used a year.
Mining to develop the natural wealth of vast reserves of important and valuable minerals. Bauxite, iron ore, tin, copper, lead, manganese and gold are very valuable, but only accessible once the rainforest is cleared.	The cheapest method of mineral extraction is open cast mining. This results in large-scale deforestation as trees and soil are stripped from the underlying rocks.	Carajas in Amazonia is the location of the world's largest source of iron ore. The bauxite mine of the Juruti Project in the Amazonian state of Para will result in the loss of 10 500 hectares of rainforest.
Road construction supports the development of rainforest for other uses.	New roads are essential for allowing people in and raw materials out. New roads are both privately constructed as well as government-funded. But they cut broad destructive swathes across the rainforest.	The Trans-Amazonian Highway extends 6000 km into Brazil's interior.
Hydro-electric power (HEP) provides cheap and plentiful renewable energy.	Once dams are constructed, the high rainfall totals in tropical equatorial rainforest environments provide great potential for HEP generation. But the reservoirs behind the dams flood large areas of cleared forest.	Power for the Carajas iron ore mine is generated by the HEP station at the Tucuruí dam on the Tocantins River.
Settlement growth reflects the rapid population increase.	Population pressure comes from both natural growth and migration. The need for relocation cannot be ignored if other areas cannot support their populations.	Brazil has over 25 million landless people who need land to settle. Migration from the poorest parts of Brazil, such as the drought stricken north-east is, therefore, encouraged. The new roads have become 'growth corridors' for many of these 'colonists'.

Figure 3.22 *Key causes of tropical equatorial rainforest deforestation* ▲

Economic, environmental and social impacts of tropical equatorial rainforest deforestation

Economic impacts

Since the start of mechanised logging in the 1950s and the development of an integrated transport infrastructure, deforestation rates have soared (Figure 3.23). Whilst logging is one solution to raising the quality of life in LICs, the economic benefits of deforestation rarely seem to trickle down to those who are most in need. Corruption, poor governance and inadequate law enforcement frequently mean that local people, for example small-scale farmers, are often exploited by those with power and money. Foreign transnationals, for example **agribusinesses**, often have a presence in areas of deforestation. Controversially, the World Bank has supported plans to expand the timber industry in areas such as the Congo Basin, Africa and parts of Indonesia.

Figure 3.23 *The smoke and fires of continued deforestation in the Amazon rainforest. The brown areas show where the land has already been cleared for logging, ranching and farming.* ▲

Environmental impacts

An estimated 20 000 tropical equatorial rainforest species become extinct each year, most without ever having been discovered.

The loss of tropical equatorial rainforest places huge pressure on the drainage basin system. Fragile soils become exposed to direct rainfall, and without the wide protecting canopy above, the resultant runoff leads to rapid soil erosion and leaching. Water courses may become silted, leading to localised flooding. Furthermore, with reduced rates of evapotranspiration, regional rainfall patterns may be altered.

A tropical equatorial tree acts as a **carbon sink**, and absorbs around 22.5kg of carbon dioxide each year. Conversely, tropical equatorial rainforest destruction, mostly through burning, accounts for around 20% of global atmospheric carbon dioxide released each year.

Social impacts

The tropical equatorial rainforest not only provides the necessities for people to survive, namely food, medicine and shelter, but is the source of beliefs, values and social structures. The Quichua Indians of Tena, Ecuador are just one example of native people who inhabit the rainforest. Deforestation clears more than just their physical environment – it destroys their social cohesion, language, traditions and culture.

The indirect effects of deforestation impact on lives far away from the rainforest, too. For example, flooding and mudslides in the Philippines in 2004, which killed up to 500 people, were widely blamed on illegal logging upstream which increased rates of runoff and silting of river channels.

CASE STUDY

Deforestation of the tropical equatorial rainforest of Amazonia

The legend of El Dorado, an ancient empire deep in the Amazon rainforest, tells of a South American Indian Chief who made offerings to the sun god by covering himself in gold dust and then washing it off in a lake. Driven by legends such as this, early European explorers ventured deep into the rainforest, searching for gold and other treasures.

El Dorado remains undiscovered, yet the riches of the Amazon tropical equatorial rainforest have been remorselessly exploited, particularly since the 1960s when large-scale deforestation has dominated.

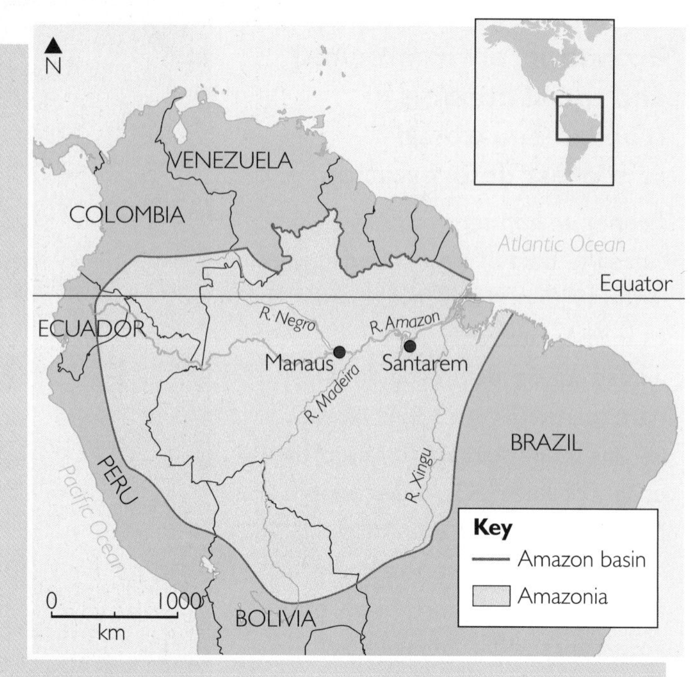

Figure 3.24 *Amazon region (Amazonia)* ▲

Opening up the Amazon

With some exceptions, such as the east-west Trans-Amazonian Highway and the north-south BR-163 'Soy Highway', the majority of the roads in Amazonia are illegal (see page 131). The initial tracks are built by loggers to help access mahogany and other valuable hardwoods. After these first cuts, the 'wounds' in the forest are opened up by an unhealthy mix of ranchers, farmers, squatters and, frequently, hired gunmen. They clear previously inaccessible areas of forest to make it appear as if the land belongs to them. Corruption and heavy-handiness then help the new settlers, or 'Grileiros', steal the land and acquire 'legal' land titles.

Economic impacts

The poorest in Brazil have rarely benefited from the economic exploitation of Amazonia. Ironically, in the 1970s, the first settlers practised primitive slash-and-burn farming, which had limited impact on the forest. Yet, they were rarely awarded ownership of the land. Instead, the Brazilian government offered those who were better-off financially up to 2995 hectares for extensive ranching, logging and other developments in order to generate a greater return.

The government has continued to sideline the smaller growers who are more likely to live in harmony with the rainforest, instead, offering ever more attractive **subsidies** to large commercial operations. Transnational organisations, frequently from the USA, are at the centre of these economic activities which have made Brazil a world leader in the exports of beef and soya beans.

Elsewhere, HEP projects (for example there are seven dams on the Xingu and Madeira rivers alone) help to deliver power to aluminium smelters and mining operations. Mineral extraction is also of great significance – the vast open-cast workings at Carajas, in north-eastern Brazil, employ over 7000 workers (Figure 3.25).

Figure 3.25 *Carajas is the world's largest iron ore mine* ▼

Environmental impacts

At its height, 5000 people every month were moving to the formerly uninhabited region of Rondonia, on Brazil's western border with Bolivia. Typical of much of Amazonia, the migrants are too great in number and lack the necessary skills and knowledge to practise **shifting cultivation** successfully. The cleared plots are too extensive and not given sufficient time to recover their fertility before being cleared again. Where migrants move into areas that are already occupied by indigenous shifting cultivators, such as the Yanomami, there is increased pressure on a system of agriculture that has worked with nature for thousands of years. In both cases, the soil is degraded, quickly loses fertility and yields decline. Inevitably, these developments have had an impact on species diversity and on the natural cycles in the rainforest.

Large-scale and capital-intensive **monoculture**, such as soya beans, demands high inputs of fertilisers and pesticides as well as lime. Leaching results in the toxins entering water courses, poisoning water and killing fish.

Social impacts

The dirt road neighbourhoods that cling to the snake-like tracks and roads of Amazonia seem to symbolise the desperate hopes of settlers. Each new track into the forest is a new opportunity for some, and quite literally the end of the road for others. Grileiros or miners sometimes try to illegally steal land from native Indians. The traditional Yanomami and Yekuana tribes in the border region with Venezuela have vigorously opposed mining (Figure 3.26).

New settlers also bring new threats to health and traditional cultures. The Panará Indians of the Pará and Mato Grosso states have all but been wiped out, killed by diseases that travelled northwards with the settlers as the BR-163 road was built in the 1970s.

Figure 3.26 *The very existence of indigenous Amerindians, such as the Yanomami hunter shown here, is threatened* ▲

ACTIVITIES

1 **a** Draw a table similar to the one below to summarise the economic, social and environmental impacts of tropical equatorial rainforest deforestation.

Economic	Social	Environmental

b Then, rank the three impacts (economic, social and environmental) in order of importance. Remember to consider impacts at all geographical scales – local through to global. Be sure to give reasons for your answer.

2 Why is so much tropical equatorial rainforest in Amazonia being cleared illegally?

3 Who do you think is to blame for the rapid rates of deforestation in Amazonia? Use examples to explain your answer.

Development issues

In this section you will learn about:

- the conflict between development, biodiversity and sustainability
- development issues of soya agribusiness in the Amazonian tropical equatorial rainforest
- the contrasting impact of tourism in Uganda and Borneo

A broad range of development issues impact on biodiversity and the potential for sustainability in tropical equatorial rainforests such as Amazonia. Plantation agriculture producing soya is one such pressure.

Figure 3.27 The states of Pará, Mato Grosso and Amazonas, Brazil ▲

Cargill: an example of an agribusiness

Cargill is a large US transnational food, agricultural, financial and industrial agribusiness. It employs 138 000 people in 67 countries, and in 2009 recorded a profit of over $3.3 billion. Cargill has published commitments to its customers and the countries within which it operates:

- We will conduct our business with high levels of integrity, accountability and responsibility.
- We will develop ways of reducing our environmental impact and help conserve natural resources.
- We will treat people with dignity and respect.
- We will invest in and engage with communities where we live and work.

Source: www.cargill.com

Cargill: a questionable example of a sustainable transnational?

Since Cargill arrived in Santarém, Brazil in 1965, growing soya has been the major driver of deforestation in the municipality. Between 2002 and 2004, annual deforestation rates jumped from 15 000 to 28 000 hectares in Santarém and the neighbouring municipality of Belterra.

Forests have started to give way to mechanised soya plantations. In Santarém, it is estimated that 10% of the land deforested in recent years is now soya plantations. Yet this figure fails to show the whole impact the soya industry and its infrastructure is having in the region. Much land is also indirectly converted – secondary forest is felled, and squatters are forcibly removed from land they have already cleared (inevitably leading to clearance elsewhere).

Source: www.greenpeace.org.

A growing demand for soya

Europe buys half the soya exported from the Amazon state of Mato Grosso, where 90% of rainforest soya is grown. The soya is used to produce animal feed, which in turn, helps to meet the European demand for cheap supermarket meat.

Figure 3.28 A soya plantation cut out of the Amazon rainforest ▲

The global trade of soya

97% of the soymeal produced worldwide is used for animal feed.

78% of UK soya beans are imported from Brazil.

The Cargill port at Santarém

Cargill built a port in 2003 on the Amazon at Santarém to export soya to Europe where the grains are converted into feed for chickens and other farm animals. Farmers from soya-growing areas like Mato Grosso moved to Santarém to be nearer the port, in order to take advantage of reduced transport costs. This influx of outsiders led to conflict with local people.

Most of the soya that Cargill exports through Santarém is brought in by trucks along the BR-163 highway from Mato Grosso (Figure 3.29). This road is in the process of being paved, which environmentalists say will lead to more deforestation along its length and more soya farmers moving to the area.

Source: www.theecologist.org

Figure 3.29 *The BR-163 'Soy Highway'* ▲

Sustainable development: a politician's point of view

'We have no problems to reach zero percent deforestation in the Amazon. We've had an agreement for four years that soya growing will not fell a single tree in the Amazon biome. That's inspected by NGOs. In the livestock sector, another programme was introduced to achieve zero deforestation in the Amazon biome within two years, where all the farms are monitored by satellite. And those who fell a single tree or who go against the agreement will be punished and will lose their production and their right to the forest.'

Brazil's Agriculture Minister, Reinhold Stephanes, 2010

Source: www.farmfutures.com

Alley-cropping: an alternative to slash-and-burn

As an alternative to the slash-and-burn approach to clear land, alley-cropping offers a sustainable farming solution in areas of tropical rainforests. Fast growing saplings of hardy species, such as the Inga tree in Central and South America, are planted in rows on previously farmed land. Within 18 months, the trees form a canopy and the resultant shaded 'alleys' create darkened spaces in which grasses and weeds, starved of light, die off. The leaves of the canopy are then pruned and placed on the ground to form a thick layer of decomposing mulch. The rotting mulch gradually returns nutrients to the soil, which may then be used to plant seeds such as maize.

Source: www.rainforestsaver.org

ACTIVITIES

1. Who are the winners and losers from the dramatic growth of soya production? Explain your answer.
2. Justify the Brazilian government's decision to allow transnationals, such as Cargill, to farm in Amazonia.
3. Outline reasons why alternative sustainable farming methods, such as alley-cropping, are less common than more traditional methods such as slash-and-burn.
4. To what extent is soya production a sustainable activity in the area around Santarém? Your answer should consider biodiversity versus development needs.

CASE STUDY

Development issues: The Batwa pygmies, Uganda

In 1992, Bwindi Impenetrable Forest and the Mgahinga forest reserves, Uganda, were upgraded to national parks (Figure 3.30). Two years later, Bwindi Impenetrable National Park was declared a World Heritage Site. The justification for this protected status is the unique biodiversity of the parks' tropical equatorial rainforest.

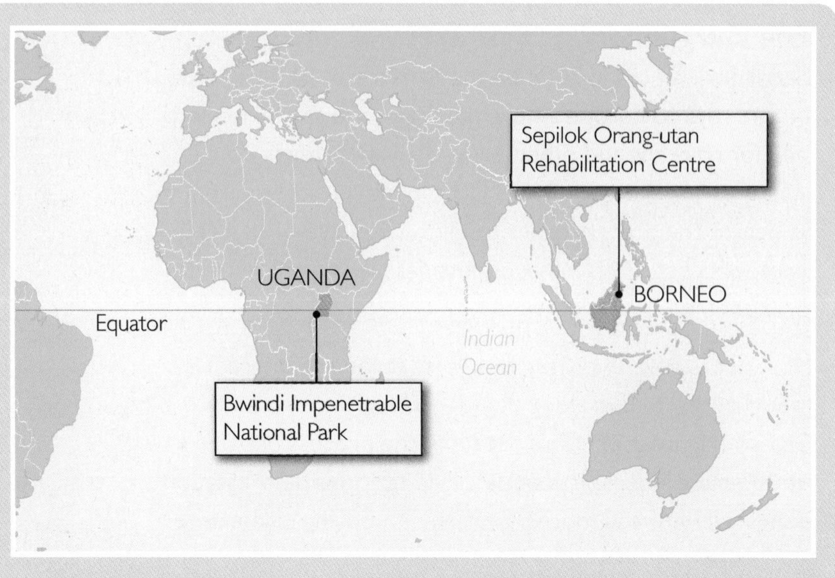

Figure 3.30 *Bwindi Impenetrable National Park and Sepilok Orang-utan Rehabilitation Centre, Borneo* ▲

As national parks, access to forest resources became severely restricted. In taking direct action, the government has protected the areas' biodiversity, including the endangered mountain gorilla. The parks have also stimulated a healthy tourist industry that in turn has encouraged economic development.

However, the success has largely had a negative impact on traditional tribes living locally. They were excluded from the government's decision-making process and some communities were evicted from their homes.

Bwindi Impenetrable National Park contains:

- 120 mammal species
- Over 50% of the world's gorilla population
- 360 bird species
- 200 species of butterflies
- 324 tree species

Human impact

The indigenous Batwa pygmies were removed from their traditional tropical equatorial forest homes in 1992 (Figure 3.31). Largely forgotten by the government and misunderstood by outsiders, the Batwa have been forced to settle on the fringes of existing settlements.

The Batwa were ill-prepared for modern life. For example, as hunter-gatherers, they hunted for food and harvested the forest resources, such as honey from hives within forest trees. Today, without any land of their own, they have to find work in order to buy food locally.

Tourism is a mixed blessing for the Batwa pygmies. The economics of tourism have clearly influenced the government, both in the decision to evict the Batwa and in marketing the colourful culture of the Batwa as a tourist attraction (not helped by tour operators who advertise 'trips to see pygmies').

The future?

However, new sustainable community-based projects, such as the Batwa Trail in the parks, offer some hope. When the new trail is complete, pygmy guides will take visitors into the heart of the tropical equatorial forest. Along the way, the guides will share some of their ancient skills, crafts, music and dance. Money raised from this project will be put back into the local pygmy community, towards services such as health and education.

Figure 3.31 *The indigenous Batwa pygmies of Uganda* ▲

CASE STUDY

Development issues: Sepilok Orang-utan Rehabilitation Centre, Borneo

Economic development

You are probably aware that in recent years orang-utan numbers in the wild have dramatically declined. This is as a result of human activities in their native environment of the tropical equatorial rainforest on the islands of Borneo, Indonesia and Sumatra, Malaysia (Figure 3.30). In these rapidly industrialising economies, the pressures of development have led to an 80% loss of the orang-utans' rainforest habitat in the last 20 years alone. The main cause for this dramatic decline in numbers is deforestation as a result of logging, mining and farming (particularly soya palm oil plantations). A further threat has been poaching to feed the appetite of the Asian pet trade.

Protection and rehabilitation

Sepilok Rehabilitation Centre in North Borneo was set up in 1964 to care for orphaned, sick or injured orang-utans (Figure 3.32). Whilst the centre lies at the edge of the Kabili Sepilok Forest Reserve, there are no boundary fences and around 80 orang-utans are free to return to the wild at any time. Several other endangered species have also been treated at the centre, including sun bears, gibbons, Sumatran rhinos and injured elephants.

Sustainable tourism

Sepilok plays an important conservation role and is a popular tourist destination. It is also a model of **sustainable tourism**. Visitor access to the reserve is restricted to walkways, and tourists are not allowed to touch or handle the apes. Sepilok is a centre for research and education and has generated global awareness of the protection laws for endangered species. For example, the Malaysian government has invested greater resources, including longer prison sentences, in the fight against illegal trading of endangered species as pets.

Responsible tourism

"Seeing so much wildlife in its natural habitat, especially the incredible orang-utans, and visiting the longhouses of the Iban tribe make this trip unique and unforgettable. Helping to preserve this spectacular part of the world and the culture and wildlife that flourish here just by visiting it is a fantastic opportunity."

Source: www.responsibletravel.com

Did you know?

The Sumatran orang-utan may be the first great ape to become extinct in the wild.

Figure 3.32 *The orang-utans of Sepilok Rehabilitation Centre* ▲

ACTIVITIES

1 Consider the impacts of tourism on the tropical equatorial rainforest.
   a What are the possible knock-on, or multiplier effects, of tourism on tropical equatorial rainforests?
   b To what extent is tourism a 'double-edged sword'?
2 Do you agree that endangered species such as orang-utans should be protected at whatever cost?
3 Explain how responsible tourism supports the principle of sustainable tourism. Use information on this page and from the www.responsibletravel.com website to support your answer.
4 To what extent does tourism support the rich biodiversity and the economic development of the tropical equatorial rainforest? Be sure to use examples in your answer.

3.10 The impact of urbanisation on ecosystems

In this section you will learn about:

- the nature of urban ecosystems
- influences on urban ecosystems
- plant succession on wasteland
- the introduction of invasive species to British ecosystems

Life maintains conditions suitable for its own survival.

the Gaia theory

An increase in the proportion of people living in towns and cities, as compared to rural areas, is the process called urbanisation. Urban areas now house well over half of the world's population. As a result, ecosystems in these built-up areas are rapidly changing, namely in response to human activities and changes to local climates, water supply, and air quality (see Figure 3.33). This creates a variety of **urban niches**.

Cause	Example	Effect
creation of watertight surfaces	• tarmac roads • concrete	limits soil moisture storage, soil depth and root growth
compaction of soil	• building foundations • transport	restricts free drainage leading to waterlogging; limited soil aeration
acidification of soil	• acid rain	reduces availability of nutrients
increase in planned areas of vegetation cover	• establishment of recreational spaces such as gardens and parks	local natural ecosystem may be invaded by non-native species from gardens or other types of landscaping
construction of residential, commercial and industrial buildings	• homes • offices • warehouses	provides shelter and shade
water, air and land pollution	• waste matter including chemicals and heavy metals such as lead	stunts plant growth; only pollution-tolerant species may survive

Figure 3.33 *The impact of urbanisation on ecosystems* ▲

- **Urban niche**: a specialist urban habitat, such as around neglected buildings and on wasteland.

Colonisation of wasteland

Urban wasteland is a general term used to describe abandoned land that has essentially been left for nature to take its course. Called **brownfield sites** they include places such as former factories or routeways such as canals and narrow strips of disused railway (see page 136). Despite the particular difficulties that these sites pose, including contaminated ground and limited soil depth, plant succession is surprisingly quick. Succession on abandoned industrial sites follows a lithosere-type succession, with adapted plant species changing the soil conditions that then favour the species which follow (Figure 3.34).

Did you know?

Some European capitals, such as Brussels and Warsaw, contain up to 50% of the flora found in their entire countries.

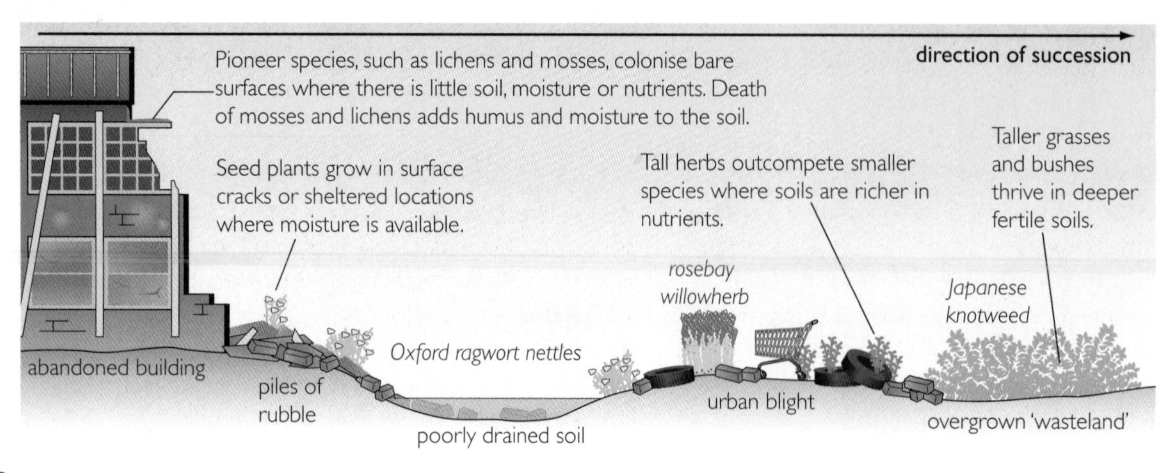

◀ ***Figure 3.34*** *Plant succession on abandoned industrial sites or wasteland*

Introduction of new species

You may not be aware, but Britain has been invaded by outsiders from all parts of the globe. And mostly, we are to blame! So-called 'invasive non-native species' are both planned and unplanned – but always not indigenous to the British Isles. Planned invaders are frequently introduced by gardeners, for example, as decoration or for erosion control. Unplanned invaders are carried by animals, birds and the wind.

While not all alien plant species are to be discouraged (some, such as tomatoes and corn, are beneficial and do not compete with native plants), without any natural predators or common ecologically limiting factors, the non-native species spread rapidly and invade local ecosystems. Frequently able to outcompete traditional species for habitat and food, they reduce biodiversity and demand careful management and conservation strategies.

Japanese knotweed

One of the most invasive weeds in Britain, Japanese knotweed, was brought into the country in Victorian times as an ornamental plant. In London alone, the weed has doubled its coverage in the last 20 years and is present along most of the capital's waterways.

Japanese knotweed is resistant to treatment – its stems re-grow, float downstream and subsequently disperse at a new site. It can only be effectively treated by spraying with strong herbicides and then completely disposing of the weed and roots, for example, by burning. Cleared sites need regular monitoring to be kept weed-free.

If left untreated, Japanese knotweed will consume local biodiversity and is a real threat to conservation areas.

Figure 3.35 *Japanese knotweed* ▶

ACTIVITIES

1. *80% of Europe's population now live in urban areas.* Describe and explain the likely impact that this has had on urban ecosystems. In your answer include reference to the following:
   - urban niches
   - plant succession
   - biodiversity.
2. For either of the species listed below, explain why it is considered an invasive species in Britain. You may need to use the Internet to help you find information for your answer:
   - Buddleia
   - Himalayan balsam.

3.11 Routeways and the rural-urban fringe

In this section you will learn about:
- wildlife corridors along routeways
- changes in the characteristics of the rural-urban fringe
- development pressures on the rural-urban fringe

Ecosystems along routeways

Road kill! This rather alarming phrase is used by many people to describe the loss of wildlife directly caused by road traffic. Conservationists argue for changes to the built environment to protect wildlife, such as road modifications to allow safe crossing points or route changes to avoid unique habitats. Yet ironically, routeways such as roads, railways and canals provide niches for distinctive ecosystems (Figure 3.36). They are unique because transport movement along these corridors brings in new, non-native species. And while these corridors are somewhat maintained, for example weed control on grass along the sides of roads or salt spreading on roads in winter, they are comparatively free from human interference.

Did you know?

The United States has 6.5 million kilometres of roadways, on which an estimated one million animals die every day.

Figure 3.36 *Characteristics of ecosystems along typical road, rail and canal routeways* ▼

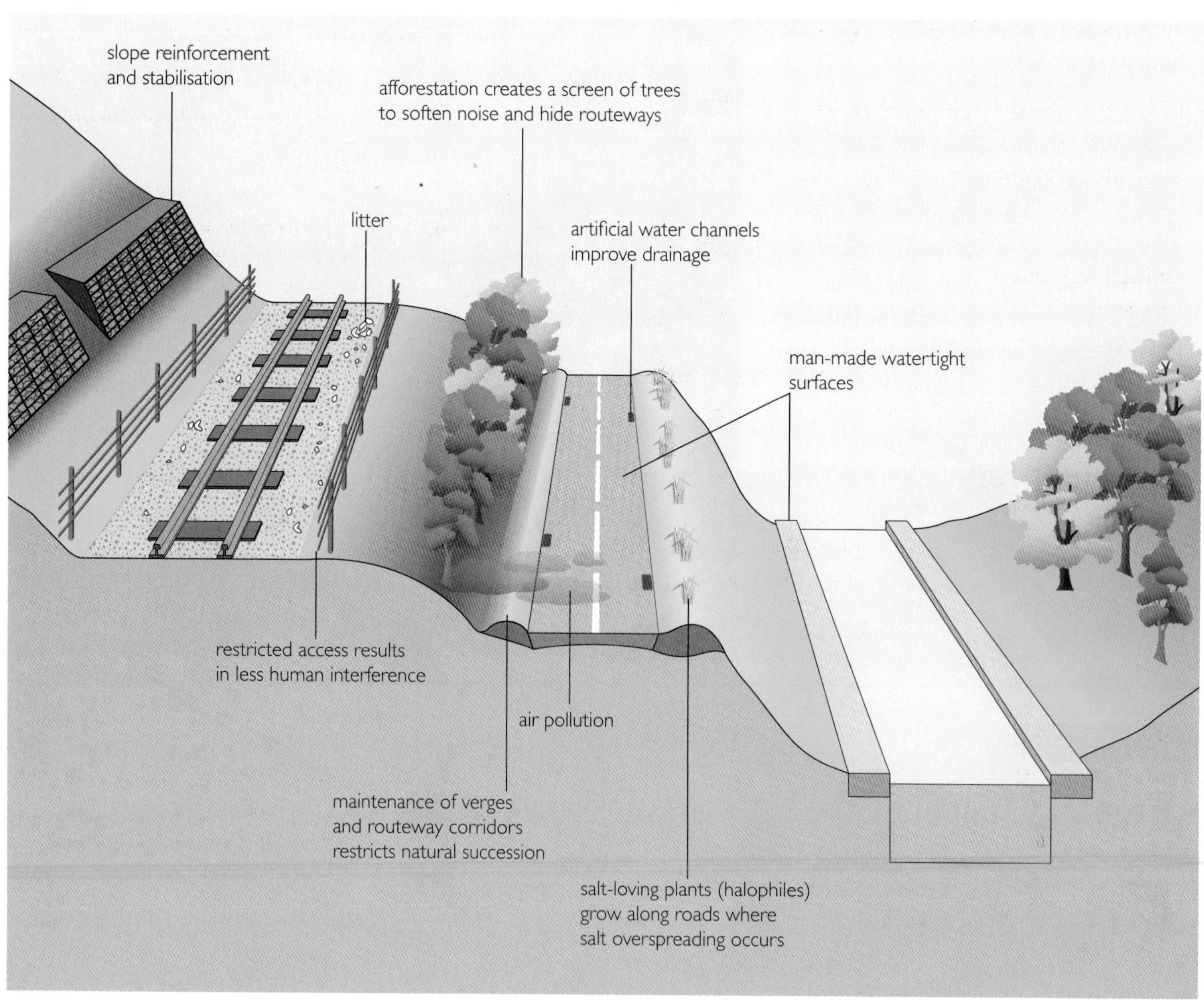

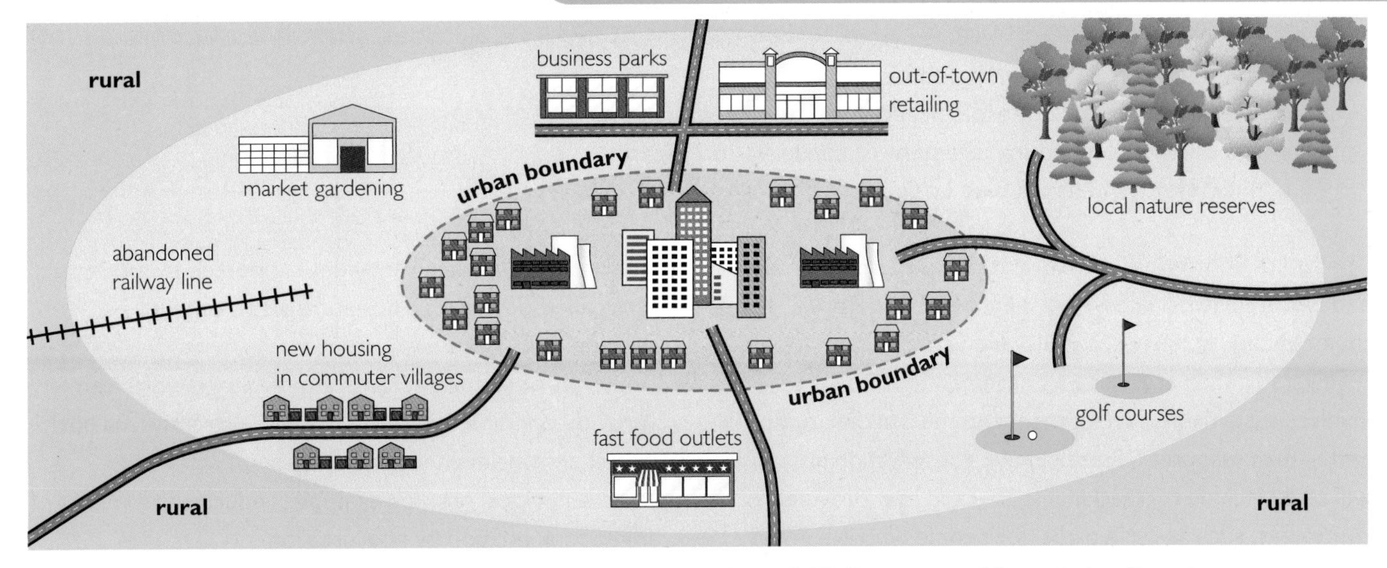

Figure 3.37 Characteristics of the rural-urban fringe ▲

Changes in the rural-urban fringe

Suburban railways and extended road networks are important influences on the extent of the rural-urban fringe, the land at the edge of an urban area that marks the boundary between rural and urban land use (Figure 3.37). As a result, it is an area of mixed land use and may include:

- intensive farming – for example, greenhouses indicating 'market gardening', the growing of high-value flowers, fruit and vegetables for sale in nearby cities
- new housing – for example, homes built on the edge of commuter villages or as a result of '**creeping suburbanisation**'
- **green belt** land – for example, patches of small woodland or local nature reserves
- recreational spaces – including playing fields, parks, golf-courses, and even theme parks
- brownfield sites – including abandoned railway lines
- industrial estates and business parks
- retail developments – for example, out-of-town supermarkets, DIY superstores or fast food 'drive-throughs' – all with extended car-parking.

Despite planning restrictions, such as classifying land as green belt, or the wishes of policy-makers to recycle existing brownfield sites, the extent of the rural-urban fringe is constantly changing in response to the various development pressures. For example, pressure to build more housing or the need to reduce traffic congestion results in a landscape that is in flux.

In addition, **urban blight** may lead to pressures on the quality of the natural environment. Fly tipping, vandalism, even farmland next to urban areas that has been neglected for a long time (such as with **non-rotational set asides**), may reduce the beauty of the landscape. Indeed, land owners have suggested that urban blight may actually help them gain valuable planning permission as any new housing or office developments would be a more attractive option.

Secondary succession is common within the rural-urban fringe as new species adapt to this changing and sometimes hostile environment.

- **Urban blight:** the associated problems of crime, neglect and lack of economic investment that lead to areas of a town or city suffering significant decay. Evidence of this decline may be abandoned and empty buildings or population loss.

ACTIVITIES

1. Study Figure 3.36. Describe the impact of routeways on changing ecosystems.
2. Using Figure 3.37, list the development pressures affecting ecosystems in the rural-urban fringe.
3. Explain why sustainable development of the rural-urban fringe is so difficult.
4. *'Green belt land comprises 13% of total land in England.' (Homebuilders Federation 2008).*
   In pairs, consider the benefits and drawbacks of such a relatively large proportion of green belt land in Britain. Before you do this, use the Internet to find out more about the history of green belts in the UK and the associated development restrictions.

3.12 Ecological conservation areas

In this section you will learn about:
- the characteristics of ecological conservation areas
- the designation and management of conservation areas
- Sheffield city centre's Blue Loop

The aim of **ecological conservation areas** is to protect and restore natural ecosystems. In protecting species and their habitats, ecological conservation areas support the biodiversity of an area. The management of these areas involves sustainable development and the use of natural and human resources. For example, conservation areas will often include recreational spaces and may provide routeways, such as cycle paths, for people who live in the area.

Other reasons for developing ecological conservation areas include:
- improving the appearance of an area, for example, brownfield sites in towns and cities (Figure 3.38)
- stimulating investment to encourage the positive **multiplier effects** of tourism or the return of industry to inner city areas
- educational, including activities supported by an education officer such as nature walks or wildlife gardening
- multi-purpose spaces, for example, a lake which can provide opportunities for a range of users such as bird watchers, anglers and ramblers
- managing flood risk, for example, reducing the impact of river bank erosion by afforestation
- maintaining rich biodiversity by protecting native species and restricting or reducing non-native wildlife.

Conservation areas tend to have very limited budgets and are typically viewed as a low-cost management solution. Their success, therefore, depends on engaging local communities and making use of volunteers. As shown on the facing page, a wide range of groups and organisations are involved in the management of conservation areas (Figure 3.39).

Figure 3.38 *A typical brownfield site* ▼

- **Natural England** replaced English Nature on 1 October 2006. Natural England is the government's principal advisor on the natural environment. It aims to both conserve England's natural environment and make it available for the benefit of everyone.

- **Sustainable development**: '*Development which meets the needs of the present without compromising the ability of future generations to meet their own needs.*' (Brundtland Report, United Nations, 1987)
In practice, sustainable development should result in a fairer society, with the benefits of economic growth shared across all parts of a population. It is not solely about conservation of the environment.

Figure 3.39 *Salmon Pastures Local Nature Reserve, Sheffield* ▲

Figure 3.40 *Designated ecological or conservation areas in England* ▼

Ecological or Conservation Area	Definition	Managing Authority or Group
Sites of Special Scientific Interest (SSSI)	Areas with strong legal protection to flora, fauna and geological or geographical features of particular 'special interest'	Government sites identified by Natural England
Special Areas of Conservation	Areas with increased protection of a variety of wild animals, plants and habitats	Government
Special Protection Areas	Areas with increased protection of rare and vulnerable species of birds	Government
National Nature Reserves	Areas that are the most important Sites of Special Scientific Interest (SSSIs)	Natural England plus organisations such as the National Trust, the Wildlife Trust and the RSPB
Local Nature Reserve (LNR) (see Figure 3.39)	Areas with wildlife or geological features that are of special interest locally	Different authorities, including local volunteer groups and county wildlife trusts
Areas of Outstanding Natural Beauty (AONBs)	Countryside considered to have high landscape value	Designated by Natural England
National Parks	AONBs given additional protections	National Park Authority, which includes members, staff and volunteers
Conservation areas	Areas of special architectural or historic interest	Protected by local planning permissions (local government)
Protected species	Legal protection given to endangered flora and fauna	Government

Sheffield City Centre Nature Walk: The Blue Loop

To walk the Blue Loop is to follow the city's industrial past (Figure 3.41). The Blue Loop Community Project is funded by Natural England and the National Lottery, and managed by a partnership of Groundwork Sheffield and the River Stewardship Company. The project aims to support and engage local communities along the waterside environment.

Sheffield's industrial past

Whilst the River Don was essential in providing water for cooling and processing in the growing industries of the 1800s, it proved difficult to navigate. The once salmon-rich waters became lifeless. The opening of the much-anticipated Sheffield and Tinsley Canal in 1819 provided the infrastructure for the mass export of coal, steel and manufactured goods. The success of the canal, however, was short-lived. In 1848, the newly developed railway reached Sheffield, significantly cutting the time taken to transport materials. This led to a marked decline in the prosperity of the canal. Whilst some commercial use continued until the early 1970s, the canal and associated industrial buildings fell into increasing disrepair and dereliction. It was not until the injection of funding and management by the Sheffield Development Corporation (1992), helped by British Waterways, that the River Don and Canal once again became a valuable resource.

Biodiversity of the Blue Loop

Whilst the Blue Loop follows a highly modified and, in places derelict, urban environment, the post-industrial landscape offers many unique habitats. For example, crevices in derelict stone walls are home to sand martins, and herons stand erect, watching for passing fish beneath one of the five weirs along the walk.

A local **biodiversity action plan** aims to protect important wildlife species and habitats. Several of the sites lie within the area of the Blue Loop.

- **Biodiversity Action Plans (BAPs)**: in response to the 1992 Earth Summit in Rio de Janeiro, the UK Biodiversity Action Plan was developed. It sets national priorities and targets for conservation. Action was also taken at local levels with the creation of local biodiversity plans.

Figure 3.41 *An annotated map of the 13 km Blue Loop, Sheffield City Centre Nature Walk* ▼

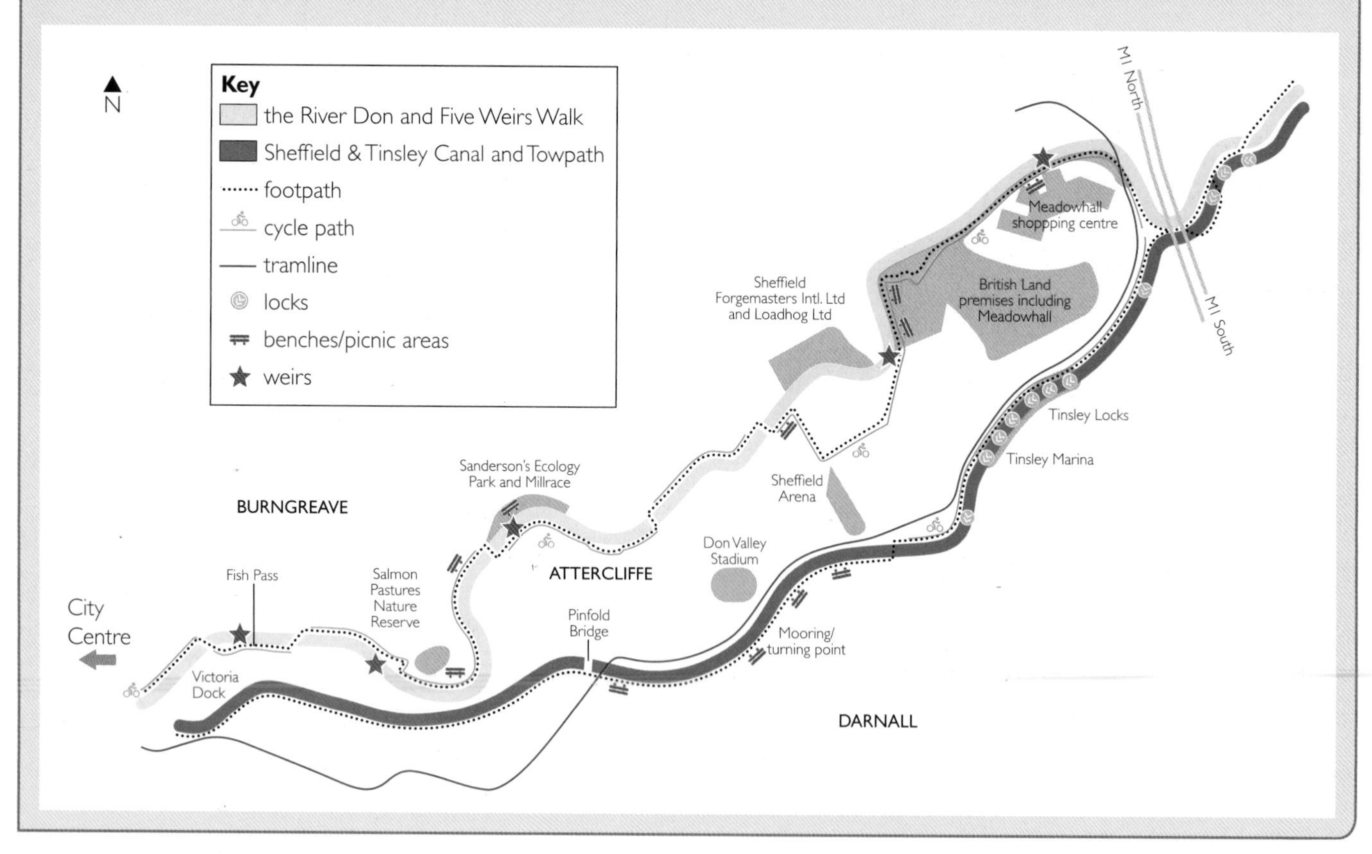

Native species include:

- fish – fish ladders have been built on the weirs to allow upstream movement of fish
- otters – careful management of river banks has provided secluded habitats for resting and breeding
- birds – vegetated river banks and islands provide habitats for urban bird species, such as song thrushes and house sparrows, that are now on the RSPB 'red list'. Water fowl, herons and kingfishers feed on the healthy populations of fish
- dragonflies – are attracted to the many kilometres of waterways to lay their eggs
- butterflies – nettles are common, particularly along the canal, and provide a perfect habitat for a variety of species.

Non-native species

Non-native species interrupt natural succession and restrict biodiversity by outcompeting native species. Examples include buddleia which damages buildings by invading cracks in the brickwork (Figure 3.42).

Wider benefits of Blue Loop

In addition to improving biodiversity, the Blue Loop has led to a wide range of local improvements and benefits.

Benefit	Example
Socio-economic	Waterside rejuvenation has encouraged tourism and private investment.
Reduced flood risk	Increased wetland areas and vegetation on the floodplain help to reduce the flood risk. New Sustainable Urban Drainage Systems (SUDS) help to better regulate the flow of water into water channels.
Recreational resource	Routes for cycling, walking and running cover most of the 13 km of the Blue Loop.
Community engagement	The Sorby Natural History Society carries out surveys to record and monitor wildlife. Themed family events, large-scale festivals and volunteering opportunities are provided for local residents by a variety of organisations.
Business partnerships	Businesses are educated in sustainable living, for example, installing **green roofs** which help conserve wildlife, improve the appearance of buildings and reduce the risk of floods.

Figure 3.42 *Japanese knotweed and buddleia outcompete native species on the banks of the River Don in Sheffield* ▲

- **Green roofs** are essentially roof tops planted with vegetation. They have existed for centuries, for example, the Vikings used grass sod on the roofs of some of their buildings. Today, modern buildings use a more high-tech approach, with waterproof base layers and drainage and irrigation systems.

ACTIVITIES

1. **a** Write your own definition of 'ecological conservation areas'. Refer to some examples from Figure 3.40.
   b Outline some of the benefits of creating ecological conservation areas in cities.
2. With reference to the Blue Loop, explain how conservation areas support sustainable development.
3. Explain why conservation areas can prove both difficult to manage and meet the needs of sustainable development.
4. Using the Internet, construct a short report on an ecological conservation area near to where you live. Use Figure 3.40 to identify some organisations that might be responsible for maintaining this conservation area.

3.13 Human activity, biodiversity and sustainability

In this section you will learn about:

- the environmental consequences of human activities
- the most serious environmental problems facing people
- the need to plan and manage environmental issues

We live in a world facing serious environmental problems. Some, such as toxic metal pollution, are long-established, while others, such as the impact of global warming, have only become understood more recently.

Whilst the human impact on land, air, water and biodiversity goes back thousands of years, the early stages of our evolution left the natural environment virtually unaffected. The use of fire, for clearing forests and stimulating plant growth for grazing by domesticated livestock, was the first advance in human technology. The last 3000 years have seen increasingly mobile, creative civilizations impact on the environment, both directly and indirectly (Figure 3.43).

Positive influences	Negative influences
The world **conservation movement** includes scientists, naturalists, resource specialists and informed, interested citizens – all studying the Earth and intervening in environmental issues whenever and wherever they can. The concepts of sustainability and environmental stewardship have developed.	All forms of **mobility** – land, air and water – have an almost unlimited potential to harm natural environments and ecosystems in all regions on Earth. Even the deepest ocean waters are now being accessed by people.
Planners at all scales, from local to national governments, understand that conservation must tie into effective urban, regional, developmental and economic planning.	World population growth results in more and more land, air, water and noise **pollution**. It also leads to soil erosion, desertification, acid rain and human-enhanced global warming (see Chapter 2).

Figure 3.43 *Indirect influences on the natural environment* ▶

Figure 3.44 *Too many people?* ▼

The most serious environmental problems facing people

The impact of humans on the natural environment attracts as much interest as speculation. It also invites as much informed comment as it does blinkered ignorance. So, then, what are the most serious environmental problems facing us today?

Figure 3.45 *The most serious environmental problems facing people* ▼

Global warming

Arguably the most pressing environmental issue of our time, covered in-depth in Chapter 2.

Destruction of natural habitats

Within the next half century another quarter of the remaining forests on Earth will be converted to other uses, such as roads, golf courses and urban areas.

Loss of biodiversity

Biodiversity has been described as '*the foundation of all life on Earth*' because it is crucial to the functioning of ecosystems. This is obvious when involving big edible animals, or plants providing fruits or timber. But so-called lesser species are just as important to sustaining ecosystems and food webs. A significant proportion of the world's wild species and genetic diversity has already been lost – and climate change threatens more.

Unsafe water supplies

Over 1 billion people in LICs lack access to safe drinking water. Some politicians predict that future world conflicts will be driven by the need for water security – so-called 'water wars'. Most of the world's freshwater in rivers and lakes is already being used for domestic and industrial consumption and farmland **irrigation**. Throughout the world, underground **aquifers** are being depleted more rapidly than they are being replenished.

Soil erosion

It is estimated that between 20% and 80% of agricultural soils globally have already been severely damaged. Erosion by wind and water is between 10 and 40 times the rate of soil formation. Leaching, **salinisation** and acidification have further degraded many of the world's soils.

Loss of wild foods – especially fish

About 2 billion people, mostly in LICs, depend on the oceans for protein. Wild fish stocks could be managed effectively, but in reality, overfishing has resulted in the decline or collapse of valuable fisheries worldwide.

Fossil fuels

Oil, natural gas and coal remain the world's most important energy resources. But burning fossil fuels is both highly polluting and a significant contributor to global warming. These resources are also finite – with reserves of oil and gas best measured in decades rather than generations.

Toxic chemicals

Toxins from industry are both manufactured and released as **effluent**. Reduced sperm counts, birth defects and mental health problems have all been associated with exposure to toxic chemicals. Despite their tiny concentrations in the air, soil, oceans, lakes, rivers and groundwater, toxins such as refrigerator coolants, detergents and plastics affect us all. We swallow, breathe and even absorb them through our skin.

World population growth

Despite varying rates of change between HICs and LICs, global totals are projected to reach 9.2 billion by 2050. More people require more land, food, water, energy and other resources. But is this sustainable?

Each of us consumes resources and generates waste, currently over 30 times more per person in so-called high-impact HICs compared to low-impact LICs. But economic development in and emigration from LICs are increasing the proportion of high-impact people. These are high-impact lifestyles in environmental terms. People in LICs rightly aspire to HIC living standards – but whether this is sustainable in environmental terms is questionable.

ACTIVITIES

1 Study Figure 3.44.
   - **a** Suggest how the environment has been affected by the urban sprawl shown in the photo. Consider vegetation, species diversity, water supplies and ecosystems.
   - **b** How might the natural environment be improved to benefit the people living there?

2 For this activity you will need to work in pairs. Look at Figure 3.45. Notice that there is information about nine different environmental problems facing the global community.
   - **a** Select one of the problems that you find particularly interesting. Be sure to coordinate with the rest of the class to ensure that you have a good range of problems.
   - **b** Use the Internet and other resources (such as textbooks and magazine articles) to conduct some additional research on the problem you chose.
   - **c** Use the information you have gathered to create a PowerPoint presentation (of up to 12 slides) and show it to the rest of the class.

3.14 Management of fragile environments

In this section you will learn about:

- the 2010 Gulf of Mexico oil rig disaster
- the UNESCO World Natural Heritage site at Aletsch Glacier, Switzerland

Ideally, environmental management needs to be proactive rather than reactive. But sometimes this is simply not possible. Environmental catastrophes have happened in the past, and many would argue that they are inevitable in the future. In such circumstances, the consequences for the communities affected can be devastating. These consequences have to be addressed and lessons learned so that similar disasters in the future can be avoided.

The *Deepwater Horizon* oil rig disaster

The 2010 *Deepwater Horizon* oil rig disaster in the Gulf of Mexico caused the world's largest accidental oil spill. It was an environmental, financial and political catastrophe which killed 11 and seriously injured 17 (Figure 3.46). That the disaster was completely avoidable has raised important questions as to the credibility of BP – the transnational oil company alleged by many to be ultimately responsible.

The explosion happened on 20 April 2010, immediately releasing oil into the Gulf of Mexico. At one point, up to 40 000 barrels a day of oil were being released. The oil first spread to the environmentally fragile wetlands of Louisiana, eventually reaching the delicate coastlines of all five Gulf states (see Figure 3.47).

Various courses of action were taken to stop the leak, including caps placed by remote controlled robots and plugs made with mud and cement. In early August, BP announced that the plugs had worked – the flow of oil into the Gulf of Mexico had stopped.

Figure 3.46 *Smoke and flames engulf the rig following the explosion* ▲

Did you know?

If all the water in the Gulf of Mexico filled an Olympic sized swimming pool, then the oil spilled from the Deepwater Horizon would amount to one drop in the pool!

The causes

Few would disagree that the world is making only slow progress towards shifting to alternative sources of energy. Crude oil, therefore, is precious. It is also becoming harder to find. Oil does not sit in a pool waiting to be extracted. It exists in the pores of sedimentary rocks and, once found, requires high pressure outside the well, and lower pressures within, to extract. The extreme pressures involved require multiple safety systems and procedures. In deep ocean conditions, the difficulties are multiplied. (The Gulf of Mexico was 1545 m deep at the site, with the oil a further 4055 m below the sea bed!) A series of mistakes, bypassed alarms, faulty equipment and inadequate, cost-cutting safety measures led to the catastrophic blowout of explosive natural gas.

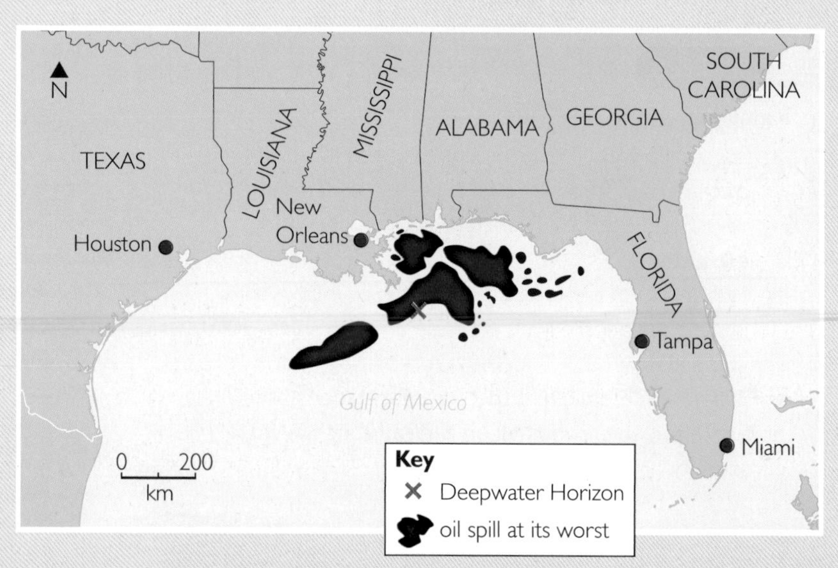

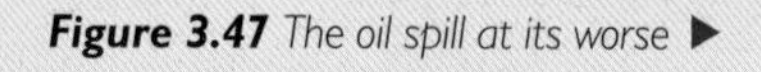

Figure 3.47 *The oil spill at its worse* ▶

The environmental consequences

Fishing and tourism provide over 80% of the jobs in the Gulf States, and much of this stretch of coast was threatened. Louisiana's delicate marshland bayou ecosystems were covered in black, poisonous sludge washed ashore by wind and tides. Fishermen and shrimpers faced financial ruin, and colonies of pelicans and others birds suffocated.

Beneath the Gulf's surface, a 48 km plume of oil, mostly below 900 m, threatened to impact on delicate ocean food webs in one of most productive ocean basins in the world (Figure 3.48). Of course, it will take several years before the final costs both economic and environmental are fully known.

Management of the consequences

The surface oil dissipated much more rapidly than anticipated. By the end of July 2010, three-quarters of the spill had dispersed, evaporated or been captured and burnt, with the remaining oil breaking down rapidly.

On-shore clean-up involved a number of public, private and voluntary organisations. BP used airboats to investigate reports of oil-affected stretches of coastline and to make clean-up recommendations. Tens of thousands of local volunteers, National Guard, Coast Guard reservists and specialized contractors were all involved in implementing coastal ecosystem management strategies:

- **Pom-poms** – collected oil around rocky areas. By the end of July, the clean-up had generated almost 40 000 tons of solid waste.
- **Containment booms** – trapped oil by causing it to 'stick' to the sides of the booms.
- **Tidal action** – high tide washed out oil from vegetation and into awaiting booms. Such non-aggressive techniques minimised damage to the fragile ecosystems.
- **Weathering processes** – oil reaching the coastline was heavily weathered, and as a result, lost some of its more toxic components. Exposed crude oil is broken down further by the combined effects of heat and water and naturally occurring bacteria.

Figure 3.48 *The consequences of the oil spill on both coastal and marine ecosystems* ▼

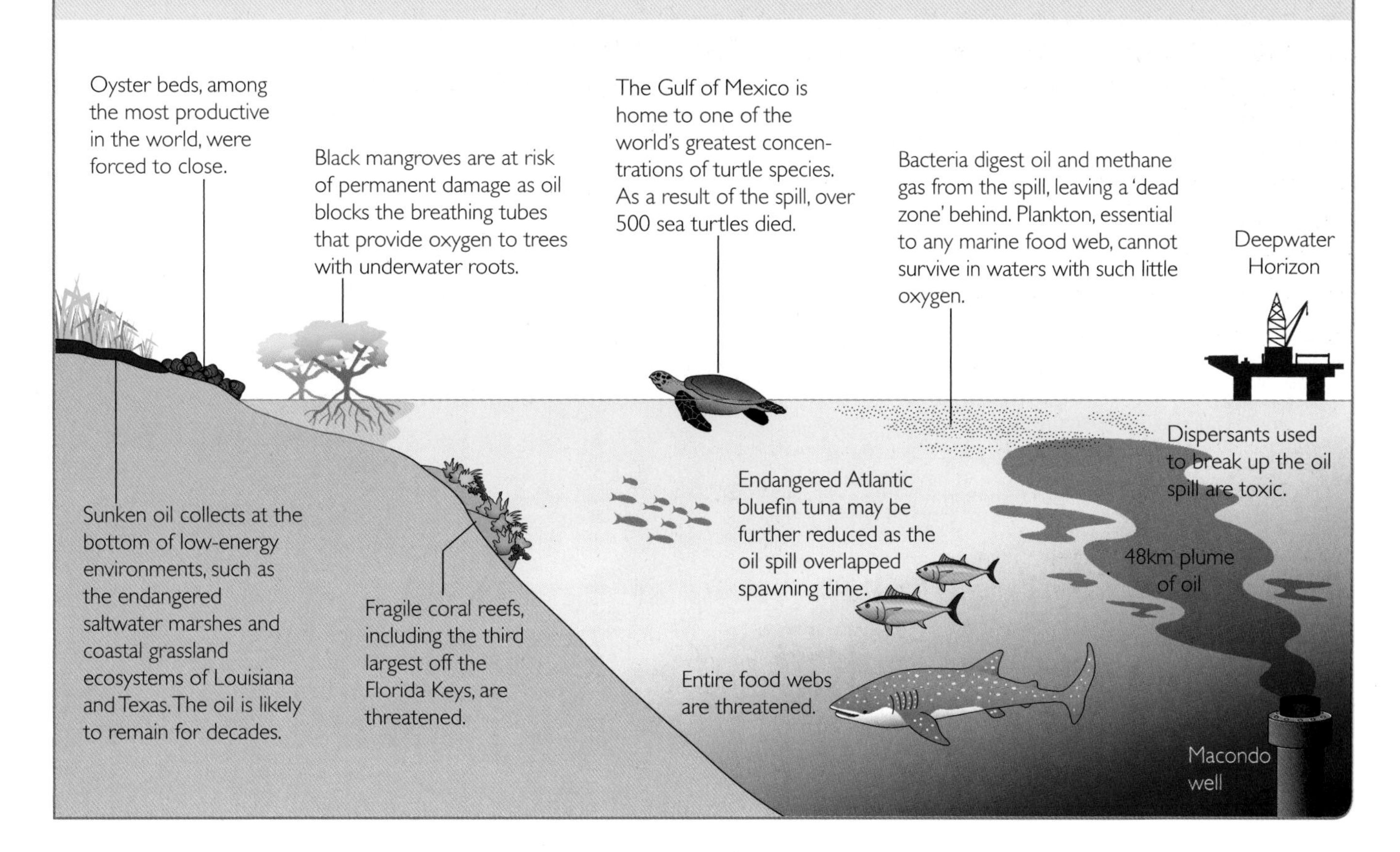

Local management: The UNESCO World Natural Heritage site at Aletsch Glacier, Switzerland

UNESCO administers an international World Heritage Programme which identifies, funds and conserves sites of special cultural and natural significance across the globe. To qualify for this status, **World Natural Heritage Sites** must meet at least one of ten criteria – number 7 of which is "to contain superlative natural phenomena or areas of exceptional natural beauty and aesthetic importance".

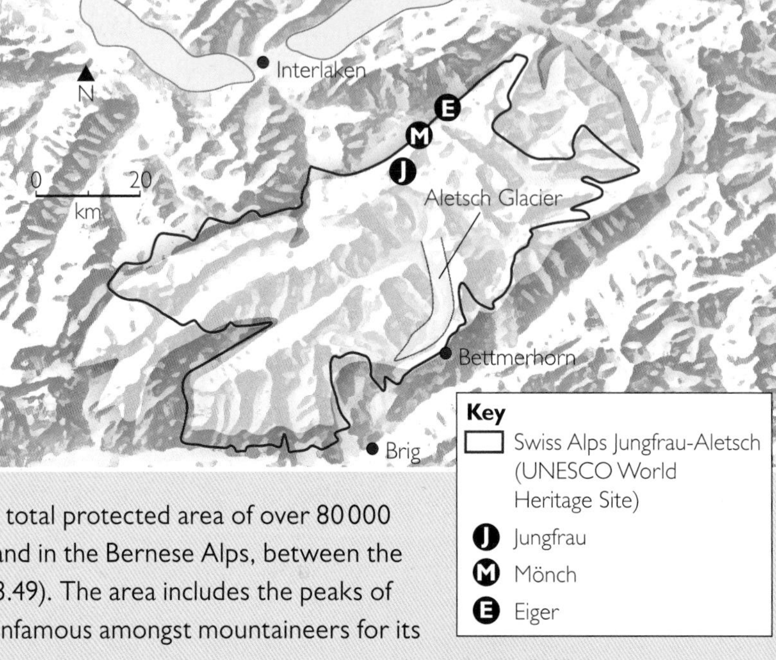

Figure 3.49 The Swiss Alps Jungfrau-Aletsch UNESCO World Natural Heritage Site ▲

There are 180 World Natural Heritage sites worldwide, and the Swiss Alps Jungfrau-Aletsch qualifies as an outstanding example. Designated in 2001, the total protected area of over 80 000 hectares is located in south-western Switzerland in the Bernese Alps, between the cantons (regions) of Berne and Valais (Figure 3.49). The area includes the peaks of the Mönch, Jungfrau and most notably, Eiger, infamous amongst mountaineers for its deadly north face.

Swiss Alps Jungfrau-Aletsch is valued for:

- its spectacular beauty and related tourist potential
- the wealth of information it contains about the formation of mountains and glaciers
- the information it provides on contemporary climate change
- its wide variety of habitats, ecosystems and examples of plant succession.

The site includes the most glaciated part of the Alps and the largest glacier in Europe – the incomparable Aletsch (Figure 3.50).

The Aletsch Glacier

Every statistic about the Aletsch Glacier challenges comprehension. At 23 km long, with a maximum thickness at its source of 900 m, it covers 120 square kms. Despite retreating 3.5 kms since 1860 (due to global warming), it weighs 27 billion tonnes – the equivalent of 72.5 million jumbo jets. It also contains enough fresh water to give every person on Earth 1 litre a day for the next six years!

It is stunningly beautiful – and with its easy access and a wide diversity of ecosystems, a very popular tourist attraction. Thousands of species of alpine plants and native mammals, including chamois, red deer and marmot, can be seen there.

Figure 3.50 The Aletsch Glacier, Bettmerhorn, Valais, Switzerland ▼

Management

The Aletsch Glacier is of exceptional importance – not just for its beauty and tourism potential, but for its educational value in understanding ecology, glaciation and ongoing climate change. Management of this fragile environment, supported by UNESCO, is essential. Management strategies include:

- **The management of tourism**. Popular viewpoints have viewing stations, cafes and toilets. But environmental degradation from tens of thousands of visitors is inevitable. Ongoing conservation initiatives include the provision of seating, wooden walkways and information boards at the most intensively used points (Figure 3.51).
- **Scientific research and management**. The Aletsch Ecological Centre in Riederalp, functions as both a research and visitor centre. It has an alpine garden, and organises exhibitions, guided walks, classes, seminars and training on environmental issues. The Aletsch and nearly all other glaciers across the Swiss Alps Jungfrau-Aletsch have been measured, some continuously since the late 1800s. Varying rates of retreat have revealed much about climate change. Other research into geology, glaciology, geomorphology, botany and zoology is also ongoing.

Figure 3.51 *Erosion, walkway management and viewpoint information where visitors concentrate at Bettmerhorn (2856m)* ▲

- **Conservation**. Forests throughout the Swiss Alps Jungfrau-Aletsch are managed for conservation rather than consumption. Fragile habitats of national importance, such as bogs and fens, are protected, and game reserves are regulated by laws on hunting. Even traditional alpine huts are maintained.

ACTIVITIES

1 *No other animal species has ever made such immense or rapid changes to natural ecosystems as people. But no other animal species has the power to regulate its own numbers. Neither do other animal species have the knowledge or means – not just to destroy the environment, but to manage and improve it to ensure a sustainable future for us all.*

In pairs or small groups examine the messages in the statement above. Why do environmentalists repeatedly insist that 'the future starts today'?

2 The beach on Alaska's Prince William Sound still holds sub-surface crude oil from the 1989 *Exxon Valdez* oil spill. The *Deepwater Horizon* well was spilling oil into the Gulf of Mexico at a rate of around an *Exxon Valdez* every four days!

a Is it fair to compare the effects of tanker disasters such as the *Exxon Valdez* oil spill with that of the *Deepwater Horizon* rig disaster? Explain your answer.

b Both the *Exxon Valdez* and *Deepwater Horizon* disasters affected the communities and ecosystems of their local areas and beyond. To what extent are the scale of a disaster and our management of it influenced by the level of economic development and the interest shown by the media?

3 The UNESCO World Natural Heritage site Aletsch Glacier, Switzerland provides an example of conservation versus exploitation issues in a fragile environment. For this, or another of your choice (such as Uganda on page 132):

a List the specific threats which identify the environment as fragile.

b Outline current management strategies.

c Discuss whether these management strategies are adequate to meet the future needs of this environment.

autotrophs Organisms (plants) that effectively feed themselves by photosynthesis using energy from the sun; autotrophs are at the first trophic level (producers) and are at the bottom of the food chain

biodiversity The number and types of organism in an ecosystem

biomass The total mass of all living organisms in a particular ecosystem, including plants, animals, micro-organisms, and people

biome Large-scale ecosystems such as tropical rainforests or savannah grasslands that are closely allied to climate regions

biota Living organisms in the soil, such as earthworms

carbon sink The term used to describe the capacity for plants to absorb carbon dioxide from the atmosphere and therefore help to reduce the concentration of greenhouse gases responsible for global warming

carr Another word for a 'fen', a low-lying area of land emerging from a freshwater pond or lake (hydrosere) as it becomes silted up

climatic climax vegetation The final stage (climax) of a vegetation succession, usually determined by the climate. In much of the UK, the natural climatic climax vegetation is deciduous woodland

decomposers Organisms such as bacteria and fungi that cause the decay and breakdown of dead plants, animals, and excrement, returning nutrients to the soil or direct to plants

detritivores Animals such as earthworms and vultures that eat dead and decaying organic matter

ecosystem The interrelationship between communities of plants and animals and the environment in which they live (soil, climate, etc.), e.g. a freshwater pond or a hedgerow

energy pyramid Pyramid-shaped diagram used to show the energy losses between each tropic level

food chain/web A diagram used to describe the links between different aspects of an ecosystem in terms of who eats who. It may take the form of a simple linear chain or a much more complex web of interrelationships

habitat The home of a single organism or a community, e.g. an oak tree or a pond

heterotrophs Organisms that feed off other organisms, otherwise known as consumers

humus Decomposed organic material that improves the quality of soils

hydrophyte A plant that has special adaptations to enable it to live in water (aquatic)

hydrosere A vegetation succession that develops in freshwater environments, such as a pond or ditch

leaching The dissolving in rainwater and downward transfer of nutrients through the soil

lithosere A vegetation succession that develops on bare ground (ploughed soil or newly exposed rock surfaces)

monoculture The production of a single crop such as wheat in an area, often leading to soils becoming impoverished

niche A term used to describe the highly localised environmental conditions in an ecosystem or the specialised plants that exist there

nutrient cycling The transfer and recycling of plant nutrients between the biomass, the litter store (dead and decaying plants and animals), and the soil

omnivores Consumer organisms that eat both plants and meat. Having two sources of food, they often exist at the highest tropic level and at the top of the food chain

pioneer community The first plant community to colonise an area such as a newly exposed rocky platform or freshly deposited sand. Pioneer plants are tough and can cope with hostile conditions, such as drought, salinity, or exposure to sun and wind

plagioclimax community Plants whose development is held in check by human intervention (e.g. by grazing livestock or burning heather moorland) preventing the natural climatic climax being reached

reducer organisms Organisms such as fungi, bacteria, and earthworms that reduce dead and decaying organic matter and return nutrients to the soil

sere A single stage in a vegetation succession

stratification Layering of vegetation, as seen in a tropical rainforest

sustainable development Development of resources that is long lasting and does not cause damage to the environment thereby compromising its use by future generations

trophic levels Levels of energy in a food chain, often represented by a diagram in the form of a pyramid, with the producers (plants) at the base and consumers (primary, secondary, and tertiary) forming the narrowing top of the pyramid. Energy is lost at each level due to respiration and excretion

vegetation succession A sequence of stages that often takes place over hundreds of years in the development of a plant community, e.g. in a salt marsh or on sand dunes

xerophyte A plant that is adapted to live in extremely dry conditions, e.g. cacti

EXAMINER'S TIPS

Section A

1 (a) Study Figure 3.12 (page 118). Outline the main characteristics of the deciduous woodland ecosystem shown in the photograph. *(7 marks)*

(a) Make sure that you focus on the photograph and use the correct geographical terminology in your description.

(b) Explain the processes responsible for the development of a vegetation succession. *(8 marks)*

(b) Focus on the processes associated with the development of a vegetation succession. You can refer to an example if you wish. Make sure you use the correct geographical terminology.

(c) With reference to one example, discuss the impact of human activities on vegetation successions. *(10 marks)*

(c) You must make use of a detailed example in describing the impacts of human activities. Ensure you make clear links between the human activities and their impact on a vegetation succession.

2 (a) Study Figure 3.36 (page 136). Outline the characteristics of vegetation along the routeways shown in the diagram. *(7 marks)*

(a) Refer to the information in the diagram throughout your answer.

(b) Explain the impact on ecosystems of changes in the rural-urban fringe. *(8 marks)*

(b) Here you need to focus on the changes that occur within ecosystems. Try to include a range of impacts, and stick to the rural-urban fringe.

(c) With reference to one example, discuss the importance of ecological conservation areas. *(10 marks)*

(c) Clearly define the term 'ecological conservation area'. Use detailed case study information to consider the importance of such areas. Remember to discuss.

Section C

1 With reference to one tropical region, discuss the impacts of human activities on natural ecosystems and local communities. *(40 marks)*

(1) Select and focus only on one tropical region (e.g. tropical rainforest). Consider a range of human impacts, using specific case study information to support your points. Make clear connections between the activities and the impacts, ensuring a good balance between natural ecosystems and local communities.

2 With reference to one fragile environment, evaluate the success of management strategies. *(40 marks)*

(2) Select and focus on one fragile environment, such as the Gulf of Mexico or the Antarctic. Describe the management strategies and, most importantly, evaluate their success.

4 World cities

Introduction

Among the world's cities is a select group whose importance is out of proportion to their population. We call them 'world cities'. They are big players, controlling much of the global decision-making and investment that affects everyone. For example, half of the world's investment capital passed through London in 2005, and the price of global foods and raw materials is decided there.

But these world cities are changing. They face problems such as urban sprawl and de-industrialisation, which they are trying to address through urban regeneration – although this seems to benefit some more than others. And the current leaders are being joined by new world cities, such as Mumbai and Shanghai.

In this chapter you will find out about the forces and processes shaping the world's great cities, and the way these cities influence our lives.

Books, music, and films

Books to read

London by Edward Rutherford
The Bonfire of the Vanities by Tom Wolfe

Music to listen to

'London Loves' by Blur
'New York City Cops' by The Strokes
'Paper Planes' by MIA
'California Dreamin' by The Mamas and the Papas

Films to see

City of God
LA Confidential
Bride and Prejudice
Slumdog Millionaire

Looking north over Manhattan

Look at the photo. Which world city is this?
Which country is it in?
What happens in the buildings?
Does this city have anything to do with you?
What might this city be like in 100 years' time?

About the specification

'World cities' is one of three Human Geography options in Unit 3 Contemporary Geographical Issues – you have to study at least one.

This is what you have to study:

- Millionaire cities, megacities, and world cities – the global pattern.
- Economic development and change related to urbanisation.

Contemporary urbanisation processes

- Urbanisation, suburbanisation, counter-urbanisation, and re-urbanisation: characteristics, causes, and effects.
- Planning and management issues.
- Contrasting case studies within countries at different levels of economic development to demonstrate these processes.

Urban decline and regeneration within urban areas

- Urban decline: characteristics and causes.
- Urban regeneration: gentrification, property-led regeneration schemes, and partnership schemes between local and national governments and the private sector.

Retailing and other services

- The decentralisation of retailing and other services: causes and impacts.
- One case study of an out-of-town retailing area.
- The redevelopment of urban centres: impacts and responses.
- One case study of an urban centre that has been redeveloped.

Contemporary sustainability issues in urban areas

- Waste management: recycling, and its alternatives.
- Transport and its management: the development of integrated, efficient, and sustainable systems.

In this section you will learn about:

- how urbanisation is occurring in different parts of the world
- how some cities become 'megacities' and 'world cities'

And still they keep coming!

Every day, coaches draw into the bus station in Kampala, the capital city of Uganda (Figure 4.1). People arrive, hoping for a job or perhaps to join a relative who moved there some months ago. They leave rural villages in which their families have lived for generations, where opportunities are limited.

Figure 4.1 *The main bus station in Kampala* ▲

They are part of a huge global migration – one that takes place every day as people move from rural to urban areas. This increasing percentage of people living in urban areas is called **urbanisation**. It plays a key role in a country's process of industrialisation – the workers who leave their rural homes in search of jobs provide cheap labour for the urban factories.

As of 2007, more people live in urban areas than rural. The United Nations (UN) predicts that by 2020 over 53% of the world's population will live in cities; other estimates suggest it could be 60%. If uninterrupted, the world's urban population, 3.6 billion in 2010, will rise further:

- to 5 billion by 2030
- at an average annual rate of 1.8%, double the rate for the global population as a whole.

The data shows that while the rate of growth of the global population as a whole slows down, the world's urban population will double in less than 40 years.

Calculating urban populations

With daily flows of people coming and leaving, the population figures of urban areas are difficult to estimate; Internet searches rarely give the same population for the same city twice! Population data for urban areas varies according to their **a**) census date, **b**) census method, or, **c**) whether the administrative urban area (called the '**metropolitan area**') or urban region is included in the calculation.

For example, London's population in 2010 was either:

- 8.2 million (the metropolitan region, with its 33 administrative areas)
- 11 million (the built-up area, including places beyond its administrative boundaries)
- 15 million (the population within a 50 km radius).

What complicates things is that as cities grow they engulf others, forming continuous urban areas or **conurbations**. Hence the population of Manchester has now become that of the Greater Manchester conurbation, including towns such as Oldham and Rochdale (Figure 4.2). This means that populations might not be comparable between dates because they might represent different areas.

Figure 4.2 *The Greater Manchester area* ▼

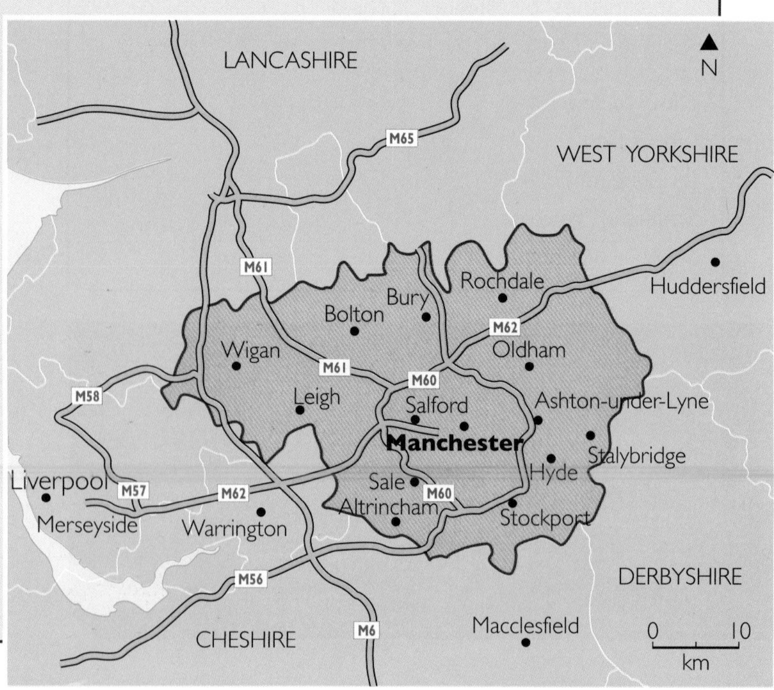

The changing urban balance

Yet, as Figure 4.3 shows, the rate of global population change is not an even one. In 1975, six of the largest metropolitan areas were located in the world's economically developed countries. By 1975, urban populations in these countries were already high as a result of the industrial revolutions of the 18th and 19th centuries. The process of urbanisation had taken place over a period of two centuries – the UK was the first country in the world to reach 50% urban population in 1861.

However, by 2025, just two of the ten largest metropolitan areas will be located in the developed world. Urban populations in developed countries are now rising more slowly, from a total of 0.9 billion people in 2003 to an expected one billion by 2030, an annual average growth rate of 0.5%, compared to 1.5% during the period 1950-2000.

Compared to the world's developed countries, which are 75% urban, less than 40% of people in developing countries lived in urban areas in 2010. However, this too is changing:

- Urban population growth there has risen dramatically by 2.3% annually since 2000 (that's a 30-year doubling time!) and is estimated to continue at this rate until 2030
- As a result, the share of urban population in developing countries is rising sharply and will reach 57% by 2030.

Just as slum conditions of the working poor challenged people in the growing cities of 19th century Britain, the numbers of those living in slums in the world's developing cities will be one of the challenges of the 21st century.

***Figure 4.3** This world map shows the ten largest metropolitan areas in the world (population in millions) in 1975, 2000 and 2025 (projected). The * shows those metropolitan areas which have incorporated new areas between the dates shown, and as such, have a large increase in population.* ▼

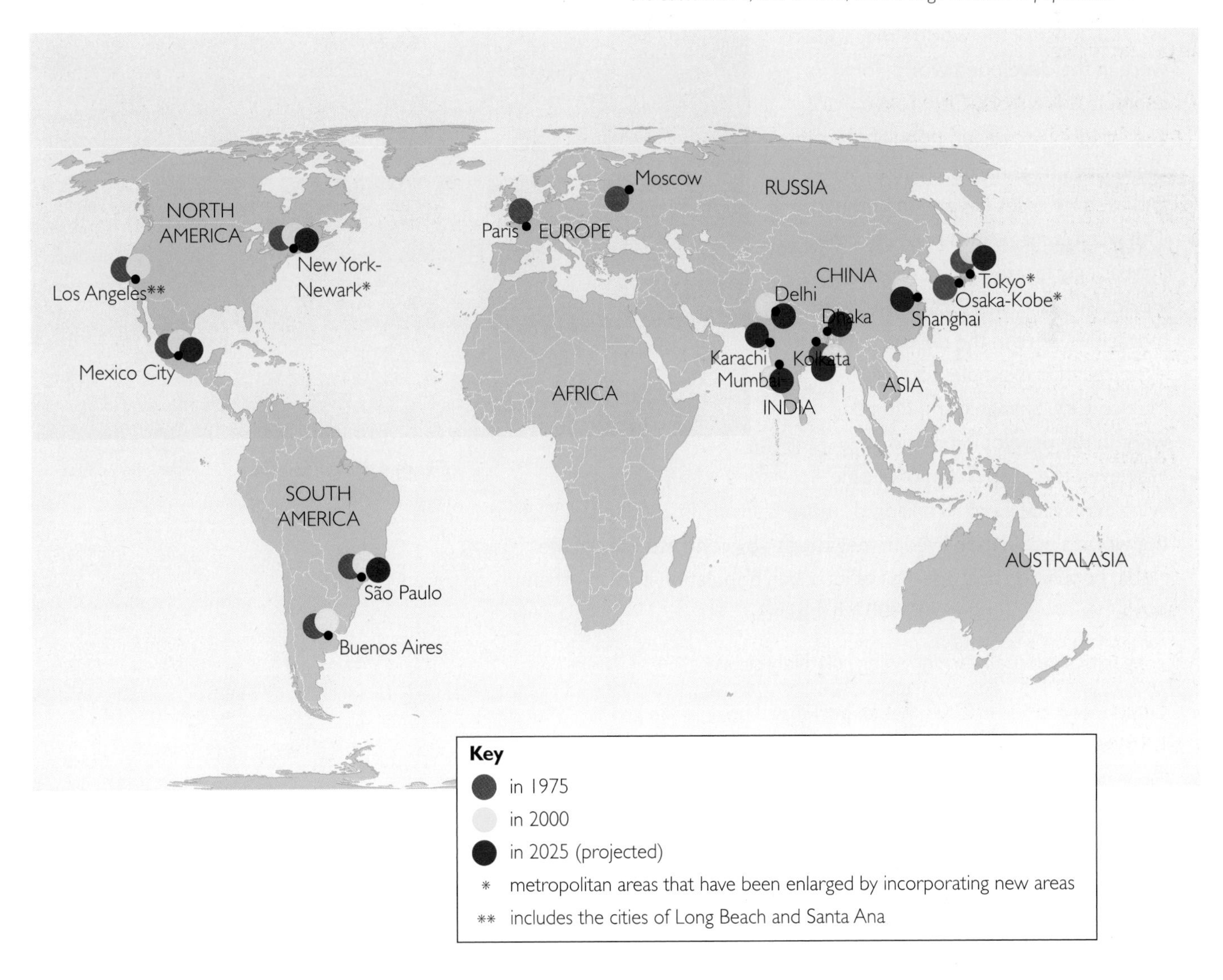

The emergence of the 'megacity'

As cities change, so do the words we use to describe them. Today, the term 'millionaire city' is used to describe any city with a population of greater than one million.

- In 1950, there were 83 millionaire cities worldwide
- In 1997, there were 285 – 106 in the developed world and a further 179 in the developing world
- By 2005, the number had grown to 336!

A few of the largest cities have grown to merge with others to form city regions, or **agglomerations**, such as Tokyo, whose region now includes the population of Yokohama (Figure 4.4). The term **megacity** is now used to describe these city regions.

- In 1980, most of the world's megacities were in the developed world, for example, New York City, Tokyo, Paris and London. However, populations of these cities have hardly changed since, and in some cases have even declined.
- The world's megacities have become increasingly located in the developing world. By 1999, only 25% of the world's megacities were in the developed world – the rest, for example, Sao Paulo, Mexico City, Shanghai and Mumbai, were in the developing world.
- However, the proportion of people living in megacities of 10 million or more is small. In 2003, just 4% of the world's population lived in megacities. By contrast, 25% of the global population and one-half of its urban population lived in urban settlements with under 500 000 inhabitants.

But megacities generally are growing relatively slowly.

- Of 20 mega-cities in 2003, half experienced population growth of under 1.5 % between 1975 and 2000.
- Only six grew at rates above 3%.
- Since 2000, 11 of the world's megacities have experienced population growth below 1.5% and five have growth rates of over 3%.

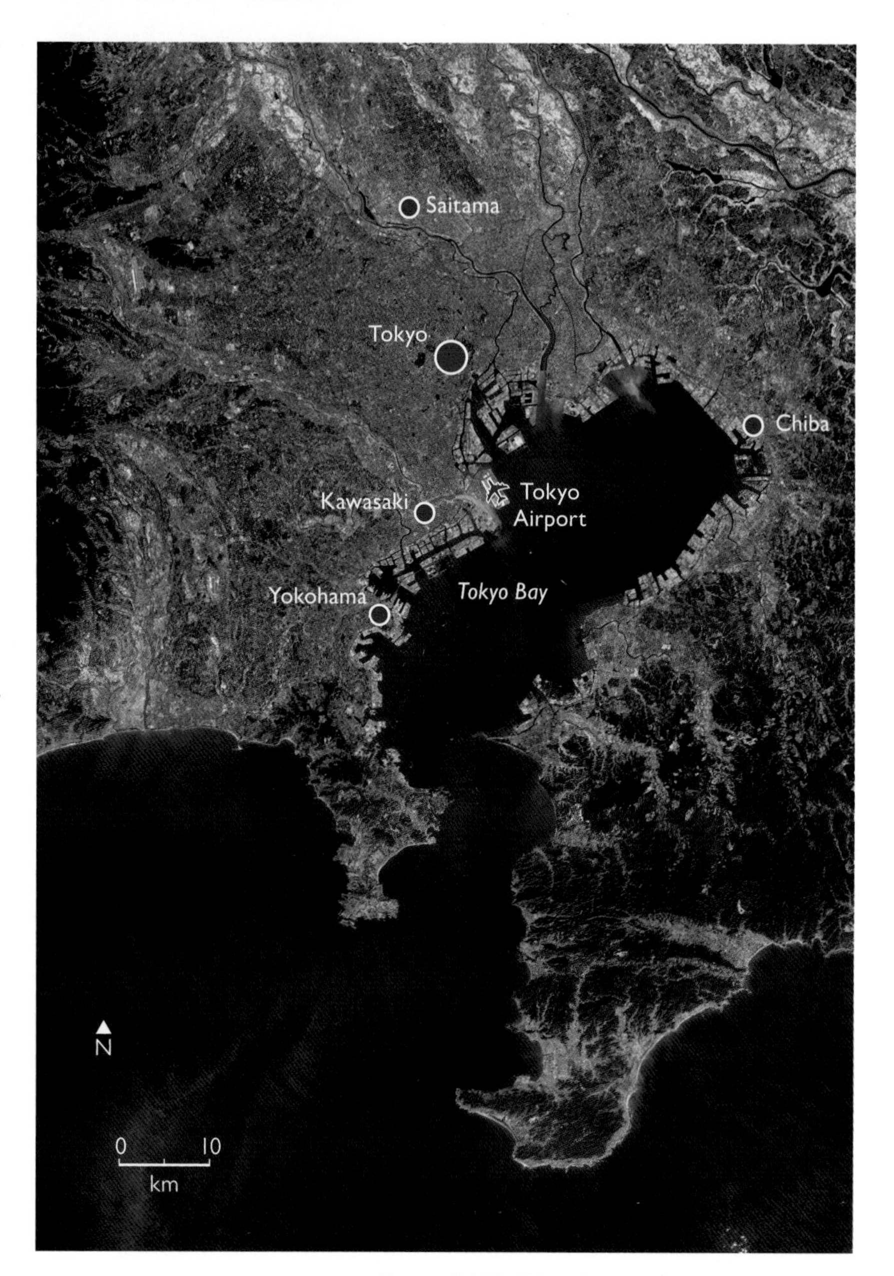

Figure 4.4 *The Tokyo-Yokohama region, the world's largest megacity* ▲

The development of 'world cities'

Among the world's largest urban areas are a select group of cities which have become critical in terms of economic activity (Figure 4.5). Just 100 cities account for 30% of the world's economy! Through these cities the workings of the global economy are channelled, with its capital, its knowledge, expertise, and – in spite of the 2007-08 banking crisis – its financial and political stability. Hence the term **'world city'** – cities which play a disproportionately important role in the global economic system. They allow global finance and trade to function in a few strategic geographic locations.

Identifying world cities

The first attempt to identify cities of global significance was made in 1998 by Beaverstock, Smith and Taylor from Loughborough University, where they set up the Globalization and World Cities Research Network. They attempted to rank cities based on four aspects of the 'knowledge economy': accountancy, advertising, banking/finance, and law. They assessed cities with offices of particular multinational corporations providing financial and consulting services, rather than those with other cultural, political and economic characteristics. Their rank order in 2008 identifies categories of importance:

'Alpha' cities:

- **Alpha++** New York, London
- **Alpha+** Hong Kong, Paris, Singapore, Tokyo, Sydney, Milan, Shanghai, Beijing
- **Alpha** Madrid, Moscow, Seoul, Toronto, Brussels, Buenos Aires, Mumbai, Kuala Lumpur, Chicago
- **Alpha–** e.g. São Paulo, Zurich, Mexico City, Jakarta, Bangkok, Frankfurt, Los Angeles

There are then significant numbers of

- **Beta cities** e.g. Melbourne, Johannesburg, San Francisco, Manila, Bogotá, New Delhi, Dubai
- **Gamma cities** e.g. Montreal, Nairobi, Panama City, Chennai, Quito, Stuttgart, Vancouver, Zagreb

Figure 4.5 *The major 'world cities'* ▼

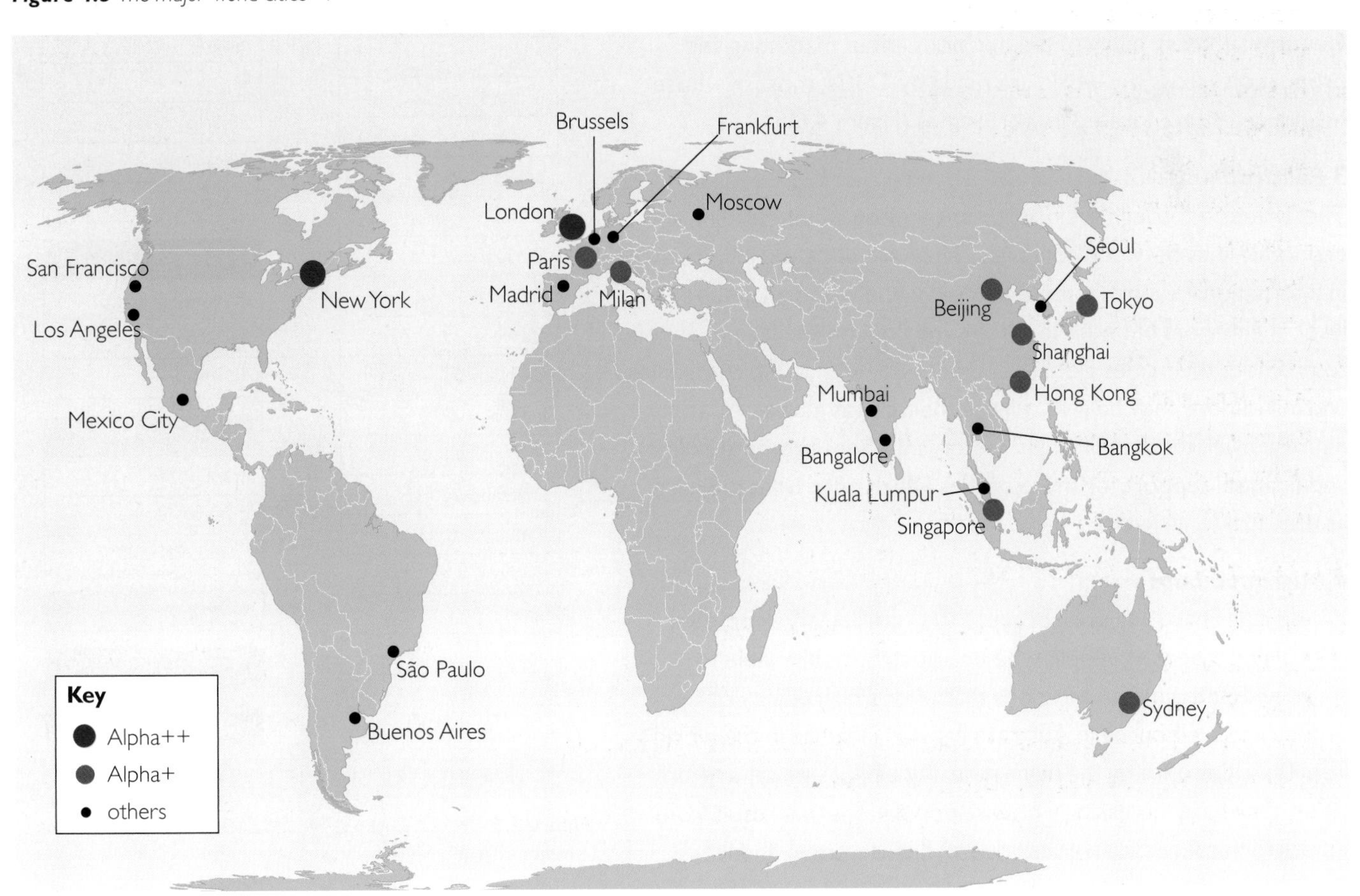

Think of a wheel with the world cities as the 'hubs' through which huge volumes of economic activity are channelled. The 'spokes' radiate out to secondary cities, regions, and centres of economic activity – these 'spokes' are flows, for example, of Internet activity, airline traffic, investment capital, and people. As hubs, world cities have four key characteristics.

1 Hubs of business, transport and trade

World cities are first and foremost 'cores' with a number of strong connections to the rest of the world, whether via the Internet and electronic communication or by air. London's Heathrow Airport, for instance, has by far the world's largest international passenger traffic. Other world cities, like Singapore, are major centres for shipping and container traffic. Yet others, such as Tokyo, dominate global trade and consumer activity. They contain:

- Corporate headquarters for TNCs, international financial institutions, law firms, industrial giants (e.g. BP) and stock exchanges that influence the world economy. To support these, a significant number of financial services, such as banking and accountancy, are also present.
- Clear identifiers of personal wealth; e.g., number of billionaires.

2 Production hubs

Although most of the manufacturing and production of goods is carried out in developing countries and Newly Industrialising Countries (NICs), many of the decisions about marketing and production across the world are made by TNCs, most of whose headquarters are located in world cities (Figure 4.6).

3 Political hubs

These cities – through their politicians – often dictate trading and economic links between countries. They influence and participate in international events (e.g. G8 Summits) and world affairs. At city level, the mayors of both New York and London see it as their role to promote their cities globally and network with those companies and institutions they believe will use their city as a base. At a national level, these individuals and organisations promote political and financial support for those parties which offer what they see as essential to growth.

4 Migration hubs

World cities have influence because they attract large numbers of qualified, talented people who are globally mobile (able to move with their jobs to locations that offer a high quality of life). Sydney and Melbourne in Australia regularly feature in the world's Top Ten cities offering the highest quality of life. Like other world cities, they have increasingly diverse populations that result from demographic flows, as well as flows of finance, trade and ideas.

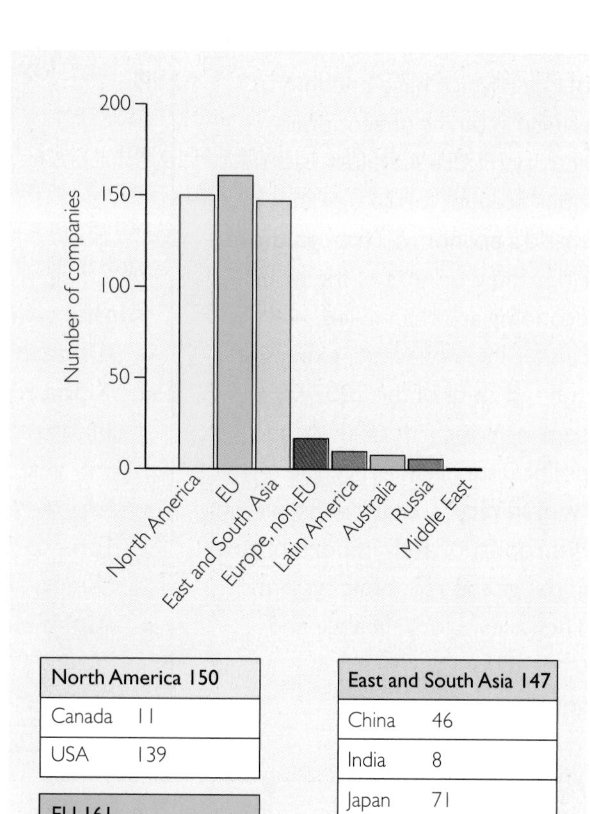

North America 150	
Canada	11
USA	139

EU 161	
Austria	3
Belgium	5
Belgium/Netherlands	1
Britain	29
Britain/Netherlands	1
Denmark	2
Finland	1
France	39
Germany	37
Ireland	2
Italy	11
Luxembourg	1
Netherlands	13
Poland	1
Spain	10
Sweden	5

East and South Asia 147	
China	46
India	8
Japan	71
Malaysia	1
Singapore	2
S. Korea	10
Taiwan	8
Thailand	1

Europe, non-EU 17	
Norway	1
Switzerland	15
Turkey	1

Latin America 10	
Brazil	7
Mexico	2
Venezuela	1

Australia 8

Russia 6

Middle East 1

Figure 4.6 *The distribution of the Top 500 companies (based on turnover in 2010) by country* ▲

World cities are hugely well connected, with communication networks drawing in flows of capital, profits and consumer behaviour like magnets. They encourage the migration of the world's wealthiest people, attracting talent and offering opportunities for innovation and change. Many world cities have economies larger than those of some countries.

- Hong Kong (Figure 4.7) receives more tourists annually than all of India.
- In 2010, Tokyo and New York had an estimated GDP similar to those of Canada and Spain, while London's GDP is greater than that of Sweden and Switzerland.
- They dominate their national economies – Paris generates 25% of the French GDP while the London region (including south-east England) with 25% of the UK's population generates 40% of the UK's GDP.

Figure 4.7 *Hong Kong, a major Alpha+ world city in south-eastern Asia* ▲

These cities dominate and influence well beyond their own national borders:

- London and New York compete to be the financial capital of the world. In 2005, London overtook New York with half of all global investment capital that year passing through London. Between them, they influence global interest rates, currency strengths, and investments, e.g. share prices and pension plans.
- They dominate culturally – London and New York are acclaimed as the 'theatre capitals' of the world.
- They dominate global languages – English is the most common language spoken internationally in film and media, aviation, finance and company law.

In coming decades, many others will join the list of 'world cities'. By 2025, the GDP produced by *each* of Delhi, Shanghai, São Paulo and Moscow will exceed the GDP in 2011 of Indonesia and Belgium. By 2020, a middle class of about 2 billion people globally will consume US$20 trillion in goods and services, twice the consumption of the USA in 2010! Imagine the global impact of their spending power on food, travel, energy, consumer goods, or greenhouse gases.

ACTIVITIES

1. Distinguish between the terms urban, city, millionaire city, urbanisation, metropolitan area, urban region, megacity, conurbation, agglomeration, and world city.
2. Identify the key trends in Figure 4.3. Suggest reasons why developed world cities are growing slowly and developing world cities are growing rapidly.
3. Draw a spider diagram to show how world cities emerge and develop:
   - in business, transport and trade
   - as production hubs
   - in political importance.
4. Explain why all world cities are located in the developed world.
5. What are the advantages and disadvantages to a country of having a disproportionately high percentage of its GDP being produced by urban regions?

Internet research

Research the characteristics of one agglomeration shown in Figure 4.3. Identify:

a its physical area
b reasons for its development
c issues that it faces (e.g. size, transport, housing, space).

Internet research

Use Google to research the term 'world cities'. How likely are London and New York to remain 'Alpha++'? Which other cities do you see as joining or replacing them by 2025?

In this section you will learn about:

- the rapid urbanisation taking place in the developing world
- a case study of the causes and effects of urban growth in Mumbai
- the management issues created by rapid urban growth

Urbanisation on a huge scale!

Imagine a city where the population equivalent of Nottingham is added every year! That's Guangzhou, China, which grew by 3.3 million people in the first decade of the 21st century, making it the world's fastest-growing city.

During that decade, the fastest-growing cities were in Asia and Africa – urban populations there grew by an average of 3.0% annually compared to a global average of 2.1%. UN data shows that, globally, 324 cities with populations of over 750 000 grew by at least 20%. 84 of these were in China, while Abuja in Nigeria grew by the greatest percentage – 139.7%!

Why are cities growing so quickly?

Many cities in Africa and Asia are experiencing **hyper-urbanisation**, that is, their rates of growth exceed each city's ability to cope in terms of housing, employment or services. There are two main causes of this:

- **natural increase**, i.e. the difference between births and deaths
- **net inward migration**, i.e. the net gain of those moving to a city.

In Africa and Asia, urban population growth is driven mostly by **rural-urban migration**. This creates both advantages and problems. On the plus side, rapidly growing cities create great opportunities for businesses and investors, and for developing infrastructure (e.g. water, transport) and services (e.g. health and education). Most migrants are young and able-bodied, so there are huge reserves of labour. But unplanned and informal urban growth in many developing cities creates equally huge environmental and social problems.

In describing Mumbai, India, economist Susan George wrote: *'Shanty towns built by desperate people fleeing rural poverty and trying to create a better life for themselves have become a common sight around Asian cities.'* These shanty towns – or informal settlements – are typical of rapidly changing economies, in which poverty or displacement force people from their rural communities into cities – cities in which governments have not provided the basic requirements of urban life such as drainage, clean water, public health and education (Figure 4.8).

Figure 4.8 *One of many informal riverside settlements in Bangkok, Thailand* ▼

Susan George was describing rural-to-urban migration, the main cause of growth for most developing cities. Figure 4.9 shows factors which 'pull' people into cities, and others that 'push' them from the countryside. There are background factors at work too, such as overseas investment which creates new jobs. However, the reality of moving rarely meets expectations. Instead of a better life, many end up amongst the millions of slum dwellers forced to squat illegally on spare land until moved on by police. Many landowners take the easier way out, charging rent and allowing squatters to stay. In this way, another shanty town is born.

Rapid urban development is part of a wider series of global economic changes. Since 1980, many companies in the developed world have relocated production of goods and services to the developing world, particularly India and China. This rapid industrialisation across southern and eastern Asia has generated employment, with new economic investment creating '**core**' regions. The theory which describes the impact of these changes is known as Wallerstein's theory of **core and periphery**.

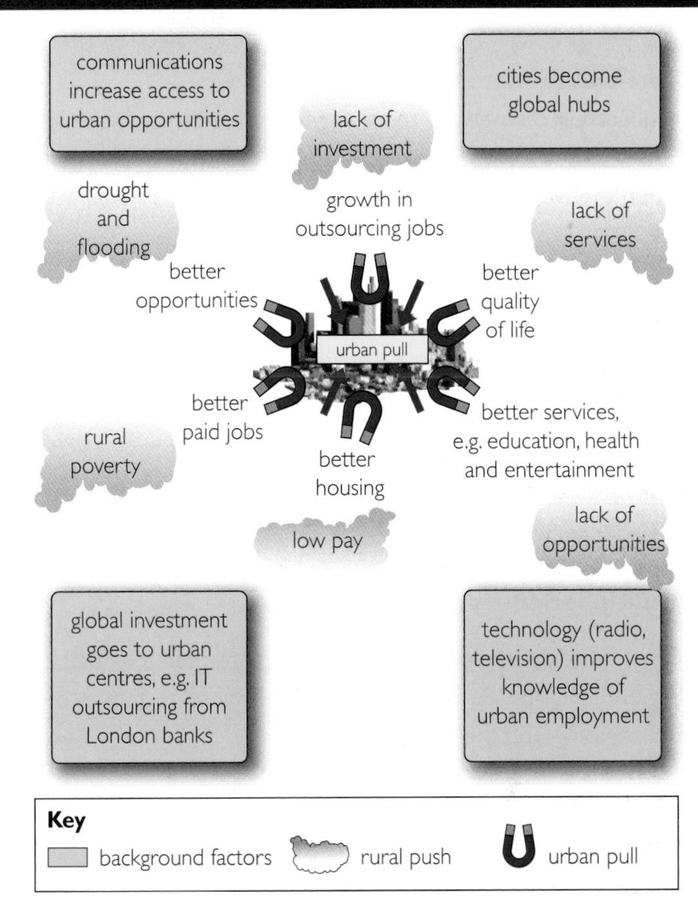

Figure 4.9 *Rural–urban migration factors* ▲

Core and periphery

In 1974, Wallerstein, an American sociologist and world-systems analyst, developed a theory about the spatial effects of economic activity. He divided the world into two types of economic areas: core and periphery (see below). Four decades later, his theory helps in understanding how centres of economic growth – what economists call '**growth poles**' – can generate further growth and lead to an upward spiral of development which produces what is called the 'multiplier effect' (see Figure 4.11 on the next page).

Figure 4.10 *The concept of core and periphery* ▼

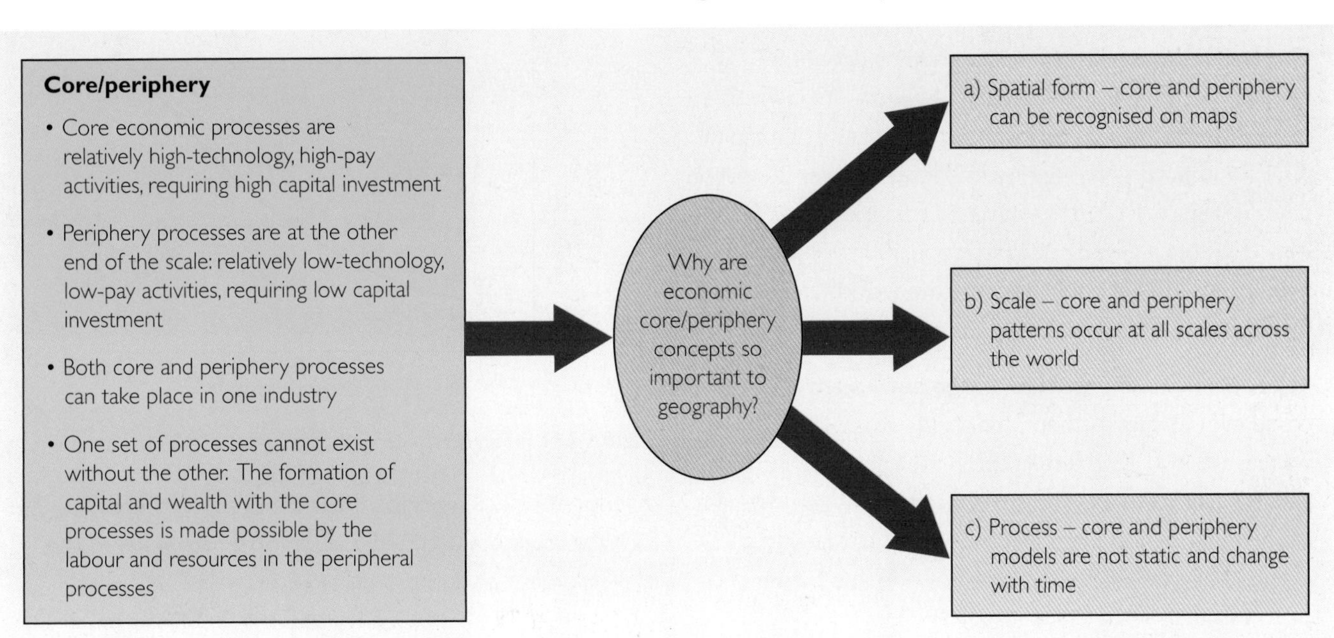

Core areas

Wallerstein claimed that core regions drive the economy. He identified a global core from the 18th and 19th centuries when the world's first industrial areas emerged – initially the UK and Europe, followed by North America and Japan. Now, economic growth in China and India is producing new 'core' areas, so his theory still applies. It also works at a national or regional scale – think of Mumbai and its urban region as one of India's 'cores'.

Core areas dominate production, investment and decision-making. They add value by turning raw materials into manufactured goods. Profits are concentrated in the region, leading to higher incomes. This process creates a '**multiplier effect**' (Figure 4.11).

Over time, the multiplier effect gets greater and a whole region develops. People seeking jobs continue to move to the region – and the spiral continues.

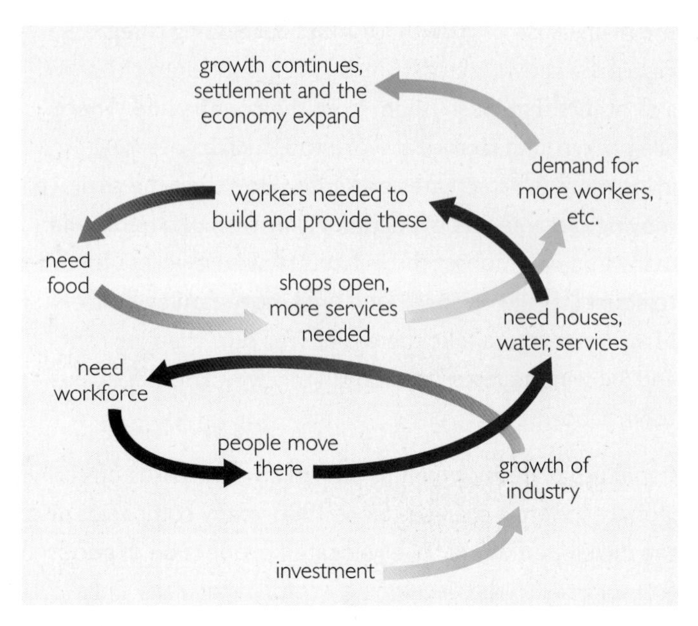

Figure 4.11 *How the economic multiplier effect works* ▲

Peripheral areas

Peripheral areas lie at the other extreme. Often lacking resources and wealth, unequal flows develop between them and the core:

- raw materials (which are of lower value than manufactured goods) are often supplied by peripheral areas
- populations in peripheral areas are drafted in as cheap labour in mines or on plantations, making those raw materials cheap to produce
- people in peripheral areas have lower wages than the core and may rely on wages sent back from the core, known as **remittance** payments.

CASE STUDY

Mumbai

Mumbai is at the heart of India's growing economy. Located on the western coast, it lies within India's richest state, Maharashtra, in terms of both total and per capita GDP. Mumbai is growing rapidly. Home to over 20 million people (Figure 4.12), it is estimated that it receives 1000 new migrants a day. By 2025 its population is likely to be over 26 million – it could become the world's largest city by 2050.

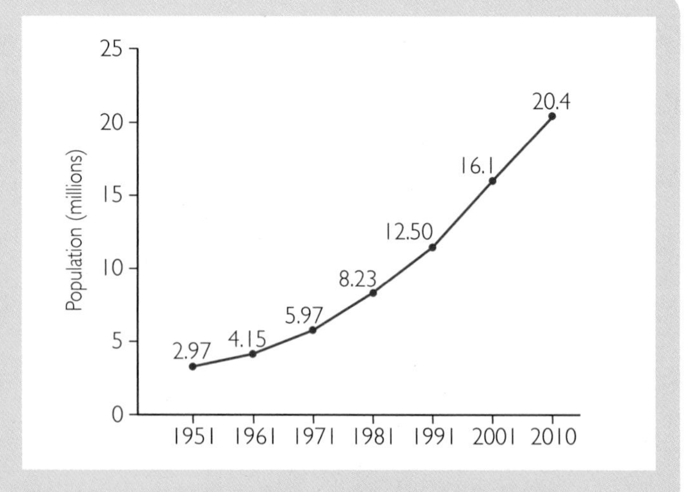

Figure 4.12 *Population growth in Mumbai, 1951-2010* ▲

In its economic strength lies a problem. To attract companies and investment, India's tax rates, like its wages, are low. As a result, companies and high income earners pay little tax. The city therefore has low revenue from which to provide public services, such as water, sanitation or public health. Most low income earners cannot afford charges for these services, so there is no private investment either, a situation which leads to the development of slums.

Mumbai is a lively, cosmopolitan city whose growth has come from:

- **services** – such as banking, insurance, IT and call centres. Mumbai's universities produce well-educated, English-speaking graduates who are employed by large western companies (e.g. BT), who contract them to provide services – known as **outsourcing**. India's wage rates are low, for example, well-paid call centre workers in Mumbai earn only US$5000 a year.
- **manufacturing** – half of Mumbai's factory workers work in the textiles industry, producing cotton textiles for export. Other booming industries include food processing, steel, engineering, cement and computer software.
- **construction** – a demand for housing, factories and offices has led to a boom in the construction industry
- **entertainment** – Mumbai has the world's largest film industry, Bollywood (Figure 4.13)
- **leisure and business services** – for example, hotels and restaurants.

Figure 4.13 *The bright lights and colours of a Bollywood film* ▲

Mumbai's wealth has had widespread impacts:

- it provides 33% of India's entire tax revenue
- 40% of international flights to India land there
- demand for property has driven rents in exclusive parts of the city higher than those in London or New York.

Globalisation has provided Mumbai with international banks, world-class restaurants, smart cars and headquarters of Indian transnational corporations like Tata Steel, Mukesh Ambani Oil and Godrej Retail – businesses that are now taking over their European and US rivals. India's middle class now numbers over 300 million people, and their tastes and preferred lifestyles are often decided in Mumbai.

Like many world cities, Mumbai is a city of contrasts. Its wealth acts as a magnet to hopefuls who arrive in the city every day, yet most live in poverty. Behind the bright lights and large middle class enjoying economic growth, huge numbers live in poverty. Unemployment and poor public health and water supplies make life both difficult and sickly for a large percentage of the population. Even those with living space often reside in cramped, expensive housing far from where they work, resulting in long commutes on crowded trains and buses. Over 3500 people die on Mumbai's suburban railway each year. Most deaths are among passengers crossing the tracks, sitting on train roofs to avoid crowds (and electrocuted by overhead cables), or hanging from doors and windows.

Dharavi – Mumbai or Slumbai?

One million people live illegally in Dharavi one of the world's largest shanty towns, located in Mumbai (Figure 4.14). They live in closely-knit communities that, along with homes, include self-help health clinics, food halls and meeting places. 60% of Mumbai's population lives in communities like this.

Dharavi is situated between two main railway lines. It provides cheap accommodation for low-skilled workers. Homes here are solid and many have electricity. Small workshops produce cheap pottery, plastic toys, clothes and handbags for export. Average incomes are low, about £40 a month, and rents about £12 a month. While over time, families can acquire building materials to improve their homes, few can afford to move out because the rest of Mumbai is too expensive.

Figure 4.14 *Dharavi* ▲

Work in Dharavi

Many people are forced to work in the **informal sector** – self-employed work that is irregular and has little security, such as street trading. Some is even illegal. Work is often 'cottage industries', for example, families making shirts in an assembly-line fashion. One cuts, another sews sleeves, another attaches the collar and so on. The product is then sold to a store buyer for 15 rupees (17p).

However, 80% of Mumbai's waste is recycled in Dharavi. Very little is considered unusable. The recycling industry in Dharavi is worth US$1.5 million a year and employs 10 000 people, including children. Barefoot child workers collect and carry plastic, glass, cardboard, wire hangers, pens, batteries, computer parts and even soap. Workshops range from small aluminium smelters recycling cans to others melting vats of waste soap from hotels and remoulding it into new bars (Figure 4.15).

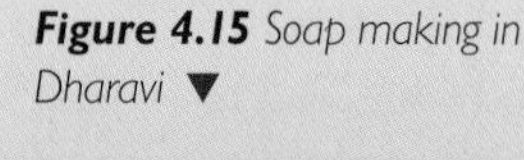

Figure 4.15 *Soap making in Dharavi* ▼

Vision Mumbai – a planning challenge

Mumbai has one major problem – its quality of life has deteriorated. Slums have multiplied and congestion, pollution and water problems have rocketed. However, it has developed a plan, called *Vision Mumbai*, to tackle these problems and transform the city into a world-class location by 2013, with state-of-the-art transport systems and higher-quality housing. Work on the plan began in 2004. If economic growth continues at 8-10% per year,

- over a million low-cost homes will be built
- slum populations will fall to 10-12% of their 2000 level
- safety, air pollution, water, sanitation, education and healthcare will all be improved.

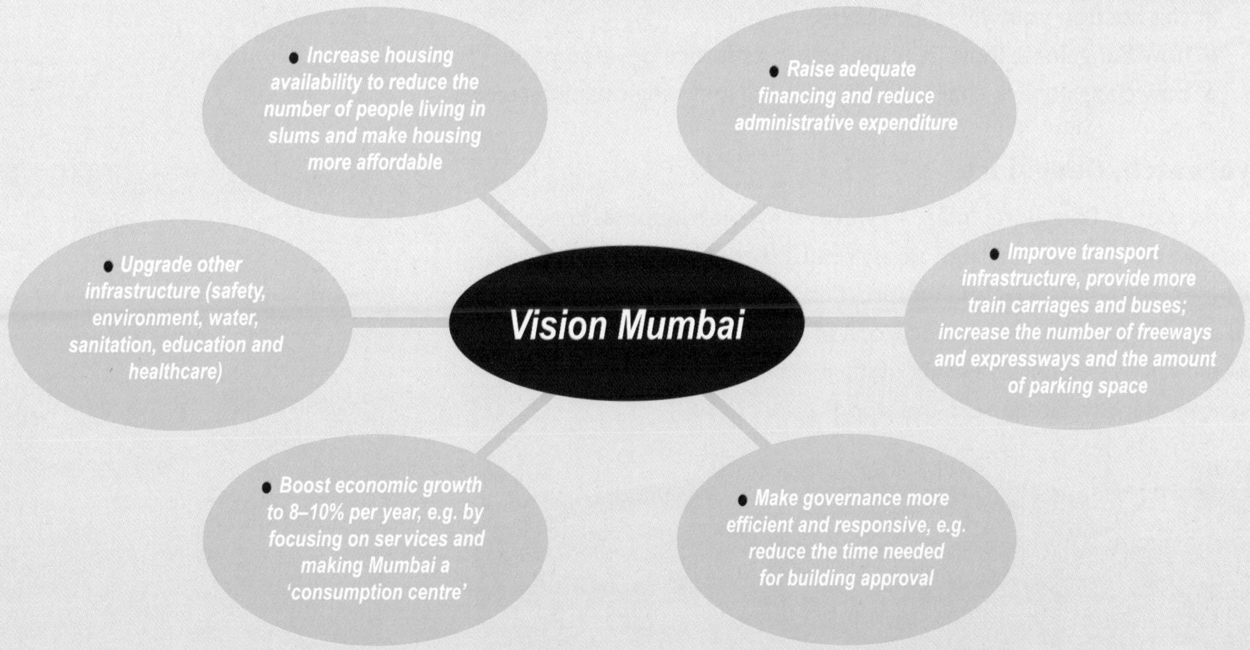

Vision Mumbai

- Increase housing availability to reduce the number of people living in slums and make housing more affordable
- Raise adequate financing and reduce administrative expenditure
- Upgrade other infrastructure (safety, environment, water, sanitation, education and healthcare)
- Improve transport infrastructure, provide more train carriages and buses; increase the number of freeways and expressways and the amount of parking space
- Boost economic growth to 8–10% per year, e.g. by focusing on services and making Mumbai a 'consumption centre'
- Make governance more efficient and responsive, e.g. reduce the time needed for building approval

Figure 4.16 *Vision Mumbai's six core targets* ▲

Part of *Vision Mumbai* means that Dharavi and other slums will have to go. The plan calls for the land, which is worth a staggering US$10 billion, to be sold and developed. Private developers will be able to buy the land for less than it is worth. In return, for every square foot of new housing they build for the poor, they are allowed 30% more for commercial development. High-rise blocks for slum dwellers will be built next to new shopping malls, offices and apartments.

Making Vision Mumbai work

Vision Mumbai is based on six core targets (Figure 4.16). Part of *Vision Mumbai* is based on 'quick wins' – things that could go ahead quickly to improve Mumbai, for example:

- clear, restore and maintain 325 open and green spaces
- build an extra 300 public toilets
- widen and 'beautify' the main north-south and east-west roads.

By 2007, 200 000 people had been moved and 45 000 homes had been demolished in Mumbai's slums, as 300 hectares of land were cleared.

ACTIVITIES

1 Distinguish between natural increase and net inward migration as factors in urbanisation.

2 Annotate a copy of Figure 4.9 with specific factors that apply to Mumbai's rapid growth as a city.

3 **a** In pairs, discuss and explain the factors that might lead to **a**) illegal dwellers in places such as Dhavari, **b**) informal work being the major type of economy in such places.

  b What might the arguments be for and against the informal economy as far as the Indian or state governments are concerned?

4 **a** In pairs, identify the main needs a city such as Mumbai has in order to function. Think of economic, social and environmental needs.

  b Explain how and why these needs are not being met at present.

  c How well might Vision Mumbai help to meet these needs?

5 Explain why Vision Mumbai might be supported by some, but not by everyone living in Mumbai.

6 In pairs, discuss whether Mumbai should consider increasing taxation revenue for public services or whether such a move might threaten the city's economy. Discuss and compare your views with those of others in your class.

In this section you will learn about:

- how Bangalore, India, is planning for its future development
- how Bangalore is challenged by cultural and religious differences

Travelwatch, Bangalore

An intersection in Bangalore, India. To western eyes it resembles chaos. A cow roams freely in the middle of a busy road, crossing where she chooses. Cars, buses, auto-rickshaws and motorcycles give way – after all, she is sacred so takes priority. In Bangalore, one of India's fastest-growing cities, traditional values sit alongside some of the most explosive urban growth and post-modern architecture in the world.

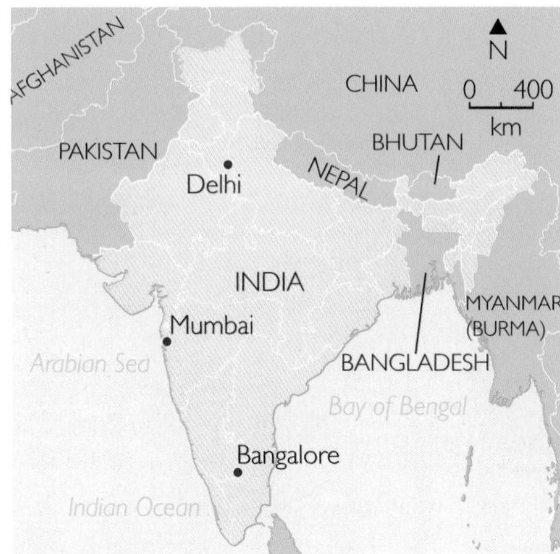

Figure 4.18 Bangalore's location ▲

Figure 4.17 An intersection in Bangalore, India. Which way cow? ▲

- **Outsourcing** is the employment of people overseas to do jobs previously done by people in the home country.

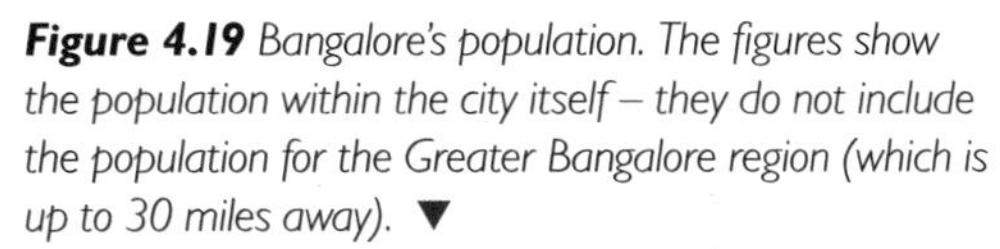

Figure 4.19 Bangalore's population. The figures show the population within the city itself – they do not include the population for the Greater Bangalore region (which is up to 30 miles away). ▼

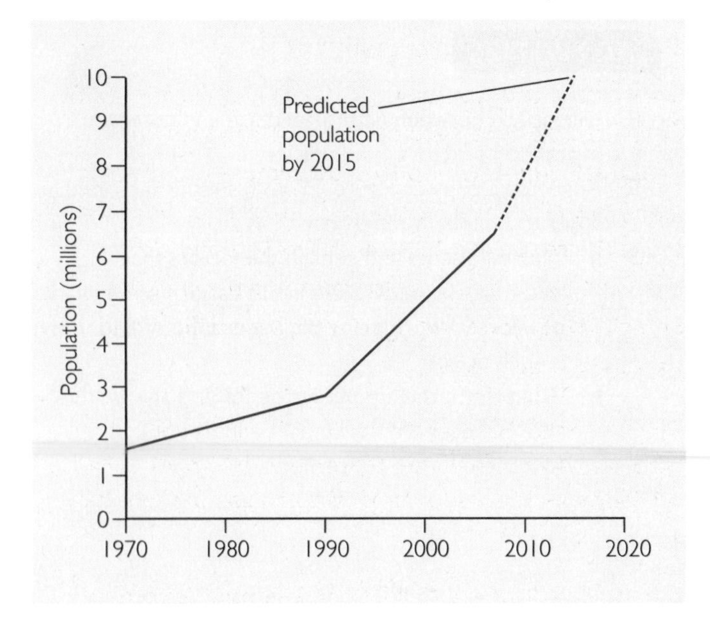

Bangalore, in the state of Karnataka (Figure 4.18), is India's hub in the '**new economy**', an economy which relies on people and their skills rather than manufactured products. Bangalore is the centre of new technology, banking, finance, and the **knowledge economy**. 40% of India's 1.3 million workers in the IT industry are based in Bangalore. The Indian government liberalised India's economy in the early 1990s, allowing large numbers of overseas companies to set up operations there. With so much employment, Bangalore's population has risen rapidly (Figure 4.19), so much so that the city has expanded beyond its boundaries. British Airways set up its accounting operations in India in 1996, setting a trend in **outsourcing**. Wages in India are only 10% of those in London so savings for companies like BA are considerable.

Now, a whole range of outsourced operations are based in Bangalore and other large Indian cities, including:

- **technical development** – e.g. companies such as Google and Deutsche Bank, one of London's biggest investment banks, have set up teams of software engineers and developers in India. In 2010, computer giant IBM alone had over 100000 employees in India.
- **support** – technical support for companies such as BT and HSBC
- **call centres** – e.g. handling queries or providing quotes for Norwich Union, Britain's largest insurer

Bangalore is well placed to take on expanding 'new economy' roles such as these. Its university provides a highly educated workforce, which, since the 1960s, has supplied technical expertise for India's defence and space research industries. Bangalore now has the highest average incomes in India, and jobs are plentiful.

The city is displaying a number of indicators of increasing affluence:

- there are six new shopping malls
- luxury car showrooms are booming
- thousands of new bars and cafes have opened since 2000, as increasing numbers of young, well-paid workers earn more and go out more often
- bar owners cannot get enough staff (staff members' salaries can be doubled if they are good enough to retain)
- taxi firms are booming because IT companies pay to transport visitors around the city and take their workers to and from the office.

Planning for future growth

To promote growth, in the 1990s Bangalore planners set up designated areas to become 'hubs' of high-tech firms. To attract companies it offered tax breaks (i.e. low tax rates) as well as cheaper labour. But where to place them? In the city centre, with its narrow streets and older buildings, traffic jams can last hours and the crowded environment makes getting to work both difficult and expensive (Figure 4.20).

The answer is out-of-town business parks. Bangalore has grown so much that IT support – which started as off-shore operations for large western companies – has spawned home-grown, Indian-owned companies which provide a range of technical and support services. Such companies need space to grow which can only be found on the city outskirts.

Further growth is certain. A report by IT company Oracle in 2008 estimated that India would need 8 million extra workers in outsourced industries! While in 2010, economic growth in the developed world was sluggish, India's annual growth was 8%! The ripple effect has started – other Indian cities, such as Chennai, are also providing similar services and attracting large companies.

Figure 4.20 *The old colonial city of Bangalore* ▶

The impact of growth

But is the wealth generated in Bangalore trickling down to all members of society? India's caste system still persists, and the disparities between rich and poor are stark. At night, jobless, low-caste Dalits scratch a living by removing human excrement from pits in the poorer neighbourhoods of Bangalore (Figure 4.21). They provide cheap labour on which India's basic services depend. They see few of the benefits of India's economic boom and explain why India's poorest count amongst the poorest people in the world.

Bangalore's 'night soil' collectors

Bangalore may be India's high-tech capital, but the practice of humans using their bare hands to clean out toilet pits still continues there. 10000-15000 Dalits are employed to clean out toilet pits in the poorer areas of the city.

The failure of the city authorities to provide a proper sewage system has forced residents to build these pits. When the pits get clogged, the call goes out for these wretched workers, who are desperate for work. After swigs of alcohol to ward off the stench, one gets into the pit and empties it with buckets or cans and the sewage is then taken away by the others. At the end of the night, they head home with earnings of 300-500 rupees (about US$6-10).

Adapted from a BBC report from Bangalore in September 2002

Figure 4.21 *The other side of life in Bangalore* ▲

India's caste system

The caste system is a religious and social class system in India where classes are defined by birth. Found mostly among Hindus, it also exists among some Muslims and Christians. Although India's Constitution makes caste discrimination illegal, barriers persist.

Under the caste system, Dalits (once known as 'Untouchables') have the lowest status. They often work in unhealthy or unpleasant jobs (see Figures 4.21 and 4.22) as well as suffer prejudice, segregation, and poverty. Prejudice is greatest in rural areas where they are not allowed to worship in the same temples as others, and must obtain water from different sources. Dalits cannot stray from their own part of the village, and Dalit children sometimes have to sit at the back of classrooms.

'Untouchability' was outlawed in 1950 and has declined, for example, President Narayanan of India (1997-2002) and the present Chief Justice both belonged to Scheduled castes (see below).

As well as the advantaged Forward castes (25% of India's population) the government now classifies people by:

- Scheduled castes – formerly 'Untouchables' (16% of the population)
- Scheduled tribes – consists of tribal groups (7%)
- Other backward classes (52%)

The classification is designed to encourage **positive discrimination** in education and jobs for the disadvantaged. Some people disagree, arguing that school and employment choice should be based on economic status as a few Dalits are now wealthier and more educated than people of Forward castes. But the caste system runs deep and is widespread. While inter-caste marriages are now more common, caste still determines some marriage choices. Some Indian news columns contain caste-based categories, and dating adverts can state caste as a choice.

Figure 4.22 *Dalit women carry breeze blocks on a building site in Bangalore* ▼

Future challenges

Bangalore's growth poses several questions for planners:

- Should more housing be built to keep pace with the increasing population and to bring prices within the reach of more people (Figure 4.23)? Rents are currently beyond the means of most workers on average pay; they can easily spend half their salary to rent one room outside the city. Construction workers often live in squalid roadside tents.
- How can Bangalore's public transport system be improved to keep pace with the city's growth? Five million vehicles currently clog its roads. Bullock carts block the road to an out-of-town business park, a road which is crammed with trucks, cars and two-wheelers – despite funding for a new expressway.
- How can enough energy be provided? Every IT company in Bangalore has to have a private generator as power failures occur daily.
- Should Bangalore International Airport be enlarged to meet the volume of international traffic? Its existing terminal is cramped, with just one luggage carousel.

Bangalore is planning for further, but different, growth. The city government plans to decentralise further by building new towns on greenfield sites around the city. It is keen to develop labour-intensive sectors like car manufacturing in Bangalore, and to disperse IT jobs to other cities in India.

Figure 4.23 *New housing in Bangalore – but for whom? The shortage of new housing has driven prices up so only the well paid can afford them.* ▶

ACTIVITIES

1. Using information from this and previous sections in this chapter, list the social, economic and environmental benefits and problems caused by Bangalore's growth.
2. What are the potential benefits and problems of an economy with growth driven by foreign companies?
3. In pairs, decide what priorities Bangalore should make about:
   a its future economic growth (what kind?)
   b how it should house its people
   c whether it should spread outwards or set up new towns on the edge of the city.
   Present your findings to the class.
4. Explain why the caste system is more likely to break down in urban than rural areas.
5. Prepare a class debate on whether:
   a positive discrimination should be encouraged to favour disadvantaged castes
   b newspapers should allow marriage adverts in which a preferred caste is stated.

Internet research

In pairs, research virtualbangalore.com and YouTube (type in 'Bangalore traffic') on the following:

a the types of companies located in Bangalore
b what Bangalore is like to live in
c the problems faced by the city.

In this section, through a case study of Los Angeles, you will learn about:

- the factors leading to suburbanisation
- how suburbanisation causes economic, social and environmental problems

Los Angeles

Los Angeles in California, USA, conjures up many different images. Disneyland, Hollywood, the Beach Boys – and sun. While it represents the American dream for some, for others it is not all glitz and glamour.

Reasons for the growth of Los Angeles

Los Angeles has attracted people in their millions to California. It forms part of the huge urban area known to Americans as the 'San-San urban corridor' (connecting San Francisco to San Diego). This huge urban area – for which Americans coined the term '**megalopolis**' – is home to 24 million people.

Los Angeles is the second largest city in the USA – but why did it develop?

- **Transport** – The arrival of the transcontinental railway from the east in 1876 stimulated rapid population growth; half a million people arrived within 40 years. The 20th century revolution in air travel created Los Angeles airport (LAX) which in 2010 was the sixth busiest airport in the world.
- **Employment** – In the early 20th century, the discovery of oil, the opening of a Ford car plant and other manufacturing industries meant continued growth. Later, the aircraft industry took advantage of the city's good weather for civil and military test flights as well as for production sites.
- **Image** – The development of the film industry in Hollywood – a suburb of LA – in the 1920s and 1930s created a glamorous image for the city. Later in the 1960s, growing affluence brought tourists to film theme parks – e.g. Disneyland and Universal Studios – all of which created further employment.
- **Greater affluence** gave people greater choice about where to live. The fastest population growth in the 1960s and 70s was in the sun-belt of California. Cold winters of the east coast cities (e.g. New York) combined with a romantic 'flower power' image from the late 1960s to make California a destination of choice; the lyrics of the Mamas and Papas' 1966 classic song 'California Dreamin'' say it all.

Figure 4.24 *This map shows the location of Los Angeles. Look at the scale and measure just how extensive this city is!* ▼

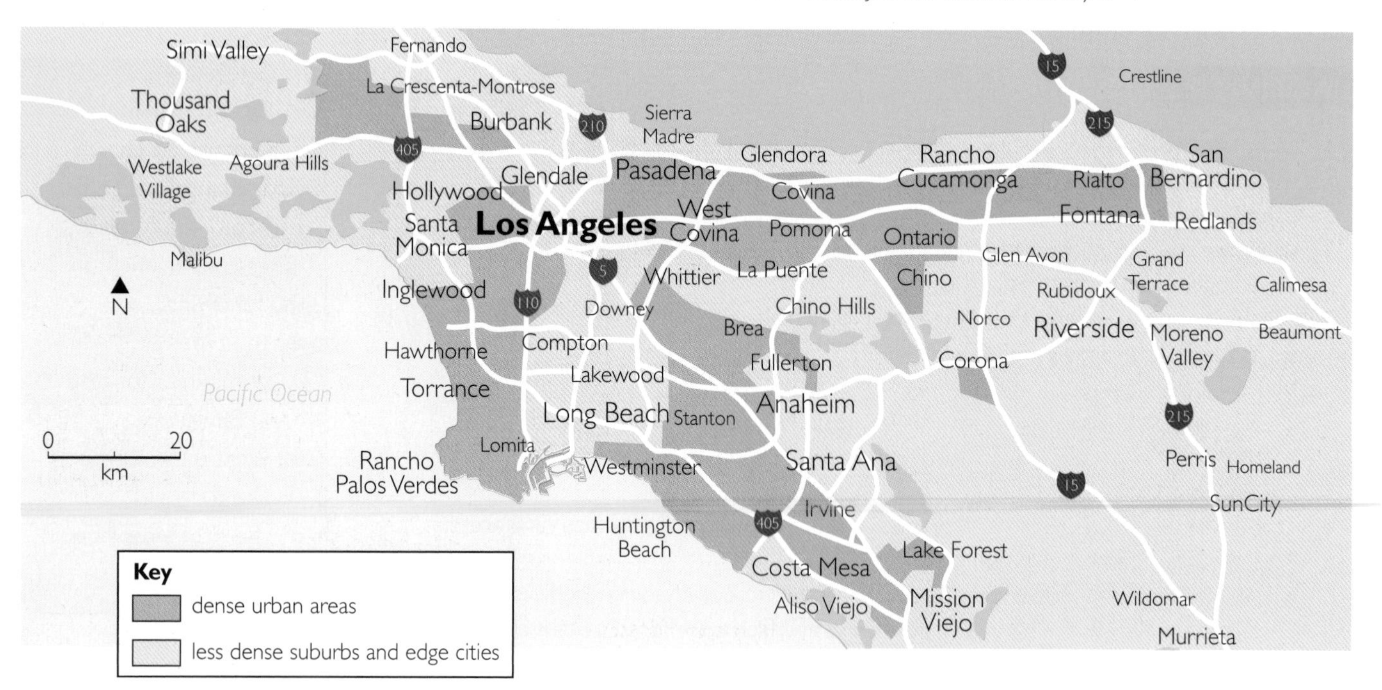

By the 1970s, Los Angeles was the USA's fastest-growing city; by 2000 14 million people lived in the metropolitan district. The fastest-growing areas of Los Angeles have been on the eastern extent of the metropolis, in Riverside and San Bernardino counties, where growth is up to 6% annually (Figure 4.25).

But Los Angeles has several problems resulting from this growth: suburbanisation, urban smog, the creation of a 'donut city', and social segregation.

Figure 4.25 *The American dream – suburban life* ▲

1 – Suburbanisation

14 million people take up space, even in a state like California where space is plentiful – it's 60% bigger than the UK. Even in its early stages of growth, the suburbs of Los Angeles spread a great distance from the city into a sprawling urban mass.

The suburbs were far from the city and offered a better quality of life – space for a house and a pool, for roads, schools, hospitals and shopping malls, everything the American dream could offer, in fact. What made them accessible was available transportation:

- The arrival of electric tramways in the 1920s and 1930s meant that people could live further away from work
- Later, freeways (motorways) spread across the city, allowing city workers to drive in from homes far away from the Central Business District (CBD). Until the 1980s, petrol prices were extremely cheap – under a dollar per gallon.
- The 1980s saw massive urban growth spreading out as far as the San Bernardino mountains and deserts – two hours travelling time from Los Angeles.

However, this only tells part of the story – the **pull** factors attracting people to the suburbs. Against these were the **push** factors. As suburbs grew the population of central Los Angeles declined, leaving a poor and often semi-derelict inner city in which some of the poorest and most crime-ridden suburbs were inhabited by those at the bottom of the socio-economic ladder. Figure 4.26 shows the varied reasons for the sprawling suburbs.

Figure 4.26 *Reasons for the growth of suburbs in Los Angeles* ▼

huge land mass
safer neighbourhoods
few planning restrictions
cheaper land for larger properties
low-density, single family housing
businesses looking for greenfield sites
crowded housing
change in economy (decline in manufacturing, increase in services)
fears for safety
urban push
congestion
high average incomes
high land rent
poor schools and services
declining jobs
pollution
accessible
more open spaces
huge investment in transport networks
large shopping centres
better schools and services
high personal mobility
cheap fuel

Key
background factors
urban push
suburban pull

However, low-density suburbanisation consumes huge amounts of land (in California, some of the best farmland has been lost to housing), and takes people further away from work. The greater Los Angeles metropolitan area (which is even larger than Figure 4.24) now extends over 88 000 km^2 (that's about 300 km x 300 km!) and includes 100 distinct towns. Time spent travelling to and from work creates social problems – stress caused by travel, and little time for partners, children and friends. Many suburban communities now exist only as places where people sleep rather than live. They are known as **dormitory settlements**.

2 – Urban smog

Suburban sprawl may mean a higher quality of life for some – but at what cost? Increasingly congested freeways and greater air pollution count amongst the greatest environmental impacts of suburbanisation. 10 million car owners create an environmental hazard – smog. High pressure systems over Los Angeles create dense cold, still air at the surface, with warmer air above. The cold dense air sits in the urban basin and traps pollution above Los Angeles (Figure 4.28). The still conditions allow the smog to build up, creating health hazards such as asthma.

The problem is a culture of car ownership and – for a city of 14 million – poor public transport. In 2005, just 10.2% of commuters in Los Angeles used public transport, compared to over 40% in London. Public transport in Los Angeles adds up to 1.7 million journeys a day, compared to 12 million in London (whose population is only about 60% of that of Los Angeles).

On July 26, 1943, in the midst of World War II, Los Angeles was attacked, not by a foreign enemy, but a domestic one – smog.

The Los Angeles Times reported that a cloud of smoke and fumes descended on downtown, cutting visibility to three blocks. Striking in the midst of a heat wave, the "gas attack" was nearly unbearable, gripping workers and residents with an eye-stinging, throat-scraping sensation. It also left them with a realization that something had gone terribly wrong in their city, prized for its sunny climate.

The following day, city officials pointed the finger at the Southern California Gas Co.'s. Aliso Street plant, which manufactured butadiene, an ingredient in synthetic rubber. Public pressure led to the temporary closure of the plant, but the gas attacks continued, proving that it was not the prime culprit.

That summer's "gas attack" was the opening shot in an epic war on smog, which now has been waged for more than half a century. From a ban on backyard trash incinerators to developing zero-emission fuel-cell electric vehicles, the fight against air pollution has inspired technological innovations and touched off heated political battles. Controlling air pollution has always ignited public controversy.

Adapted from a brochure produced by the South Coast Air Quality Management District

Figure 4.27 *The beginning of Los Angeles's smog problem* ▲

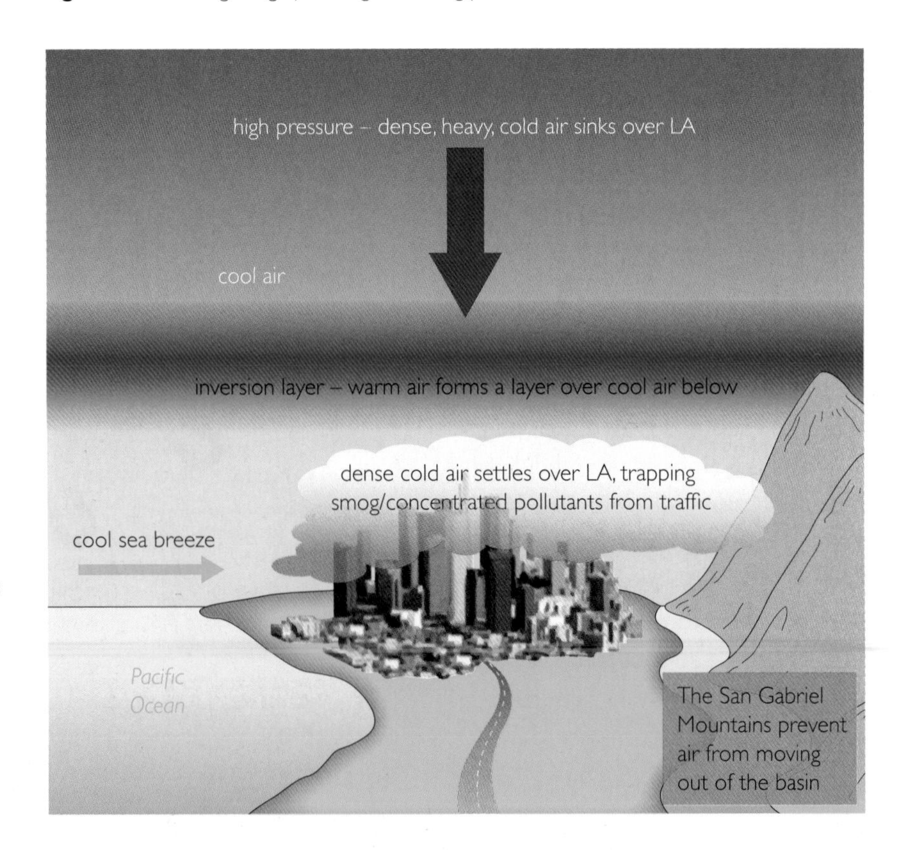

Figure 4.28 *Los Angeles sits in a basin. Exposure to pollution is known to cause both short- and long-term health problems* ▶

3 – 'Donut' city and edge cities

Los Angeles has been described as a 'donut city' – a city with a hole in the central (downtown) area. Why?

- Long-established car, tyre, steel and aircraft factories closed due to competition from overseas and changing technology. Most were located in the inner suburbs.
- Businesses followed people out of central Los Angeles to the suburbs, which offered more space, cheaper land and – in many cases – lower local taxes.
- Modern high-tech electronics, aerospace and light-manufacturing industries wanted large sites with car parks on the edge of the city.
- Those who could, mainly middle and high income earners, moved out to the suburbs. Large retail malls followed the population and their spending power out to the suburbs (Figure 4.29). The further away suburbs developed, the less inclined people would be to travel to central Los Angeles for shopping.

As a result, the inner and central areas declined, leading to dereliction and concentrations of poorer people living in sharply segregated areas. The Long Beach and Santa Ana freeway areas lost a million jobs, leaving many migrants unemployed.

As new industrial sites developed on the edge of the city, Los Angeles was named 'the city that turned itself inside out'. In the centre – decline. On the edge – expansion. Increasingly large suburbs known as **edge cities**, such as Anaheim, Irvine and Ontario, developed along the freeways (see Figure 4.24). There, concentrations of new industries, office developments, giant shopping malls and leisure zones built up. For those living in the suburbs, work suddenly developed on their doorstep. However, the people living in the edge cities have very little contact with the major city of which they are technically a part. From Anaheim into central Los Angeles is 2 hours by bus, and there are few public transport networks.

4 – Social segregation

One impact of suburbanisation has been that it has acted as a kind of sieve; those who can afford to leave inner Los Angeles for the suburbs do so, leaving an increasingly deprived population behind. There are high income earners in central Los Angeles, but they tend to live in small clusters, in secure tenement blocks guarded and protected by concierge and security services. Often these blocks are surrounded by lower income, deprived areas with many social problems and crime hotspots.

Figure 4.29 *A large retail mall in outer Los Angeles* ▼

As a result, inner Los Angeles has developed like a patchwork – not like the land use models of Burgess and Hoyt (Figure 4.30). Today the downtown 'hole' of the 'donut city' has been filled and is dominated by the headquarters of TNCs. Close to these live some of the well-paid executives. Wealthy areas are found in the inner city next to poor areas, and the same happens on the edge of the city.

Many of the migrants to Los Angeles – especially Mexicans from a few miles south across the border – settle there for work. Like the ethnic communities described in section 6.5 of this book, they live in **ethnic enclaves**, which are communities of people from similar cultural or ethnic backgrounds e.g. nationality, language or religion. Some enclaves are wealthy, while others are poor, making the land use pattern more complicated. Figure 4.31 shows the distribution of different ethnic enclaves across Los Angeles. Many migrants in these areas suffer exclusion because they cannot afford to pay for services such as health and higher education.

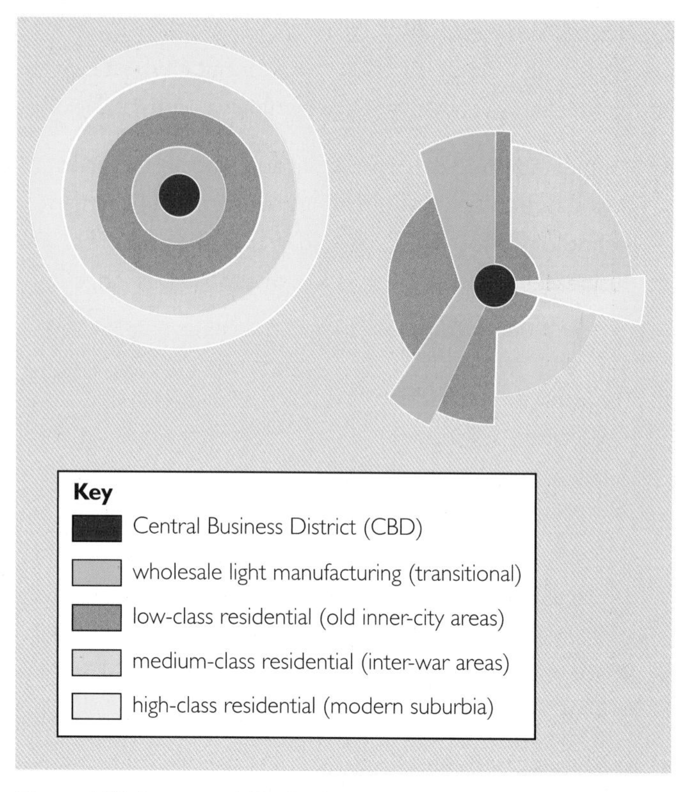

Figure 4.30 *Burgess and Hoyt's urban models* ▲

Did you know?

Some of the most significant Civil Rights riots took place in the LA district of Watts in 1965, in which 34 people died.

Figure 4.31 *Ethnic enclaves in Los Angeles and Orange counties* ▼

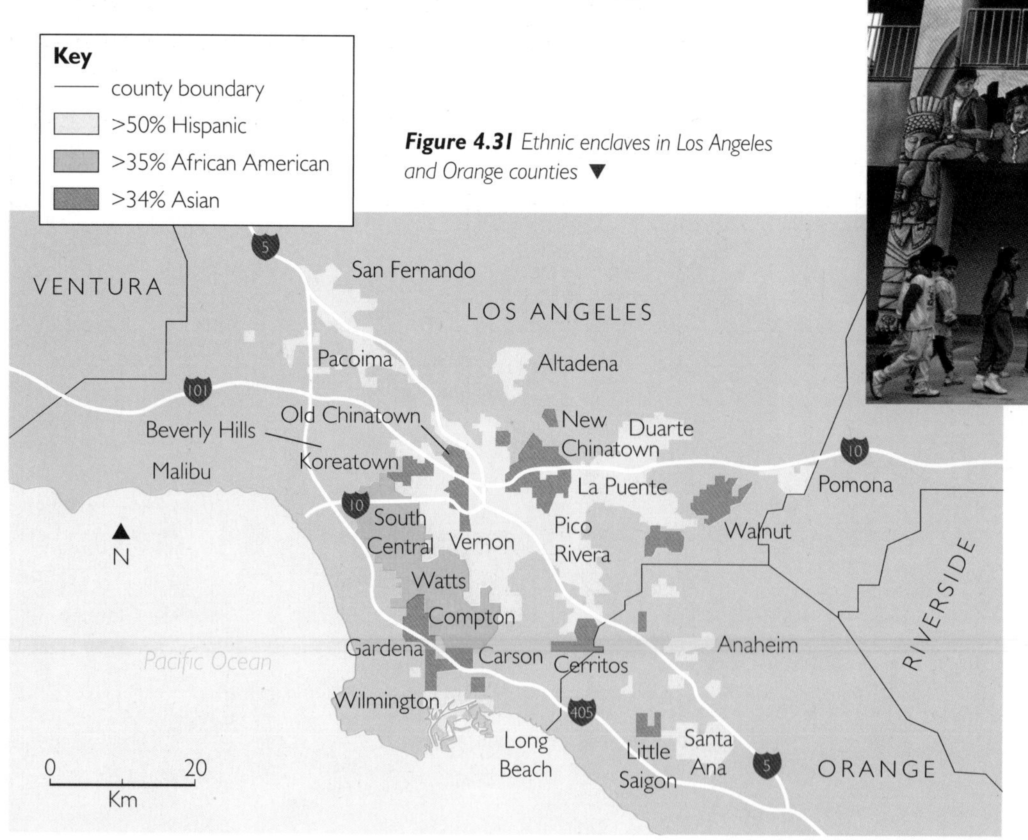

Figure 4.32 *Leo Politi primary school in East Los Angeles where 92% of the students are of Hispanic origin. The mural is representative of many found in Hispanic enclaves in Los Angeles and Orange County.* ▲

Other problems facing Los Angeles

Historically, Los Angeles's growth has relied on the assumption that cheap natural resources, such as water, would always be available. But this assumption is proving to be damaging – the rising cost of energy and the shortage of natural resources are causing problems for the city.

- **Water** – Water is piped to Los Angeles from 350 km away; continued demand for irrigating domestic gardens and filling pools is causing disputes with neighbouring states. Up to 50% of water is wasted from evaporation before it even reaches the city.
- **Waste** – 24 million people in the Greater Los Angeles area produce 50 000 tonnes of waste every day. However, a deposit scheme now allows for a 25 cent refund for any liquid container, from a coffee cup to drinks can, and has dramatically increased recycling.
- **Energy** – In August and September 2010, record heatwave temperatures (up to 45°C) caused power blackouts when power stations were unable to cope with power demands for air conditioning.

Did you know?

California has 883 golf courses, using around 90 billion gallons of water each year. Irrigated pastures for beef herds use as much water as all 38 million Californians, even allowing for all their swimming pools and lawns!

Planning a more sustainable future for Los Angeles

Los Angeles has its problems – but the Progressive Los Angeles Network (PLAN) proposes an agenda for sustainable living. The emphasis is on issues ranging from transport, the urban environment, food and nutrition, to economic development, housing and workers' rights. It really means giving residents more control over the future of their city.

Some of PLAN's proposals for a sustainable city

- Increase urban parks and clean up contaminated brownfield sites
- Promote clean fuel vehicles and green energy
- Require developers to build affordable housing in all new residential developments
- Attract food markets, farmers' markets and community gardens
- Improve public transport with 'clean' buses and new rapid bus lines
- Create a universal low fare card that allows easy transfer between buses and trains
- Promote safe, walkable and bikeable neighbourhoods
- Promote smart land use where people can drive less and live nearer to where they work, shop, study and play
- Ban new retail developments which undermine local retailers and steer future development to locations near public transport

ACTIVITIES

1. How are the following terms linked: megalopolis, suburban sprawl, suburbanisation, 'donut' city, edge city, ethnic enclave?
2. Draw a spider diagram to outline the reasons for the growth of Los Angeles.
3. Copy and complete the table to show the impact of suburbanisation in Los Angeles on people and the city environment.

	Positive impact	Negative impact
Social		
Economic		
Environmental		

4. Outline the process by which ethnic enclaves can develop in a city. You will find pages 279–281 helpful. Why are many enclaves likely to be much poorer than the city as a whole?

In this section you will learn about:

- the effects of counter-urbanisation and de-industrialisation
- the process and impact of regeneration in East London

Did you know?

The population of Ashwell, a village in Hertfordshire expanded by 20% during 1961-1981 largely as a result of people moving to it from London.

London – back from the edge

Imagine a major capital city with run-down industries and derelict land stretching along the river some 20 km eastwards from its centre out to its edge. Its population has been declining since World War II as people leave for more rural counties and a quieter life. Imagine some of those left behind, living in communities where unemployment is as high as 60% among adult males.

That city was London in 1981, one year after its docks closed. 10 000 people were directly put out of work in the docks and up to 100 000 more in companies that depended upon them. London seemed fated with decline:

- A decade earlier, it had been overtaken by Rotterdam as Europe's largest port
- The Lea Valley, to the north of London's docks, had been the UK's largest manufacturing area in the 1970s, but once the docks disappeared many companies closed or moved production overseas to countries where labour was cheaper
- It had lost 16% of its population as a result of counter-urbanisation (see below) between 1939-81 (Figure 4.33).

Now, that former port is the site of what is still, after three decades, the world's single largest urban regeneration – London's Docklands. It has enabled London to become a world city, and, in 2005, to overtake New York as the world's No. 1 financial centre. More importantly, it has helped to create a period of **re-urbanisation** in which London's population – like that of most other world cities with similar projects – has once again increased.

Causes of the decline 1: Counter-urbanisation

After World War II (1939-45) and until the 1980s, London experienced net out-migration. This reversed the urbanisation process, and is known as **counter-urbanisation**. It is similar to the 'donut city' effect in Los Angeles (see page 171), but, unlike suburbanisation, people moved out of London altogether to smaller towns and villages.

Counter-urbanisation resulted from several factors such as:

- the clearance of slums and bomb-damaged areas after World War II
- the creation of new towns (e.g. Basildon, Stevenage and Harlow, each about 30-50 km outside London) designed to absorb London's 'overspill population' from slum clearances
- the creation of London's green belt in 1947. This was a broad belt of land around London in which development was restricted or banned, and was designed to prevent further suburban expansion. People therefore had to move further away. Although the green belt protected countryside around London (and other cities with such areas), the result was that towns and villages grew rapidly beyond it. Further expansion of the city simply 'leap-frogged' the green belt.

Rapid population increase occurred in the new towns and in towns and cities beyond the green belt. Most of those who left were commuters and their families, seeking cheaper property and better environmental quality in which to bring up their children. Most continued to work in London (which offered higher salaries) but took advantage of lower house prices further away, thus adding to London's daily commuters.

Figure 4.33 *London's population during the 20th century* ▼

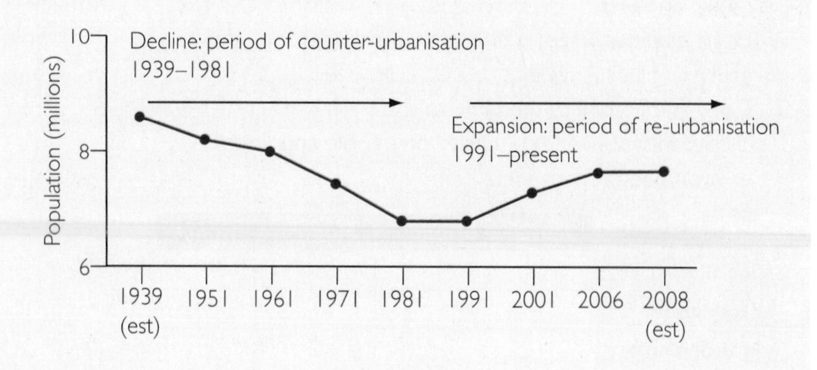

Causes of the decline 2: De-industrialisation

In the 1980s, goods produced by primary (e.g. mining) and secondary (manufacturing) industries in the UK were more expensive than those produced overseas. UK wages were higher than overseas, making British products more expensive. The growth of manufacturing in Asia, with its cheaper workforce, led to cheaper imported goods.

As a result, large numbers of mines and manufacturing companies closed during the 1980s, causing **de-industrialisation** (Figure 4.34). Closure caused huge unemployment in cities in the north and the Midlands and declining industry in east London's Lea Valley. It was a bleak period for Britain's youth – just listen to the rock group The Specials' song 'Ghost Town' from 1981. Until the mid-1960s, London was a major centre of light industry – 33% of its labour force worked in manufacturing. However, in 2010, manufacturing only accounted for 10% of the UK's GDP and 6% of employment.

To replace lost employment, the government under Margaret Thatcher planned deliberate changes to the UK economy by encouraging a '**post-industrial**' economy, directing investment towards:

- **the 'knowledge' economy** – developing high-value expertise, management and consultancy sectors. The biggest contributor to this was to be banking and finance, where international banks would use the UK as a base to finance global investment. The other significant sectors would be media and advertising, law, and IT. Such employment can locate anywhere – it is '**footloose**' – with decisions about where to locate being made largely on the basis of financial incentives such as tax breaks.
- **tourism** – the world's fastest-growing industry, resulting from greater prosperity in HICs, cheaper air travel and increased car ownership
- **property** – many cities sought to **re-brand** their past by creating a new image based on industrial heritage in former ports, canals, and old factories (Figure 4.35).

Figure 4.34 *De-industrialisation in east London* ▲

Relic feature	Regenerated as ...	Examples in the UK
Industrial past	• 'industrial heritage' environments	Beamish Museum (County Durham), Ironbridge (Shropshire)
Derelict land and docks	• brownfield sites for the expansion of office space and housing • waterside locations for new housing	Canary Wharf (London's Docklands); also Bristol and Liverpool, and canal-side in Manchester and Leeds
Old factories and warehouses	• loft apartments • shopping centres	Docks in Liverpool, Tyneside, and London, canal-side in Leeds and Birmingham Fort Dunlop (West Midlands)

Figure 4.35 *Different ways of regenerating industrial sites* ▲

Causes of the decline 3: The docks close

In 1981, the last of London's docks closed. Until the mid-1970s, they were the biggest docks in the UK, worked by over 10 000 dockworkers loading and unloading ships. However, two things happened:

- the increase in the **size of ships** meant deeper water was needed, so Tilbury, 20 miles downstream, provided a better site (Figure 4.36)
- **container ships and computerisation** replaced the need for people to load and unload ships by hand. Felixstowe, 70 miles from London, with cheaper land and a deeper harbour, developed into the UK's largest container port, and replaced London as the UK's biggest port (Figure 4.36).

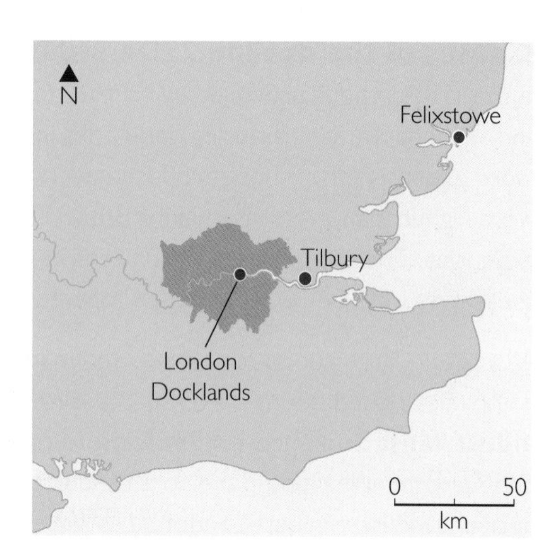

Figure 4.36 *London Docklands and the ports at Tilbury and Felixstowe* ▲

Closing London's docks had devastating effects:

- Between 1978 and 1983 over 12 000 jobs were lost. In some parts of East London, over 60% of adult men were unemployed in 1981.
- The docks were left derelict (Figure 4.37 and Figure 4.38). Dereliction in some parts was so severe that the costs of regeneration were considered too high and, therefore, the area was unattractive to investors.

The rapid succession of industrial and dock closures in East London in the 1970s and early 1980s led to huge unemployment. Economic decline caused population decline as people left to find work elsewhere; east London's population declined by 18.5% between 1971 and 1981. Half of the remaining population worked in lower-paid manual occupations. Businesses closed, including local shops and pubs.

Figure 4.37 *London's busy docks at their peak* ▼

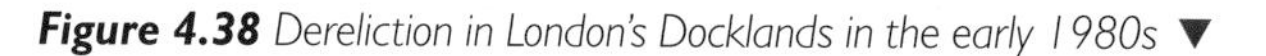

Figure 4.38 *Dereliction in London's Docklands in the early 1980s* ▼

Changing the image of inner cities

High levels of unemployment in Britain's major cities, e.g. London, Leeds, Newcastle and Liverpool, gave inner-city areas a poor image and one in which there seemed little economic future. Deprivation, lack of opportunity, and ethnic tension led to inner-city riots in Liverpool (Toxteth), Leeds (Chapeltown) and London (Brixton) in 1981. The Thatcher government reacted by attempting to rebrand the image of inner cities. During the 1980s and early 1990s, a number of 'Garden Festivals' were held, attempting to generate a 'green image' which might serve as a useful basis from which to regenerate the land further (Figure 4.39).

Garden Festival City	Year	Current use following the Garden Festival
Liverpool	1984	a mixture of housing and derelict sites
Stoke-on-Trent	1986	mostly maturing garden parkland with some retail and offices
Glasgow	1988	Glasgow Science Centre and a digital media village beside the River Clyde
Gateshead	1990	a housing estate
Ebbw Vale (South Wales)	1992	a shopping centre, housing, parkland and woodland

Figure 4.39 *Garden Festivals in the UK* ▲

Since 1980, a variety of government regeneration programmes have been launched. Some, like Garden Festivals, have been dropped. Others have tried to boost cities by attracting tourists. The European 'Capitals of Culture' (Glasgow in 1990 and Liverpool in 2008) attempt to focus on cultural regeneration in European cities.

CASE STUDY

Regenerating Canary Wharf

London's biggest asset was undoubtedly its Docklands, a 21 km^2 stretch of land parallel to the Thames and available for redevelopment. No other city in the world had such a vast amount of land close to its financial heart, the City of London. Regeneration there has been spectacular. Key features include the Canary Wharf complex (Figure 4.40), with No 1 Canada Square as its flagship building. Tenants in the Canary Wharf complex include headquarters of global banks (e.g. Barclays and HSBC), investment banks (e.g. Morgan Stanley), newspapers (Trinity Mirror) and news agencies (e.g. Reuters), and professional services firms which carry out business consultancy and accounting (e.g. KPMG). 100 000 commuters now travel to work in these office towers each day, adding to the 325 000 who work in the City of London.

Figure 4.40 *The Canary Wharf complex, with its tallest building No 1 Canada Square* ▲

Transport developments (known as **infrastructure**) have been completed to ensure that commuters can get to work. These include:

- extending the London Underground network, e.g. the Jubilee Line
- developing the Docklands Light Railway, a surface rail network which covers most of the Docklands area
- new road links, such as the Limehouse Road Link, taking traffic through the Docklands
- London City Airport (located about 6 miles from Central London), which provides easy access to the City and Canary Wharf from European cities such as Paris and Geneva.

Understanding the process of regeneration

The London Docklands Development Corporation was set up in 1981 to manage the **regeneration** of the docks. Unlike most development projects, in which developers seek planning permission through local authorities, local government was removed from the equation and developers were free to do as they wished. The Docklands regeneration involved many **key players**, including:

- landowners
- designers and developers, with ideas about how land could be redeveloped
- investors, who provided finance
- local people, who objected to many of the redevelopment plans
- central government, which created policies designed to encourage development and investment.

This kind of development, whereby private companies make the decisions and are given benefits (such as tax breaks), is known as **'market-led' regeneration**, and, sometimes **property-led regeneration**. In Docklands, economic regeneration was seen as a priority and the government felt that property developers would know how to develop the land in ways that would attract business. Jobs would be created, and wealth, it was argued, would 'trickle down' to poorer communities. It focused on

- the creation of employment
- the regeneration of existing housing stock
- the creation of new housing stock from old dockside buildings (Figure 4.41).

Figure 4.41 *Refurbished dockside wharf buildings along the Thames* ▲

Government action also supported urban regeneration.

- **Tax breaks** were critical. As long as planning permission was granted by 1991, companies could obtain huge tax breaks on any new building. These incentives were designed to attract companies to invest, and still exist.
- **Deregulation**. In 1986, the Thatcher government deregulated the financial City of London, where business had previously been restricted to UK banks and insurance companies. This meant that any bank could locate its offices in London. Hence the presence of international banks e.g. HSBC (Hong Kong and Shanghai Banking Corporation), Bank of America, and Deutsche Bank.
- **Working hours**. In 1998, the government of Tony Blair refused to sign-up to the European Working Time Directive, which limited working hours to 48 per week. The banks in Docklands argued that this would restrict trading in the financial and investment markets of South-East Asia (by arriving at work early) or those of the USA (by staying late).

Did you know?

Many cities have followed London's example by regenerating their own dockland areas, such as Melbourne and New York.

The impacts of regeneration

In three decades, the Docklands regeneration has transformed east London.

- The creation of a vast area of commercial space has enabled London to become one of the world's major financial centres. Canary Wharf is now full, but there are plans to double its space by 2025 by building further upwards. As well as business space, there is high quality conference and exhibition space at ExCel.
- It has restored the number of jobs lost to manufacturing (Figure 4.42). London lost 600 000 jobs in manufacturing between 1970-2000, but gained a similar number of jobs in business services.
- Without so much expansion for business, it is unlikely that London would ever have reached its 'Alpha++' World City ranking (see section 4.1).
- Together with London's 2012 Olympic and Paralympic Games, it has created a much-improved transport network so that east London is readily connected to other areas of the city and to overseas via City Airport. City Airport is 15 minutes by taxi from Canary Wharf, compared to one hour – at least! – from Heathrow.
- It dramatically improved the urban environment.
- It has re-vamped the image of east London by creating large housing spaces out of the former warehouses and derelict land along the Thames. The desirability of housing there has **gentrified** the area – that is, middle class high-income earners have replaced the dockworkers who used to live there. Corner shops and local convenience stores have been replaced with up-market high-spend shops and restaurants (Figure 4.43).
- As such, it has been one of the factors which have contributed to the population increase of London since the 1980s.
- Once a CBD with two centres – the West End and the old City of London – London now has three with Canary Wharf.

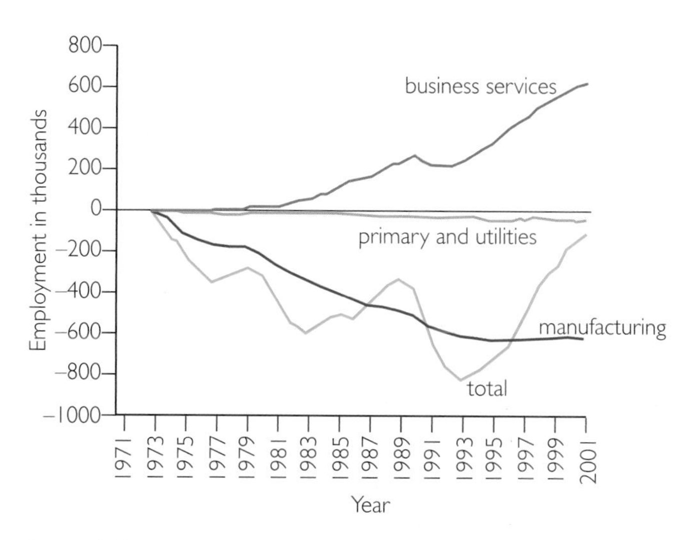

Figure 4.42 *Comparing sectors of employment in London 1971-2001* ▲

Did you know?

Technology has allowed much higher buildings on London's skyline. Steel piles now hold up tall buildings such as The Shard, built into London's soft gravels and clays.

Figure 4.43 *Restaurants in Canary Wharf* ▼

Nonetheless, the Docklands regeneration and gentrification has been criticised by many because:

- It failed to bring wealth to local people, benefiting instead well-qualified, high income earners from elsewhere. Canning Town (Figure 4.44), one of the poorest areas of London, is just 2.5 km from Canary Wharf, yet incomes **per household** (£23 000 in 2011) are less than 25% of the average income **per person** (£100 000 – before bonuses) in Canary Wharf. It is ironic that the very boroughs of London in which most Docklands regeneration has taken place (Newham and Tower Hamlets) are actually its poorest.
- It drove a social wedge between the 'new Eastenders' (i.e. financiers) and the former dockworkers, between the social housing estates which lacked investment and the private security-fenced apartments and houses overlooking the river.
- East London still suffers deprivation, poor health and poor environmental quality (Figure 4.45). In fact, health statistics show that on a journey along the Jubilee Line between Westminster in Central London and Stratford, its terminus in East London, life expectancy drops by nine years – one year for every station. Wealth and poverty sit almost side by side.
- There was no investment in the environment, no open spaces for people.
- It drove house prices well beyond the reach of local people. Two-bedroom flats in new apartment blocks in Canary Wharf in 2011 cost between £400 000 and £1 million.

Figure 4.44 *Canning Town in east London, just 2.5 km from Canary Wharf, but one of the UK's most deprived areas* ▲

The biggest change has occurred in the inner London boroughs, creating three localised trends in population:

1. Many well-qualified young adults come to work in London from the rest of the UK and the world.
2. Many inner boroughs now have markedly younger populations compared to the UK as a whole.
3. A population increase without any increase in house-building due to the division of houses into flats and the trend among young adults towards sharing houses rather than purchasing their own properties.

Re-urbanisation

Since 1985, London's population has grown, reversing counter-urbanisation and bringing a period of population increase, or **re-urbanisation**. Even though over 50 000 people each year move elsewhere in the UK, London's population is projected to reach 8.1 million by 2016.

Figure 4.45 *Map showing deprivation in east London* ▼

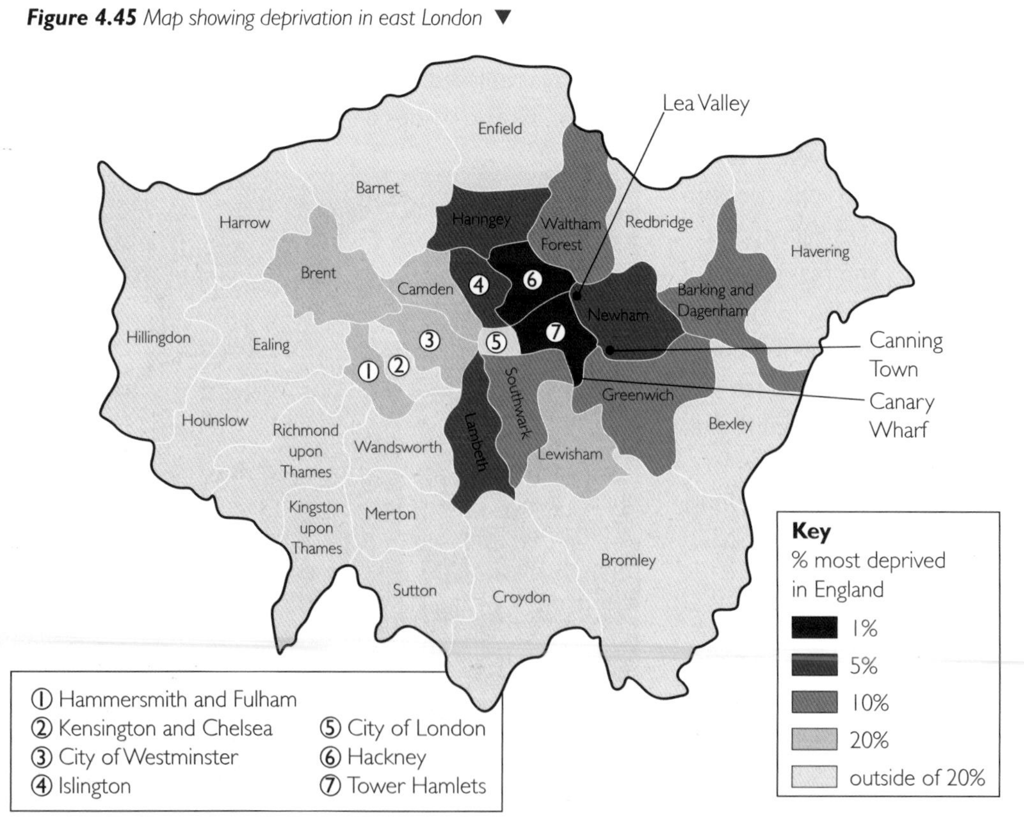

Re-urbanisation of inner London has two other elements:

- a **natural increase** of about 40,000 a year. Some inner London boroughs (e.g. Newham and Tower Hamlets) have the UK's highest birth rates. By contrast, outer boroughs have lower birth rates and a higher percentage of elderly people.
- **in-migration** is increasing rapidly, especially from EU and non-EU migrants. Some are temporary (e.g. students and people on temporary visas) whilst others are permanent.

A new era of regeneration – Urban New Deals

Whereas the Docklands regeneration focused all investment into a tightly defined area, policy since 1997 has been to spread the benefits beyond Docklands elsewhere into east London through the New Deal initiative. One example is the regeneration of Shadwell in east London (Figure 4.46). Instead of centralised top-down decision-making as in Docklands, regeneration here followed a 'bottom up' approach focusing on the communities affected, with funding drawn from both government, private companies and charities such as housing associations, who were invited to share the costs and reap profits from regeneration. Hence the term '**community-focused regeneration**'. Its achievements included:

- renovation of housing – new windows, doors and brickwork
- training and apprenticeships of local young people by building companies
- the creation of affordable homes for local people
- an expanded range of shops
- refurbishment of some housing blocks into secure, concierge-monitored blocks for the elderly.

Figure 4.46 *Renovations to housing in Shadwell – a) housing in 2003 before renovation, b) the same housing after renovation in 2007* ▼

ACTIVITIES

1 a Draw a table to show the advantages and disadvantages of **a**) deliberately running down and closing London's industries and docks, **b**) the 'knowledge economy' as a replacement for the industries and docks.

b In pairs, discuss whether you believe that the deliberate closure of London's port and industries was the right thing to do.

2 Use a spider diagram to evaluate the strengths and weaknesses of using tax breaks, deregulation and longer working hours as ways of attracting companies to locate in the UK.

3 Copy and complete the table below to show the economic, social and environmental benefits and problems arising from London's Docklands regeneration.

	Benefits	Problems
Economic		
Social		
Environmental		

4 One Newham local councillor is on record as saying 'The Docklands regeneration is not the sort of thing we'd want to repeat.' In 1000 words, discuss whether you agree or disagree with his statement.

In this section you will learn about:
- the nature of sports-led regeneration
- how valuable the 2012 Olympic and Paralympic Games have proved to be for London

Shares in British construction companies soar as London wins the Olympic bid

London 2012 will leave a lasting legacy for this country

The fact that the Olympic Games are coming to London will have positive repercussions right across these islands

London wins the bid!

6 July 2005 is a day that many Londoners remember. Expecting that Paris might win, but hoping it wouldn't, people cheered long and loud when the International Olympic Committee (IOC) announced that London was to host the 2012 Olympic and Paralympic Games. The competition had been keen. London won because it planned to use the 2012 Games as a way of bringing regeneration to east London – a concept known as **sports-led regeneration**.

What does sports-led regeneration involve?

Sports-led regeneration is normally based around major events such as the 2012 Olympic and Paralympic Games, and is a major challenge for host cities. London had seven years to prepare for the 2012 Games, with tight schedules and substantial cost – nearly £10 billion! More than any Olympic and Paralympic Games so far, London set out to provide a **legacy** of sustainable impacts, lasting well beyond the departure of the last athlete.

Hosting the Olympics is a huge commitment for any city – and the same is true for London. The 28 days duration of the 2012 Games involves:
- the creation of a new Olympic Park – London's first major park built in over a century
- in the Olympics, 11 000 athletes competing in 300 events, supported by 5000-6000 coaches and officials, and attended by 4000-5000 other members of the Olympic community
- in the Paralympics, 4000 athletes and 2500 officials
- 20 000 of the world's newspaper, radio, television, and Internet journalists
- the sale of over 9 million tickets, with half a million spectators a day travelling to events in and around London
- 63 000 operational personnel (47 000 will be volunteers, many as stewards, marshals, and drivers).

The Olympic Park	Near Stratford, East London.
What's there?	The main stadiums and facilities: • Olympic Village • Olympic Stadium (all track and field events) • Aquatic centre (all swimming and diving events) • Velopark (track cycling and BMX) • Two-stadium hockey complex (hockey, Paralympic football) • Basketball arena (basketball, handball, wheelchair basketball) • Eton Manor (Paralympic archery, wheelchair tennis) • Handball arena (handball, modern pentathlon, Paralympic goalball)
And outside the Olympic Park?	Other London-based events include: • ExCel (Exhibition Centre for East London) (contact sports e.g. boxing, judo, taekwondo, plus weightlifting, wrestling) • O2 Arena (gymnastics, basketball, wheelchair basketball) • Earls Court (volleyball) • Wembley stadium (football final) • Wimbledon (tennis) • Royal Artillery Barracks, Woolwich (shooting) • Lord's Cricket Ground (archery) • Greenwich Park (equestrian and modern pentathlon) • Hyde Park (triathlon, swimming – 10 km open water) • Regents Park (road cycling) • Horse Guards Parade (beach volleyball)
Elsewhere in the UK?	• Bisley in Surrey (shooting) • Hadleigh Park in Essex (mountain biking) • Broxbourne White Water Canoe Centre, Hertfordshire (canoe/kayak slalom) • Eton/Dorney rowing lake in Berkshire (rowing) • Weymouth in Dorset (sailing) • Various football grounds (football heats)

Figure 4.47 *The locations of Olympic and Paralympic venues* ▲

The cost of building the facilities listed in Figure 4.47 needs to be weighed against benefits:

- **costs** – include **tangible** costs (things that cost money, e.g. building a stadium), and **intangible** costs (things that cannot be measured but are important such as the effects of building disruption on people)
- **benefits** – again, either tangible (e.g. receipts from tickets, sponsorship deals, higher tourist spending) or intangible (e.g. the gains provided by having a new park).

The Olympic bid was made on the basis that benefits would exceed costs although many of these benefits depend on the eventual value of the regenerated land. Already, land values in Stratford have risen sufficiently to recoup half of the cost of the 2012 Games.

Who are the key players?

The 2012 Games involved a number of organisations (Figure 4.48). The impact of the Games on the local economy, people, and environment is affected by the decisions which these organisations make.

Key player	What they do
International organisation	
The International Olympic Committee (IOC) with officials from many countries in the world	decide who will host the Olympics
UK central government, which has appointed:	
The London Development Agency (LDA)	responsible for economic and urban development in London
The London Organising Committee of the Olympic Games (LOCOG)	responsible for organising the Games, e.g. events, athletes, drug-testing, etc.
The Olympic Delivery Authority (ODA)	responsible for building venues, from planning and design to land purchase to the construction and completion of the stadiums and other facilities
Olympic Games Legacy Company	a company established to ensure that the 2012 Games leave a legacy of venues and housing
Regional government – the London Assembly	
The Mayor of London and leader of the London Assembly	the London Assembly has control over transport and influences policy and planning decisions across the city
Transport for London (TfL) which is part of the Mayor's Office	responsible for transport across London, ensuring that half a million people a day can get to and from the Olympic sites effectively

Figure 4.48 *Key players involved in the 2012 Olympic and Paralympic Games* ▲

Impact on the local economy

The new Olympic Park is located in Stratford, on the site of a former industrial estate known as Marshgate Lane. 380 companies employing 11 000 people had to be moved and paid compensation by the Olympic Delivery Authority (ODA). Most stayed within east London.

However, the new Olympic Park is part of a new vision for east London in which IT and new technologies will be developed. In 2010, Google and Facebook signed up to be part of a new vision for the Media Centre which the UK government hopes will create 3000 high-tech jobs and see the UK as a new Silicon Valley. The debate is whether this is what east London needs as it is the lack of employment for those without qualifications (and therefore not suited to the high-tech jobs being created) that is the problem. For example, in Canning Town, within Newham Borough, the 2001 census showed that over 43% of the adult population had no qualifications.

Figure 4.49 *The new Olympic Park under construction. Note the Olympic Stadium and Aquatics Centre in the centre of the image.* ▶

Impact on the local environment

The main focus for the 2012 Games in east London is the Olympic Park, which lies along the Lea Valley in east London (Figure 4.50). The collapse of manufacturing in the 1970s and 1980s led to widespread dereliction in this area (see section 4.5).

Creating the new Olympic Park has involved re-landscaping the area into a huge leisure facility for local people, containing all of the largest Olympic venues. Having the main stadiums on one site helps to prevent traffic congestion – during the 1999 Olympic Games, the host city Atlanta was brought to a standstill as athletes and officials joined spectators trying to travel to the main Olympic venues. The 100 metres men's final had to be delayed for over an hour as athletes tried to make their way to the event.

However, the main legacy of the new Olympic Park is its transformation of an area in which environmental quality was poor. It is this legacy that helped London to win the Olympic bid.

- Derelict factories and contaminated land had long created a toxic waste problem in the area (Figure 4.51). Parts of the new park had been used for dumping rubble from bombed buildings in the post-World War II clearance.
- Electricity pylons had blighted environmental quality in the area. These were the first things to be demolished as construction began and the power lines were put underground.
- All soil was stripped away and cleaned in a 'soil hospital' before being returned ready for landscaping.
- A major clean-up of the River Lea and its canals has created a new waterside park for local people.

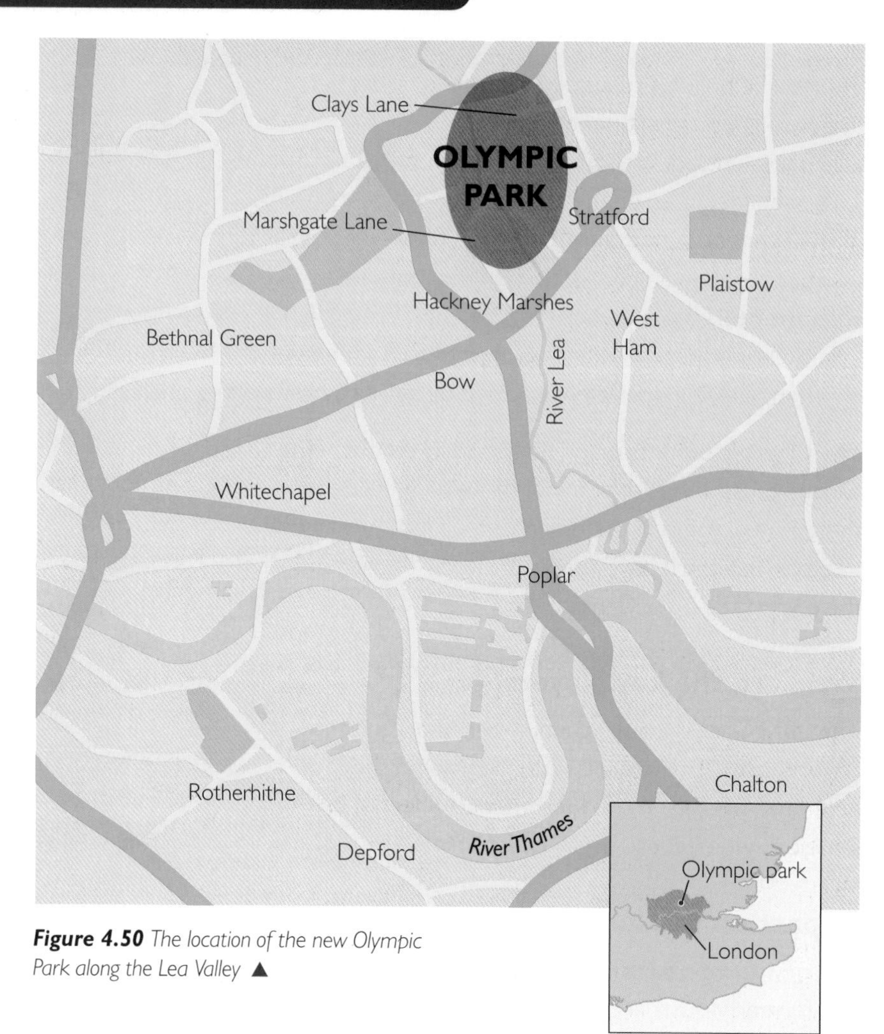

Figure 4.50 *The location of the new Olympic Park along the Lea Valley* ▲

Figure 4.51 *Dereliction along the Lea Valley* ▶

Impact on local people

The biggest resource the 2012 Games will leave local people is housing. Housing needs in east London are severe, particularly affordable housing in low-income areas such as Newham. Rapid price increases since the 1990s have affected London as a whole, and particularly east London since development of the new Olympic Park.

Between 2008-2011, the most rapid price rises were in areas around the perimeter of the Park (Figure 4.52). The Olympic Park has re-branded the area in people's minds – high-salaried City workers are now interested in buying property here where they might not have been in the past. This contributes to increasing prices, which can be disastrous for those trying to purchase or rent in the area. There is a serious need for **affordable housing** which low wage earners can either buy, rent, or part-purchase.

Figure 4.52 *Icona, Stratford, a former brownfield site now new housing overlooking the new Olympic Aquatic Centre. 87 of the 249 units here are affordable, low-cost homes.* ▲

The new Olympic Village should provide 2800 units, 35% of which will be affordable housing (i.e. about 1000 homes). However, the original plan provided for 4000 homes, of which 75% would be affordable (i.e. 3000). Although 9000 more planned houses around the Park will include affordable units, the London borough of Newham is in particular need of family housing for rent, for part-purchase and for shared ownership with housing associations. With the high costs of the Games, the greatest temptation is to maximise returns by selling property on the open market.

Did you know?

Most Olympic venues are never used again after the last athletes leave. Even Beijing's spectacular Birds Nest stadium is now just a tourist attraction; no sports events were held there between late 2008 and mid-2011.

ACTIVITIES

1 Copy the table below. In pairs, read through this unit and list the costs and benefits (both tangible and intangible) of staging the London Olympics in 2012.

	Costs	Benefits
Tangible		
Intangible		

2 a Compare the quality of the environment of the Lea Valley area before and after the Olympic-related development.

b How far will the 2012 Games leave a legacy of environmental gain? Use examples to support your answer.

3 a Identify the economic and social gains and losses of the Olympics for the local area.

b Overall, would you say that the Olympics provide more gains than losses? Explain your answer.

4 In 1000 words, consider this statement about the Olympics: 'Good for the country, good for London, not so good for local people.' Do you agree or disagree? Use examples to support your answer.

4.7 Retail-led regeneration – is the high street dead?

In this section you will learn about:

- reasons for the expansion of out-of-town shopping
- the nature of retail-led regeneration in east London

High streets are under attack. Figure 4.53 shows St Austell's White River Place, a new retail development in the centre of a small town in Cornwall. It has been designed to attract shoppers back into a depressed area of the county. One problem is that the centre has largely attracted businesses which were already there in the town, rather than bring in new trade. So, despite the re-development, nearly 30% of shops in St Austell town centre are empty (Figure 4.54) putting the town fifth in the top ten towns with the most vacant retail space in 2010. In the UK, the hardest-hit town centres are in northern England: 22% of shops are empty in Yorkshire and The Humber, and 19% in the North-East, North-West, East Midlands and West Midlands.

St Austell is hardly a world city! But it does illustrate how hard town and city centres are working to attract shoppers. Larger cities have developed schemes such as traffic-free pedestrian shopping precincts (e.g. Manchester's Arndale Centre) and making retail areas more pedestrian-friendly (e.g. London's Oxford Street). They are each fighting threats from **out-of-town shopping**.

Figure 4.53 *The White River Place in St Austell, Cornwall* ▲

Figure 4.54 *Empty shops in St Austell's main shopping street* ▲

The growth in out-of-town shopping

There are sixteen major out-of-town shopping centres in the UK, as distinct from open-air retail parks found in most towns. Brent Cross, north London, opened as the UK's first out-of-town shopping centre in 1976. In some cases, out-of-town shopping has been used to help regenerate run-down areas, such as Sheffield's Meadowhall, and Gateshead's Metro Centre. The trend towards **de-centralisation** of retailing – where retail companies move to more spacious out-of-town locations – has been the most significant change affecting city centres over three decades, and has had major impacts.

Three factors are important in their development:

- **population change** – both suburbanisation and counter-urbanisation have taken wealthy shoppers from inner cities to the outer suburbs and towns beyond. Shops are therefore 'chasing the money'.
- **push factors** – high street shops face high rents, high running costs (e.g. business taxes) and a shortage of space in which to expand. This pushes many companies towards cheaper options, either on city outskirts or establishing online shopping.
- **pull factors** – land is available and cheaper in out-of-town locations. Out-of-town shopping depends upon access by car, so space is taken up with car parking (Figure 4.55). Cheap land is critical. In Oxford Street, London, retail rents were £4400 per square metre per year in 2009! Compare that with £217 per square metre in Sheffield's Meadowhall shopping centre and it is easy to see why all except premium shopping brands are increasingly avoiding large city centres.

The issue is complicated by other threats to city centres:

- **competition from supermarkets** (or superstores) usually located on city outskirts. These offer increasing ranges of goods such as clothes and DVDs, which have traditionally been sold in specialist shops. Superstores also offer free parking as well as 'one-stop-shopping' that allows consumers to get everything under one roof.
- **internet shopping** accounted for 10.5% of UK retail sales in 2010, up from 7.9% in 2009
- **out-of-town leisure parks** typically contain a multi-screen cinema, bowling alley and restaurants, and offer free parking
- **high city centre rents** and falling sales result in more empty shops and make city centres a poorer shopping experience for customers. The percentage of empty shops in UK city centres rose from 12% to 15% between 2010-11. However, this figure is far higher in some places – in Kent, 38% of Margate's shops stood empty in 2010, the UK's highest figure.

So which way will high street shops in city centres go? They offer benefits such as convenience, service and a chance to see products and compare brands. But city centre parking charges, crowds and traffic congestion make shopping in out-of-town centres easier – provided you have a car. For increasing numbers without cars in inner city areas, the high street remains an attractive option.

Figure 4.55 *Aerial view of Bluewater Shopping Centre in Kent. The 240-acre site has parking for 13 000 cars and 50 coaches.* ▼

CASE STUDY

The Westfield Centre, Stratford

In 2011, the Westfield Shopping Centre opened in Stratford, east London, on the edge of the new Olympic Park (Figure 4.56). It is the UK's – and Europe's – biggest shopping centre. It covers a shopping area of 176 000 square metres and is owned by The Westfield Group, an Australian property development company. It resembles American shopping malls more than British high streets. It is part of a larger commercial centre, known as 'Stratford City', which will also in time develop office space.

Figure 4.56 *The Westfield centre, Stratford* ▼

The core of the Westfield centre is its shopping complex. 80% of the 300 stores are up-market fashion outlets specialising in luxury brands such as Louis Vuitton and Prada. Mainstream chain stores include Marks & Spencer, Debenhams, Next, and Waitrose, bringing luxury goods, high street and supermarket functions to one location. That is unusual for London, which normally differentiates high-spend stores in Chelsea and Knightsbridge, compared to mass-market stores in Oxford Street.

However, the Centre aims to offer more than shopping. 'Stratford City' also includes:

- 50 restaurants and a 14-screen multiplex cinema. To maintain its image, it does not offer KFC or McDonald's, and offers more upmarket choices such as the Square Pie Company.
- 120 000 square metres of hotel space
- 640 000 square metres of commercial space – half the space of the Canary Wharf complex about 5 km away!

Retail-led regeneration

Stratford City and the Westfield Centre form a regeneration project focused on retailing which is designed to improve the area close to the new Olympic Park.

- The Westfield Centre will help to regenerate Stratford by bringing in business and creating jobs. Westfield is in the heart of the London borough of Newham, one of London's two most deprived boroughs. The Centre will create 8500 new jobs, many of which will go to local residents. This compares with the Docklands regeneration (see section 4.5) which provided jobs for highly qualified people, rather than locals.
- Westfield is opening a Retail Academy designed to raise qualification levels in the retail industry, helping to address poor levels of educational qualifications among Newham residents.

- It will redevelop Stratford's town centre which had become run-down because of competition from out-of-town shopping.

Its location

The Westfield Centre is located 5 km east of the City of London and about 9-11 km from London's main shopping areas of Oxford Street, Knightsbridge and Chelsea. It has a wide catchment area; Westfield estimates that 4.1 million people live within easy catchment of Stratford (Figure 4.57). Incomes in London are the highest in the UK – in 2010, average per capita weekly incomes were over £600, more than 40% above the UK average – generating high spending power. Although east London is deprived, access from affluent areas of Essex and Suffolk is excellent by eight surface, tube, Docklands and international rail connections, which is critical since there are only 5000 parking spaces in the Centre.

- By surface rail, it is 25 minutes by train from Chelmsford, and 40 from Colchester
- It has direct access by the overland line to some of London's wealthiest suburbs, e.g. Hampstead (25 minutes away) and Richmond-on-Thames (50 minutes)
- After 2012, it should have a huge market via Eurostar train services to Stratford International station from France, Belgium and Germany.

Figure 4.57 *The catchment area of Stratford Westfield centre* ▼

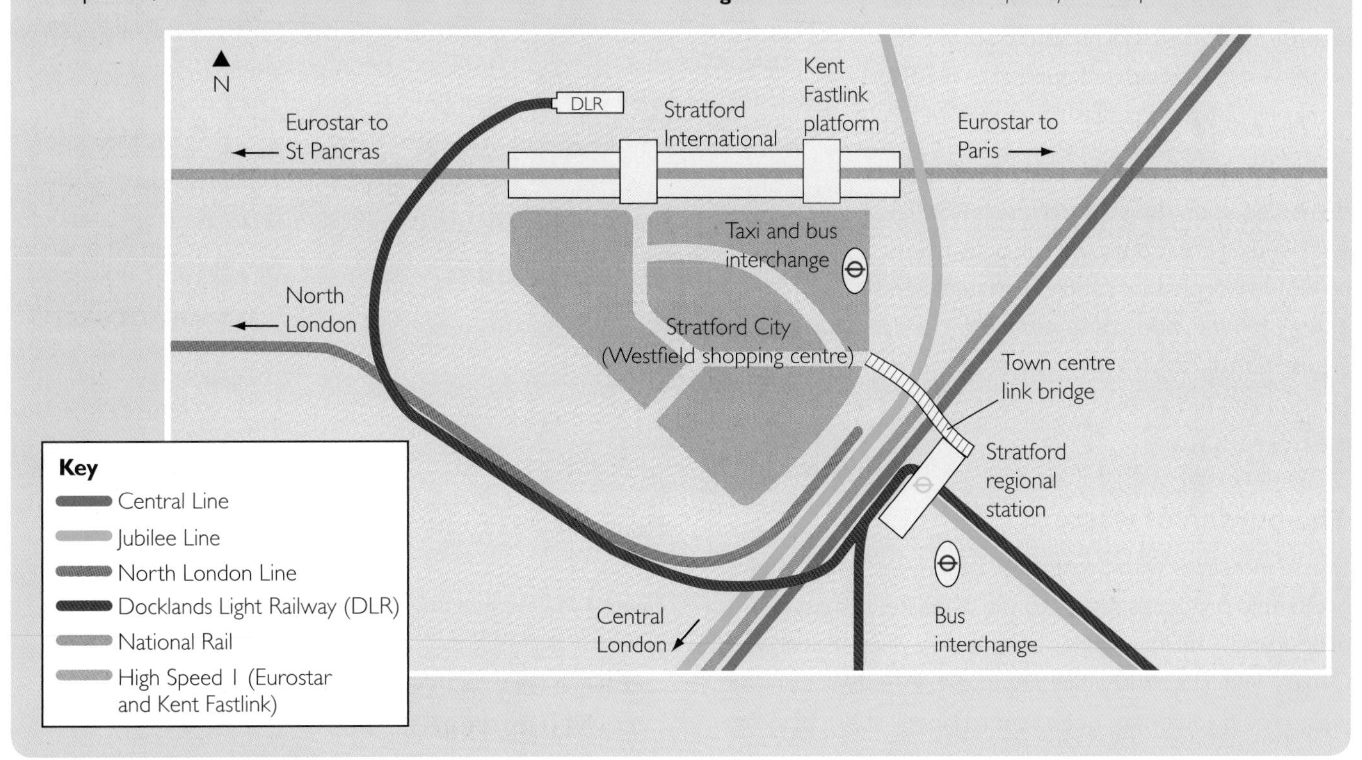

ACTIVITIES

1. Draw a table to compare city centre and out-of-town locations and shopping experiences. How would you conduct a field study to assess which of the two was the better for a town or city?
2. In pairs, discuss and present a SWOT analysis about how you see the future of city centres. (SWOT analyses identify **S**trengths, **W**eaknesses, **O**pportunities and **T**hreats).
3. Now do a similar SWOT analysis for Stratford's Westfield Centre. Which of the two has the greater number of **a**) Strengths, **b**) Weaknesses, **c**) Opportunities and **d**) Threats?
4. Discuss why transport is such a key factor in retail regeneration.
5. In 1500 words, discuss the merits of market-led (section 4.5), sports-led (section 4.6) and retail-led regeneration as ways of improving cities.

In this section you will learn about:

- the challenges of waste management in cities
- the challenges of transport and how it might be managed

1 – What a waste!

The 'world cities' are becoming increasingly important. For example, London's financial core has helped to make it the fourth largest urban economy in the world, in spite of the 2007-08 banking crisis. This comes at a price and places stresses upon the environment both within and beyond the city. As its prosperity increases, more people live and work there, consume more goods, energy, food and transport, and generate more waste. London needs 125 times its own surface area to supply the resources it consumes! 90% of all products bought are waste within 6 months of purchase. In 2010, managing London's rubbish cost £600m.

Like most cities, London's early development depended upon the ability of the immediate surrounding area to provide all of its needs. It requires a constant supply of material and energy to feed, house and service its needs. Figure 4.58 illustrates the resource flows into London and the amount of waste produced by the city each year.

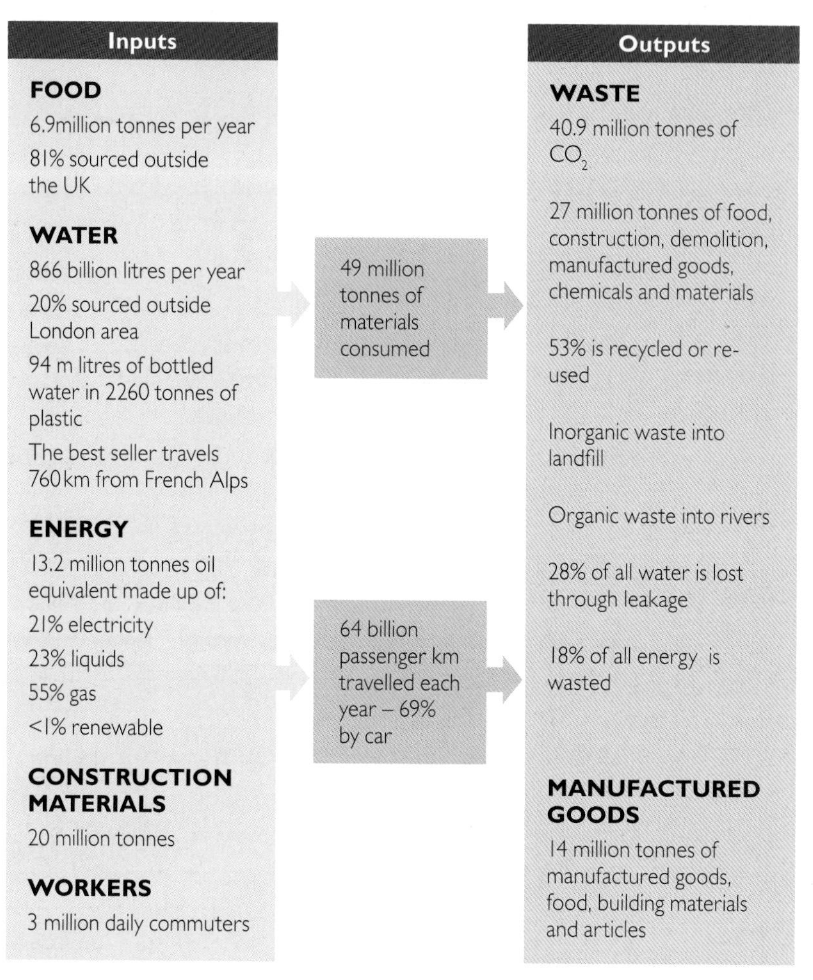

Figure 4.58 *London's resource flows* ▲

The burden of waste

Waste increases with prosperity. Every year, each London household produces over a tonne of waste; the city produces 3.4 million tonnes of rubbish a year, enough to fill Canary Wharf's main office tower every 10 days. London's waste has traditionally been sent to landfill sites. Burying waste was cheap, but difficult in a densely built up city.

At present, only one-third of London's waste is dealt with within the city, and most of this is incinerated. The remaining two-thirds is sent outside the city for landfill. Since the 1960s, London's waste has been taken to the surrounding Home Counties (e.g. Essex) by barge, road and rail. Now, councils increasingly encourage recycling, but the issue is complex as shown in Figure 4.59. Change has to come. Landfill sites are almost full and 2009 Environment Agency data shows that all **landfill** sites in south-east England will be full by 2013.

Figure 4.59 *London's recycling secret* ▼

The dirty secret of where your recycling really goes

Every item of paper placed in Camden's recycling bins is sent to Malaysia, Indonesia, India or China. 10% of newspapers and pamphlets are sent to Malaysia and 90% to Indonesia, while a fifth of mixed papers are sent to China with the remaining 80% ending up in India. Only steel, aluminium and glass are recycled in UK processing plants.

Many recycling experts hold that this is no bad thing. The ships that bring China's vast number of imports to the UK would return home empty were it not for the fact they are stuffed with used British plastics, which in turn make the next generation of plastic goods.

– Adapted from the *Camden News Journal*

Figure 4.60 shows the components of London's waste. At least 80% of what is shown could be reused, recycled or composted. Recycling is increasing – but only slowly.

- In 2001–2, 9.3% of London's household waste was recycled, rising to 13.2% in 2003-4, compared to a UK average of 17.7%. In 2010, London Borough Councils recycled an average of just 25% of waste, the lowest rate in England.
- Part of the problem is London's housing stock, half of which is flats. These account for just 10% of recycled waste because recycling facilities are often not provided by management.
- The UK government hopes to increase recycling of household waste to 33% by 2015.

The environmental pressure group Friends of the Earth believes that London should recycle 50% of its waste as cities in Switzerland and the Netherlands already manage this. By comparison with London, Berlin recycles 41% and New York 34%.

The EU has now set a limit on the amount of biodegradable waste that can go to landfill, and a £32 per tonne tax imposed by the government aims to reduce this amount by two-thirds. If products were made so that they could be recycled, this number could even reach 100% – it could also create jobs.

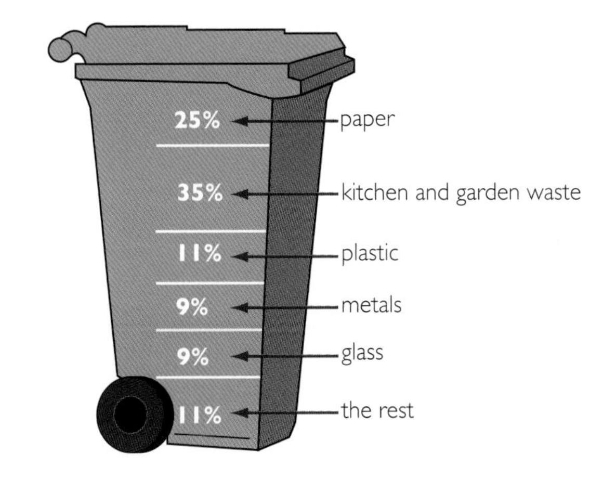

Figure 4.60 *The components of what Londoners throw away* ▲

Raising London's game

In 2010, London Mayor Boris Johnson outlined plans to turn London into a zero-waste-to-landfill city by 2025. Part of the initiative is to increase household recycling and composting rates to 60% by 2031. The report – describing waste as a 'Wasted Resource' – predicted that London's greenhouse gas emissions could be reduced by two million tonnes of CO_2 equivalent each year by reducing waste and increasing recycling. The report showed that London's landfill produces over half a million tonnes of greenhouse gas emissions annually, and that cuts in carbon emissions of 1.5 million tonnes each year could be achieved by:

- reducing the use of materials in unnecessary packaging
- creating waste-to-energy generation plants.

Figure 4.61 *What we in the UK throw away* ▼

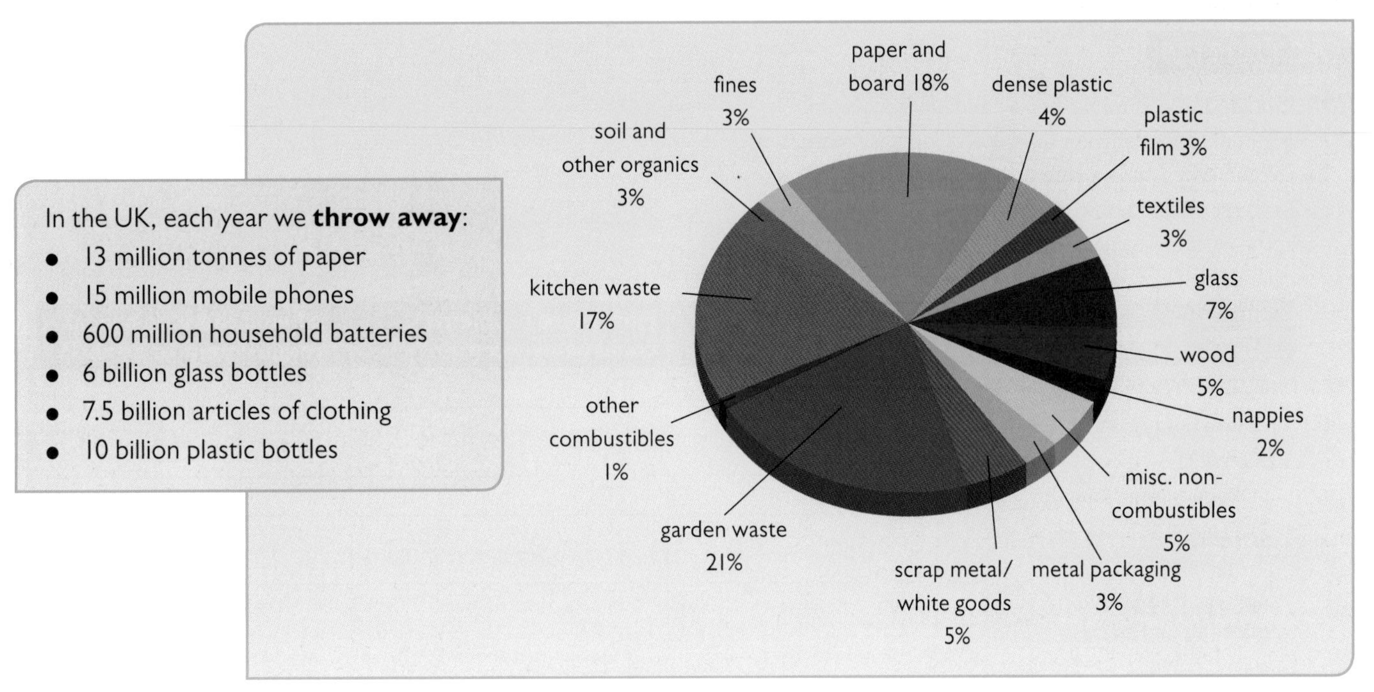

To achieve improvements, London boroughs need to:

- provide easy-to-access waste collection, recycling and composting services, especially for people living in flats
- trial recycling rewards schemes, such as the 'Recycle Bank' initiative which rewards households with shopping vouchers based on volumes recycled, and has proved effective in the USA
- introduce new waste-to-energy systems. For example, anaerobic digestion helps organic waste to ferment, producing methane (which in turn is used to generate electricity) and compost with a high nitrogen content for agriculture.
- create links with London businesses and manufacturers to reduce packaging.

As things stand, the debate is between landfill – which is increasingly difficult to sustain – and incineration, which actually destroys many products that could be recycled. A number of strategies exist by which councils can increase recycling in their areas:

- doorstep collection, rather than expecting people to take recycling to collection points
- charging for waste beyond a certain level, e.g. number of bags or bins
- charging households who do not recycle
- offering collection points for all waste including free exchange schemes for unwanted but functioning electrical goods.

Each of these has potential problems, and relies on councils, rather than private companies or individuals, to take responsibility for waste.

Landfill	
For	**Against**
• makes good use of old quarries • easily managed and safe • produces methane that can be used as a fuel source • can be sealed and landscaped for parkland or building sites	• attracts birds and pests • loose materials blow around • gives off smells of methane • waste materials seep into surrounding soils and contaminate groundwater • can suffer from subsidence • generates heavy traffic

Incinerators	
For	**Against**
• long life span • cost effective • safe disposal of toxic substances • production of energy from the burning waste • ability to reclaim metals such as aluminium • usefulness of the residue for road building	• gives off toxic gases (carcinogenic and oestrogenic) • fine particulates escape into food chains • CO_2 (a greenhouse gas) is emitted • not all waste is combustible (25% of original waste remains after burning) • residues used in path and road surfaces seep into the ground as they weather and decay

Figure 4.62 *A summary of the arguments for and against landfill and incineration* ▲

ACTIVITIES

1 Outline the economic, social, and environmental arguments for reducing landfill and increasing recycling.
2 Explain why waste incineration is a controversial issue.
3 In pairs, discuss and complete the table below showing some of the alternative strategies for reducing waste.

Strategy	Problems in the way of achieving this	How these problems could be overcome	Should there be a legal requirement to do this?
a 'Zero Waste' policy			
reduce non-essential packaging			
make recycling easier			
end landfill as a way of dumping waste			
use waste to produce materials and energy			

2 – Managing transport

Managing transport in Bangkok

Bangkok's rapid population growth has been unplanned (Figures 4.63 and 4.64). The city has one of the lowest proportions of road surface to area of any major city (around 8%, compared to London's 22% and New York's 24%) making traffic density an issue. As the city grows, longer journeys to work or school lead to traffic congestion, a major issue. Most of the city lies at about two metres above sea level, and flooding during the monsoon regularly disrupts traffic flow.

Bangkok's prosperity compared to the rest of Thailand means that it has 55% of the nation's cars with only 15% of its population. 2.3 million cars and 2.5 million motorcycles fill the roads every day. Journeys now average just 5 km per hour, depending on time of day. During early morning, journeys are quick along elevated 6-8 lane freeways, 50 metres above ground level, and accessed by ramps with toll booths. Roads face several problems.

- **Congestion**. During rush hours and monsoon seasons, roads become congested and flooded at ground level, causing traffic chaos. Road development in Bangkok is financed by toll charges – at rush hours, traffic queues build up at the toll booths.
- **Accidents**. Many freeway junctions have converging cross-overs which drivers negotiate at high speed. Bangkok accounts for 40% of Thailand's traffic accidents. Though road deaths have declined sharply, they are still four times that of the UK.
- **Air quality** is poor, though improving. The phasing out of lead petrol has had a positive effect – by 2000, only 3% of schoolchildren in Bangkok suffered lead levels above World Health Organisation safety levels compared to 11% just five years earlier. Bangkok's pollution is still above EU limits but below that of some US cities. Motorcycles now have 4-stroke engines, a change from the late 1990s. And, there are campaigns to persuade owners of Bangkok's iconic (but polluting) 'tuk-tuks' to make a similar change.

Figure 4.63 *A typical traffic jam in Bangkok* ▲

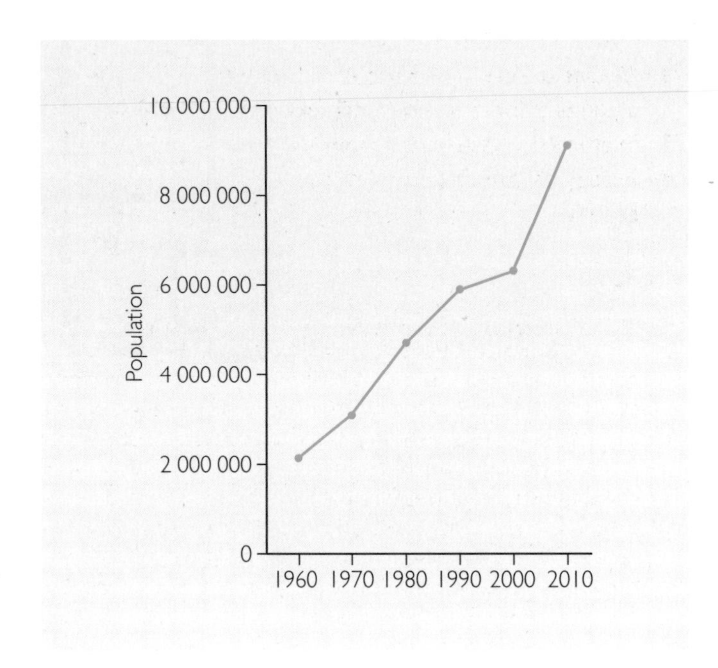

Figure 4.64 *Bangkok's population growth. The rapid rise between 2000-10 is partly due to the expansion of the administrative area of the city, but growth is rapid nonetheless.* ▶

Solving Bangkok's traffic problems

Bangkok's Mass Transit Authority (BMTA) is responsible for managing traffic and solving traffic problems by expanding bus and rail transport. But public transport has not kept pace with Bangkok's population and economic growth. After a military coup in 2006, Bangkok's military government promised to solve traffic jams, but little has been done. Instead, it cut spending on public transport and increased defence spending by over 30% instead.

Over 80% of Bangkok's working population travels to work by private transport, such as:

- motorbike taxis, which are fast and can overtake traffic jams, but are dangerous
- private buses and vans linking the city with the suburbs. Fares are more expensive than public buses, but these use expressways and take short cuts to beat the jams.

1 – Developing rail systems

Rail networks are much needed in Bangkok. There is an existing system but it is slow and at present covers little of the city (Figure 4.65). In the 1990s, the Thai Government developed a rail system, including an elevated 'Skytrain', a new light-rail tram system, and an underground network, all aimed at reducing road congestion. However, the 1997 Asian financial crash brought work to a halt and the system has never been fully built as planned.

- The 'Skytrain' opened in 1999, linking northern and eastern suburbs to the CBD, with 23 stations (compared with 52 on London's Piccadilly Line). The Skytrain target was 600 000 users daily, but it has never reached this, partially due to the fact that there are no escalators so passenger access to elevated stations is difficult.
- A new elevated train links the CBD with Bangkok's new Suvarnabhumi Airport.
- Further extensions were approved in 2007. Five lines are under construction, six more are planned together with four additional monorails. These were due to open by 2013, but progress is at best slow and funding is limited.
- Bangkok's first underground system opened in 2004 between the central railway station and northern parts of the city, the same area as the Skytrain! It is more expensive than the Skytrain, so passenger numbers have fallen from 400 000 daily to 150 000.

Figure 4.65 *Bangkok's limited rail network* ▼

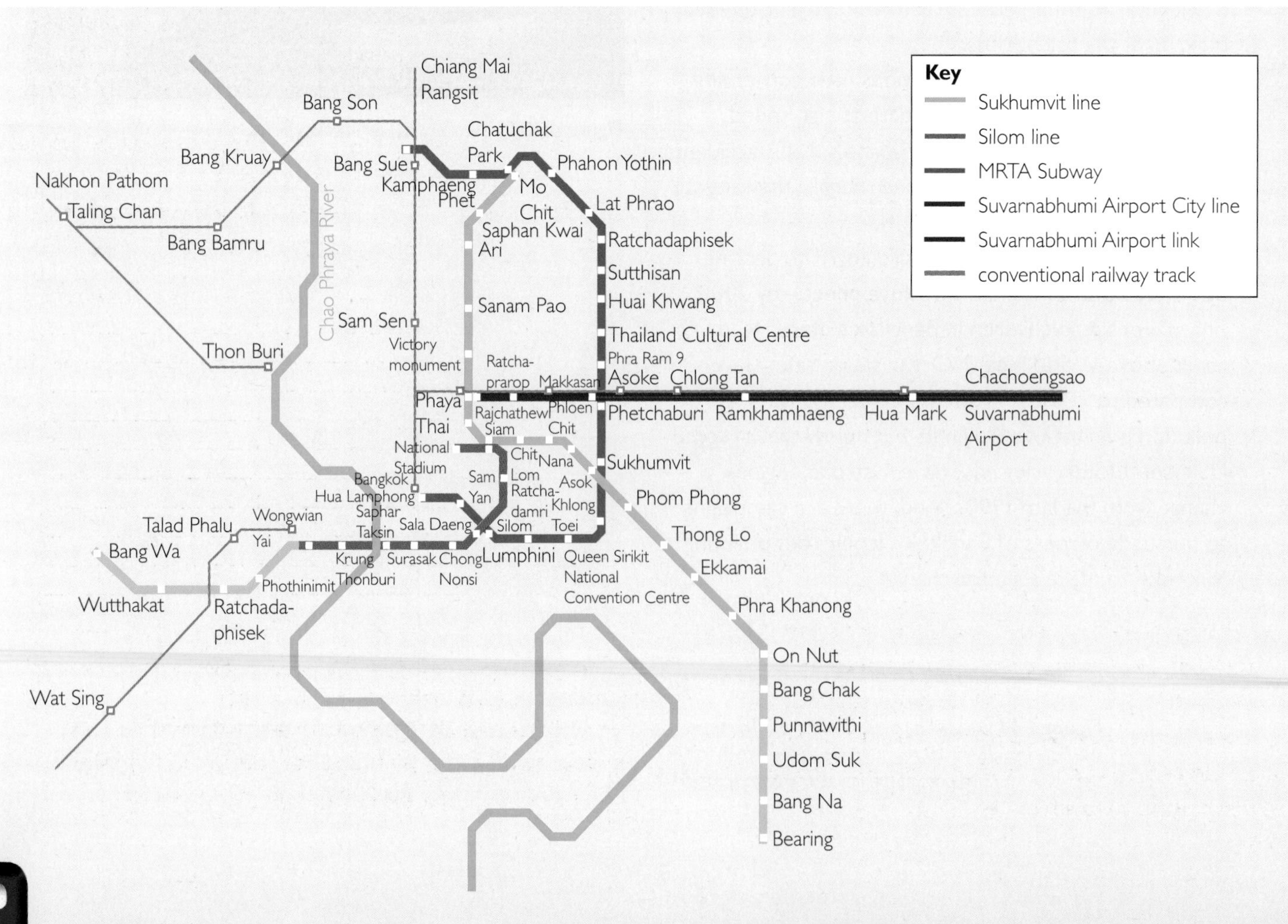

2 – Bus transport

Bus services provided by Bangkok's Mass Transit Authority carry 3 million passengers each day. But ageing vehicles and rising oil prices make it difficult for the authority to provide reliable services. Bus services are certainly frequent and cheap, with fares of about 7 baht (14p) to most destinations. The fares for air-conditioned buses are more expensive at 11-24 baht (22p-48p).

In 2010, the government announced a 'Bus Rapid Transport' system, a network of dedicated bus lanes separated from traffic. The first of five routes opened in 2010, covering 16 km and 12 stations in the southern part of Bangkok's CBD. It has not been well received – by constructing bus lanes within existing roads, one road lane or more has been lost, making congestion worse.

3 – Boat and Ferry Services

Travel by ferry is attractive but can never carry high volumes. Capacity is low and travel times slower than rail or bus. Nearly 500 000 passengers travel by ferry services, including the fast Chao Phraya Express (Figure 4.66) along the main Chao Phraya River. Other smaller ferries cross the river at points where roads meet the water.

Can Bangkok's transport system be made more sustainable?

Compared to many Asian cities, Bangkok is extreme in its transport characteristics. Private vehicle use is too high for its density of people and roads, and public transport infrastructure is inadequate for the population, especially rail. Government policies are biased towards private transport, and planning and construction of transport infrastructure is slow.

How far should Bangkok go in pursuing policies to remove cars from the road? London's 'congestion charge' has provided one kind of solution – funding public transport expansion using charges levied each time private vehicles enter central London. Meanwhile, greater numbers of buses and bus lanes have led to a doubling of passenger traffic on London's buses between 2004 and 2009. Should Bangkok do the same?

Figure 4.66 *The fast Chao Phraya Express ferry which runs along the Chao Phraya river* ▲

Internet research

Research one example of a sustainable transport strategy in a world city, e.g. London's congestion charge. Research the following questions:

a What kind of public transport is there and how wide a coverage does it have?

b What traffic problems does it face?

c What strategies are there to manage traffic now and in the future?

d How effective are current policies?

ACTIVITIES

1 In pairs, discuss and complete the table below about transport by different means in Bangkok.

Transport type	Advantages	Disadvantages
Private transport		
car		
motorbike		
motorbike taxi		
buses and vans		
Public Transport		
Skytrain		
underground		
bus travel		
dedicated bus lanes		
river ferries		

2 Construct a table showing the economic, social and environmental costs of road traffic congestion.

3 In pairs, discuss and list arguments for and against the adoption of policies to charge for private vehicle entry into Bangkok in order to raise capital for public transport improvements.

agglomeration A city which has grown so large that it merges with others to a form city region, e.g. Tokyo, whose region now includes Yokohama

CBD Central Business District or city centre

community-focused regeneration Where regeneration focuses on the needs of communities, e.g. housing, schools

conurbation The growth of a city such that it engulfs other towns or cities, to form a continuous urban area

core and periphery A theory that shows how different economic development between regions leads to a prosperous 'core' region and a poorer 'periphery'

counter-urbanisation The processes that lead to people moving away from cities, which in net terms may lead to a decrease in a city's population

de-centralisation The processes by which the retailing industry in cities moves to more spacious out-of-town locations

de-industrialisation The decline of industry

donut city A term used in the USA to describe cities where de-population of the inner core and movement of businesses away from the city centre has left behind a 'hole' or inactive city centre

dormitory settlements Suburban communities in which people sleep between workdays; with lower populations by day than by night

edge cities A term used in the USA to describe suburban growth away from major cities. These areas take on the functions of cities as more and more people and businesses move out of traditional urban areas

ethnic enclaves Communities of people from similar cultural or ethnic backgrounds, e.g. nationality or religion

gentrification The improvement of urban areas by individual property owners, which usually leads to increased commercial activity in local retail areas

growth poles Centres of economic growth around which further growth occurs

hyper-urbanisation Where a city's growth greatly exceeds its ability to cope in terms of housing, employment, or services

informal employment Unofficial employment, unregistered for tax; it is usually irregular, with little security, e.g. street trading

infrastructure The structures and services that allow places to function effectively, e.g. transport, water, etc.

market- or property-led regeneration Regeneration led by property developers and financiers with little input from government except the creation of opportunities or tax breaks

mega-city A US term for any city over 10 million people

megalopolis A huge urban area, used in the USA to describe areas such as or larger than mega-cities

metropolitan area The administrative urban area of a city

millionaire city A city with over 1 million people

multiplier effect An upward spiral of development, so that one economic activity leads to the development of others

new economy Also known as the 'knowledge economy', which is based on creativity and specialised expertise in finance, media, and management, rather than manufacturing goods

out-of-town shopping centres Major retail centres built away from city centres (as opposed to retail parks which are retail outlets found in smaller towns or suburbs)

post-industrial economy Economic development which focuses upon services and the knowledge economy, rather than manufacturing

re-branding Ways in which regeneration deliberately aims to improve perceptions of an area's status or reputation

regeneration The process of urban or rural improvement, which may be economic, social, or environmental in nature

retail-led regeneration Regeneration brought by improvements in retail space designed to attract shoppers

re-urbanisation The processes by which urban populations increase following a period of decline

rural-urban migration The movement of people from rural to urban areas

shanty town or favela Any informal settlement on spare land which may be legal or illegal

social segregation The way in which suburbs of different income and social status emerge apart from each other

sports-led regeneration Regeneration which focuses on the benefits brought by sports events

squatters Illegal occupants of property

suburbanisation The growth of suburbs

urban Describing a town or city

urban region The area over which a city exerts influence within its country, which is larger than its metropolitan area, e.g. the influence of the London region over south-east England

urban smog A mix of naturally-occurring fog and air pollution or smoke in urban areas; may cause health problems, e.g. asthma

urbanisation The increasing percentage of people living in urban areas

world city A city which plays a disproportionately important role in the global economic system, e.g. global finance and trade

Section B

1 (a) Study Figure 4.6 (page 156). Referring to the diagram, explain how some cities become 'world cities'. *(7 marks)*

(b) Explain the rapid increase in urbanisation in the developing world in recent decades. *(8 marks)*

(c) Referring to examples, explain the impacts of either inward migration or outward migration on major cities. *(10 marks)*

2 (a) Study Figure 4.33 (page 174). Assess how far the trends shown are typical of urban areas in developed countries. *(7 marks)*

(b) Explain the impacts of de-industrialisation on cities in developed countries. *(8 marks)*

(c) Explain the impacts of out-of-town shopping on urban areas. *(10 marks)*

Section C

1 Referring to examples, explain why the growth of huge 'mega-cities' can create challenges. *(40 marks)*

2 Assess the different impacts brought by different approaches to urban regeneration. *(40 marks)*

EXAMINER'S TIPS

(a) Refer only to the diagram – look for connections between hubs and spokes.

(b) You don't need examples, but refer to Mumbai to explain the process.

(c) Consider different factors, e.g. inward migration (push-pull factors resulting in slum growth, enlargement of cities) or outward migration (impacts of suburban growth, loss of green belt, growth of edge cities). Use examples.

(a) Look at the trend – is London typical or not? Discuss the decline (is this typical? – think LA!) and then recent growth.

(b) Think about London's industries and ports, and the impacts.

(c) Think about what effects it has locally and on the CBD.

(1) It would be easy just to describe the story of LA or Mumbai here. But the command is to 'explain why' there are challenges – so you need to focus on explaining the challenges for LA (e.g. why suburbanisation, edge cities, donut effect etc. are all challenges) and Mumbai (e.g. poverty, slum areas).

(2) Think about different approaches to regeneration, e.g. market-led, sports-led, etc. You need to show the impacts of these – try planning what you will say about the economic, social, and environmental impacts for each approach. Remember that to hit the top levels you must assess – i.e. weigh up which impacts are greatest and least for each approach to regeneration.

Development and globalisation

'Ask me! Ask me!' School-children in China

Look at the photo. How important is education, for them?
How important is your education, to you?
How much does education cost?
Is your education important to this country?
Can education play a part in development?

Introduction

In 2011, China overtook Japan as the world's second-largest economy. In 2006, it had overtaken the UK as the fourth-largest. China's rise through the rankings of global economies is remarkable; it is predicted to be the largest by 2025. In three decades, the world order has changed, with southern Asia rivalling the economic dominance of Europe and North America.

But what of Africa? The world's least developed countries are still in sub-Saharan Africa, just as they were in the 1970s. The problems of debt are being resolved gradually, but in the end trade will be the means by which they develop. However, global trade is controlled by the world's dominant economies. For Africa, long-term struggles to develop continue.

In this chapter, you will learn about the patterns and processes of development. You will learn about the forces of globalisation. And you will find out about some of the great development issues.

Books, music, and films

Books to read

No Logo by Naomi Klein
Fugitive Denim by Rachel Louise Snyder
Confessions of an Eco-Sinner by Fred Pearce
Capitalism as if the World Matters by Jonathan Porritt

Music to listen to

'Kid A' (album) by Radiohead
'Arrogance Ignorance and Greed' by Show of Hands
'Natwest-Barclays-Midlands-Lloyds' by Manic Street Preachers
'Radio Africa' by Latin Quarter

Films to see

Capitalism: A Love Story
Cyclo
Tsotsi

About the specification

'Development and globalisation' is one of three Human Geography options in Unit 3 Contemporary Geographical Issues – you have to study at least one.

This is what you have to study:

- Development: the economic, demographic, social, political, and cultural changes associated with it.
- The 'development continuum'.
- Globalisation: factors and dimensions – flows of capital, labour, products, and services; global marketing; patterns of production, distribution, and consumption.

Patterns and processes

- Newly Industrialised countries (NICs): their initial growth, with particular reference to the 'Asian Tiger' economies.
- The further growth of NICs, with reference to China.
- The globalisation of services, with reference to India.
- Growth in the 21st century: the impact of new markets and new technologies (e.g. in Brazil, Russia, and oil-producing countries).

Countries at very low levels of economic development

- Very poor countries: characteristics and issues – quality of life, social problems, and debt.

Global social and economic groupings

- The concept of the North-South divide, and its relationship to the development continuum.
- Reasons for the social and economic groupings of nations, with particular reference to the EU.
- The consequences of the groupings of nations.

Aspects of globalisation

- Transnational corporations (TNCs): characteristics and spatial organisation.
- TNCs: reasons for their growth and spatial organisation.
- A case study of one TNC.
- The social, economic, and environmental impacts of TNCs on their host countries, and their countries of origin.

Development issues

Each to be studied with reference to contrasting areas.

- 'Trade vs. aid.'
- 'Economic sustainability vs. environmental sustainability.'
- 'Sustainable tourism – myth or reality?'

Introducing development and globalisation

In this section you will learn about:
- what development and globalisation are
- how the 2010 football World Cup is an example of globalisation

The 2010 World Cup

In the summer of 2010 not many people could escape the noise of the vuvuzelahs. Why? The FIFA World Cup took place in South Africa and hit television screens around the world. 32 countries competed for football domination, and at the same time some of the world's largest companies were also seeking global dominance. Six major global brands had been selected by football's governing body (FIFA), as official premier league 'partners' (see Figure 5.1) and their logos were displayed around the pitches of all 64 matches. And what's interesting is this:

- The World Cup took place in South Africa. It was the first time a developing country had ever hosted the football World Cup.
- The World Cup final was watched by an estimated 2 billion people around the world. It was an example of **globalisation** in action.
- Only large multinational companies like Adidas and Sony could afford the sponsorship fees needed to get them into the premier league of 'partners', but companies such as South Africa's Prasa made it into the third tier.

Premier league 'Partners'	Company headquarters	Product
Adidas	Germany	sportswear
Coca-Cola	USA	beverages
Emirates	UAE	airline
Hyundai/Kia	South Korea	vehicles
Sony	Japan	electronics
Visa	UK	finance
Second Tier Global Sponsors		
McDonald's	USA	foods
MTN	South Africa	telecommunications
Mahindra Satayam	India	IT consultants
Seara	Brazil	foods
Yinglis Solar	China	solar power
Third Tier Local Sponsors		
Ultimate	BP Africa	oils/lubricants
NeoAfrica	Africa	IT and business consultancy
Prasa	South Africa	national railways

Figure 5.1 *FIFA's tiered system of sponsorship brought in over $1 billion* ▲

From local to global marketing

Compare the marketing in 2010 with 1966. In that year England won the World Cup and the advertisements at the side of the pitch were dominated by local brands. As countries developed and the global economy changed, so football's sponsors and the advertisements have changed.

- In the 1980s major American and Japanese companies dominated football advertisement.
- In the 1990s companies such as Daewoo from countries like South Korea sponsored football in order to advertise their products.
- By 2010 sponsors from Brazil, India and China, and the products they advertised, showed that the global economy had changed.

- **Globalisation** is the way that people, cultures, money, goods and information 'move' between countries with few or no barriers.
- **Development** means 'change' and implies that change is for the better. It usually means economic change, which improves people's standard of living.

A global market

716 million people watched England win the World Cup in 1966 – a fraction of the 2 billion who watched the 2010 final. Online television, smart phones (like the iPhone), YouTube and other digital platforms means that more people than ever – anywhere in the world – can watch major sporting events, communicate with each other and see the sponsors' advertisements.

Sport and development

Major sporting events such as football's World Cup can help a country's **development** by stimulating economic growth. Projects can be planned which aim to attract investment and close the gap between rich and poor, but they can also divert investment away from the real needs of local people.

- South Africa spent $2.2 billion on new infrastructure and attracted half a million visitors during the tournament.
- A 1% increase in GDP was anticipated.
- The cheapest game tickets, 'priced for locals', were $50 – a month's wage for many of Soweto's 2 million residents (Figure 5.2).

Figure 5.2 *Soweto's children play in the shadows of Johannesburg's new Soccer City Stadium. Many South Africans did not benefit from the 2010 World Cup.* ▲

Divided world

For the month the World Cup was on, Africa's image changed. It was portrayed as a sophisticated, successful and modern urban environment. But for millions of Africans nothing changed – many still walked miles every day to gather basic necessities such as water and wood.

Contrasting lives, Uganda and the UK

John Rwaburingamais is a primary school teacher in Uganda. He can't afford to build his own home, but his life has its benefits – fresh food is plentiful and cheap so he doesn't need a fridge. Thousands of miles away, British Dan Brittain also teaches in a primary school, in London, UK. But that's where the similarities end.

John Rwaburingamais		**Dan Brittain**	
income: **£60.00** per month,		income: **£1720.00** per month, after tax and national insurance	
after tax	**£51.00**		**£1290.00**
• rent	£6.00	• rent	£750.00
• food	£13.00	• food	£200.00
• transport	£4.75	• transport	£40.00
• bank charges	£1.50	• bank charges	£0.00
• health/hygiene	£4.50	• health/hygiene	£15.00
• utilities (gas, electricity etc)	£2.50	• utilities (gas, electricity etc)	£120.00
leaves **£18.75** for luxuries (Internet, mobile phone, leisure)		leaves **£165.00** for luxuries (Internet, mobile phone, leisure)	

Figure 5.3 *Contrasting incomes (given in GB pounds), Uganda and the UK* ▲

ACTIVITIES

1. In your own words, define globalisation and development.
2. In pairs draw a spider diagram to show how the following are all examples of globalisation:
   a) major global companies,
   b) TV and Internet,
   c) major sport events
3. How different and similar are the lives of the two teachers shown in Figure 5.3?
4. In your opinion how far is personal income a good indicator of development? Explain your answer using examples.
5. Discuss as a class whether major sporting events can a) help countries in Africa to develop, and b) suit the purposes of the rich countries rather than the poor.

Worlds apart

In this section you will learn about:
- the development continuum
- how globalisation began
- who gains and who loses from globalisation

The development continuum

John Rwaburingamais' and Dan Brittain's lives (see page 201) are worlds apart. In the past the way we looked at the world in terms of development was simple – we described countries as rich or poor, and England and Uganda would have fitted neatly into those categories. But the way we look at development is different now – countries might be richer than some, but poorer than others. Development is measured in terms of a **continuum** – there is no clear cut-off point between rich and poor.

John, in Uganda, may have a low income but his short-term needs are met.

- His food is locally sourced, his clothes and furniture made by friends and family in his village.
- His rural life is sustainable at this level of development and his impact on the natural environment is low.

The same can't be said for Dan in London. His food, clothing, energy and water are all shipped into London where his urban life depends on a complex global system of trade. Uganda and the UK represent opposite ends of the **development continuum**.

Three types of continuum

There are three types of development continuum:

- between countries such as Uganda and the UK
- within countries, e.g. between urban and rural areas, as the photos in Figure 5.4 show
- within local areas, such as cities, which have poorer and richer districts.

Development and change

A range of different changes affect a country's development and progress along the development continuum.

- **Economic change** – including an increase in a country's level of wealth. It can also be affected by things like trade and debt.
- **Demographic change** – improvements in healthcare, sanitation and education can increase life expectancy and reduce birth and death rates. These improvements are also examples of **social changes**.
- **Political and cultural changes** – these can include greater political freedom, an increase in democracy and equality for women.

Figure 5.4 *Kampala (top) Uganda's capital, looks like many modern cities, but many people in rural Uganda (below) live in poverty* ▲

The beginning of globalisation

The year 1492 was a major turning point in the world's history. In that year Columbus first landed in America and established European control. He began a pattern of trade and development that shaped the global economy and made the world the place we recognise today where richer countries have control over poorer ones. But it wasn't always like that.

- The **development continuum** is the range of levels of economic development, from the poorest to the wealthiest countries.

The story of Tenochtitlan

Spanish soldier Bernal Diaz wrote the passage on the right about Tenochtitlan, the Aztec capital and site of the future Mexico City, when the Spanish first conquered it in 1519.

' we arrived at a broad causeway and saw so many cities and villages built in the water ... we were amazed by the diversity of trees and gardens, planted in order and with tanks of fresh water into which a stream flowed in at one end and out at the other. All the roads have shelters made of reeds or straw or grass so that they can retire when they wish to, and purge their bowels unseen and also in order that their excrement shall not be lost.'

When the Spanish arrived Tenochtitlan was doing well. In 1519 it had 300 000 residents, five times more than London. It was an agricultural city divided by canals and was self-sufficient in food. Using mud from the canals as fertilizer, three harvests were possible each year.

But within a generation, Tenochtitlan had changed beyond recognition as the Spanish imposed their own systems on the local people.

- The old farm systems fell into ruins. By 1538 the first food shortages occurred.
- By the eighteenth century large Spanish-owned estates had increased production and the valley was self-sufficient again. But most of the native population had lost their land and were pushed into areas of poorer soil.
- Clearing forests for large estates caused environmental destruction.

By 1984, the Mexican Government reported to the United Nations that: 'land use problems are linked to the lack of soil nutrients. Overgrazing and deforestation have led to extensive resource depletion in the valley. There are serious pollution problems, due to the waste from homes and industrial zones. Water shortages mean that 17 million people depend on transfers from the surrounding mountains.'

The arrival of the Spanish back in 1492 meant that thousands of people had their way of life changed. While the Spanish gained, the inhabitants of Tenochtitlan lost.

Establishing patterns

Like the fate of Tenochtitlan, today's patterns of wealth, trade and development have been shaped by the past. European countries became wealthy by trading with the countries they controlled which supplied them with raw materials, food and labour. Once these supply lines were established, European nations became manufacturers and eventually shipped more expensive, finished products back to their colonies.

The power of the core

The patterns of trade that emerged from the 1500s onwards created a wealthy 'core' of European countries. The nations supplying the Europeans remained poor and on the edge, or periphery. Geographers use core and periphery theory to explain the process by which some countries become wealthy and others poor.

1 Core and periphery – theory

The core is where most wealth is produced. Global core areas include North America, Europe and Japan. This core:

- owns and consumes 80% of global goods and services
- earns the highest incomes
- makes most decisions about the global economy, e.g what goods are produced
- provides most global investment.

The poorer periphery is usually distant from core markets. Poorer countries:

- own and consume 20% of global goods and services, despite having 80% of the global population
- earn low incomes – 2.5 billion people live on under $2 per day
- make few decisions about the global economy
- provide little global investment.

Recent global shifts of industry have meant that:

- Manufacturing has fallen in the old core and risen in 'new' peripheral areas because of cheaper labour. But core countries still profit because they dictate to the new production lines.
- Now, flows of finished and semi-finished goods from peripheral countries are added to the traditional flows of commodities and raw materials (Figure 5.6). But investment and decision-making remain in the core.

In the graph below, economists Fisher and Clark show how employment patterns shift as countries change and develop over time.

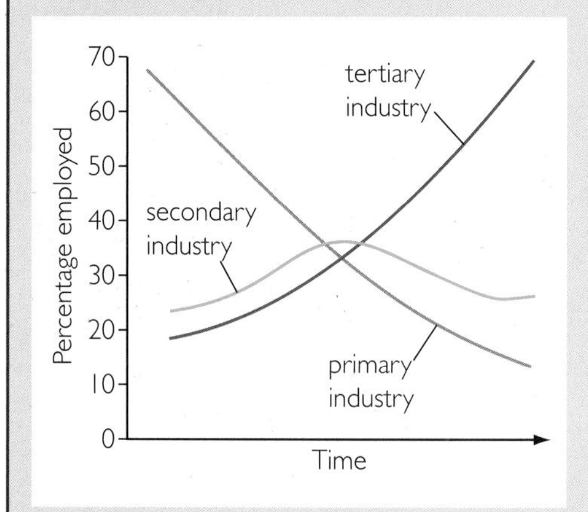

Figure 5.5 *Fisher Clark model of employment change over time* ▲

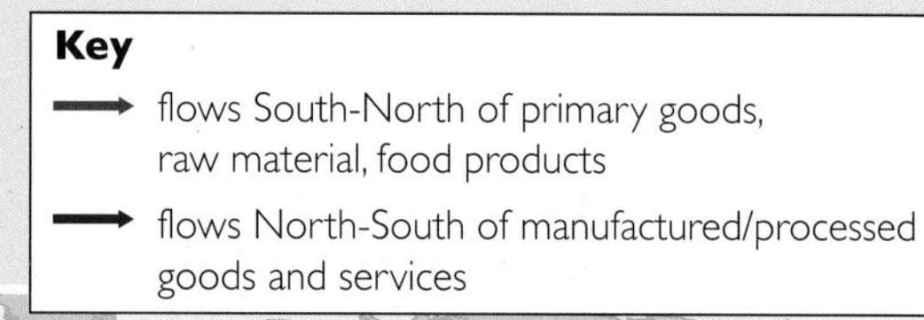

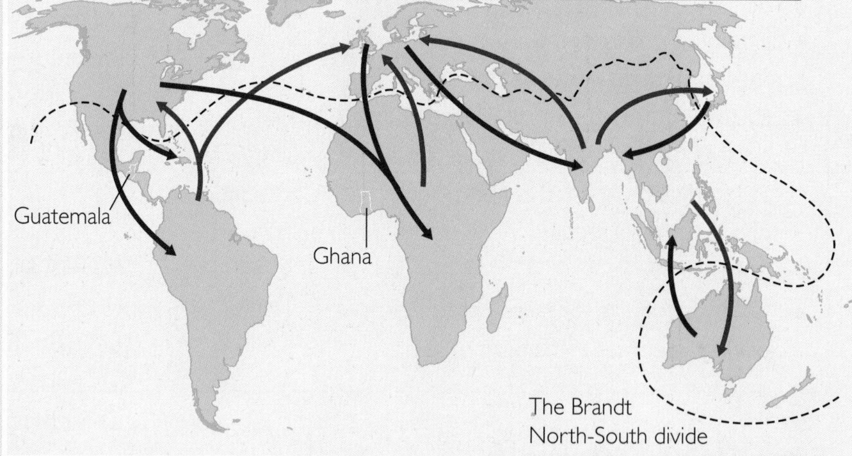

Figure 5.6 *Traditional flows of commodities and finished products* ▲

2 Dependency theory

Development of the core countries came at the expense of the peripheral ones, according to economist André Frank. Core countries depended on their investments in peripheral countries for increased wealth. The periphery countries depended on the core nations for their development, and grew export crops like cotton and tobacco instead of food crops for their own use. Each depended on the other – known as dependency theory, see the map in Figure 5.6 above. Today, most African states and many Caribbean plantations still depend on Europe for their trade.

Core periphery – reality

Just as the richer core countries controlled trade in the past, they are still doing it today. The two case studies below provide examples of the control that richer countries have.

CASE STUDY

Guatemala's cotton

In the 1980s 75% of Guatemala's cotton crop was exported. The money earned was used to buy pesticides, machines and equipment for the next year's crop.

However, if Guatemala had processed the raw cotton into finished clothes, and then exported these, the country would have earned far more. More significantly, only 1% of the land devoted to cotton would have been needed to produce the same income, and Guatemalans could have grown other crops, and grown enough food to feed themselves.

Instead, trade arrangements made this difficult and Guatemala was tied to exporting raw cotton until competition from other suppliers and materials killed off its production altogether by 2005. Transnational companies now import cotton into Guatemala so that the workers in clothing factories there can produce T-shirts for overseas markets.

CASE STUDY

Ghana's cocoa

Before independence from Britain in 1957, Ghana was the world's largest producer of cocoa. The British government used to dictate the price that Ghanaian farmers received for their cocoa crops. Even after independence three things have prevented real change:

- **Commodity traders** based in London and New York buy cocoa for large companies like Nestlé and Cadbury's. To guarantee supplies and the best prices they buy in the futures markets (this means they buy now to ensure delivery in 3-6 months). If Ghana's prices are too high, traders buy from other countries such as Ivory Coast, and that forces prices down.
- **Overseas tariffs** set by the European Union (EU) are much higher for processed cocoa than for raw cocoa beans. Ghana would gain more income by turning raw beans into processed cocoa powder or chocolate and exporting those. But the EU tariffs force Ghana to export raw cocoa beans instead.
- **The World Trade Organisation (WTO)**. Ghana joined the WTO in 1995, in an attempt to increase its global trade. Until then the Ghanaian government had subsidised its farmers to encourage them to stay on the land and grow food for Ghana's growing cities. But the WTO imposed the condition that when Ghana joined its farmers should no longer be subsidised. As a result, some farmers in Ghana can't sell their own produce because imported subsidised food is cheaper, so some tomato and rice growers have given up altogether.

Figure 5.7 *Farmers in Ghana can't sell their crops because cheap subsidised imports from the EU have flooded the market* ▲

- **Tariffs** are taxes or customs duties paid on imports.
- The **WTO** deals with the rules of global trade with the aim of easing trade and getting rid of anything hindering it. It negotiates trade agreements and makes sure that members keep to the rules.

A modern world

In the second half of the twentieth century the richer core countries continued their control over the poorer peripheral countries in two ways – through economic and population modernisation.

Economic modernisation

By the end of the Second World War some poorer parts of the world were devastated. Communism spread through Eastern Europe, China and Korea and caused panic in the USA. By the mid 1950s, American money was pouring into India and South-East Asia in an effort to prevent further expansion of communism. W.W. Rostow (an American) had produced an economic model aimed at promoting economic development and reducing poverty through **modernisation theory**. The idea was that aid, loans and investment would prevent countries turning to communism and enable them to proceed on a smooth development path.

- Countries such as Singapore, Thailand, South Korea and Taiwan experienced rapid economic growth due to the money that poured in, and their exports grew rapidly. They became the Newly Industrialised Countries (NICs) of South-East Asia (see page 208).
- India introduced the Green Revolution's modern farming techniques during the 1940s-70s. It became dependent on the USA for machinery, fertilizers, and seeds but development was boosted and it probably helped to avoid a communist revolution.

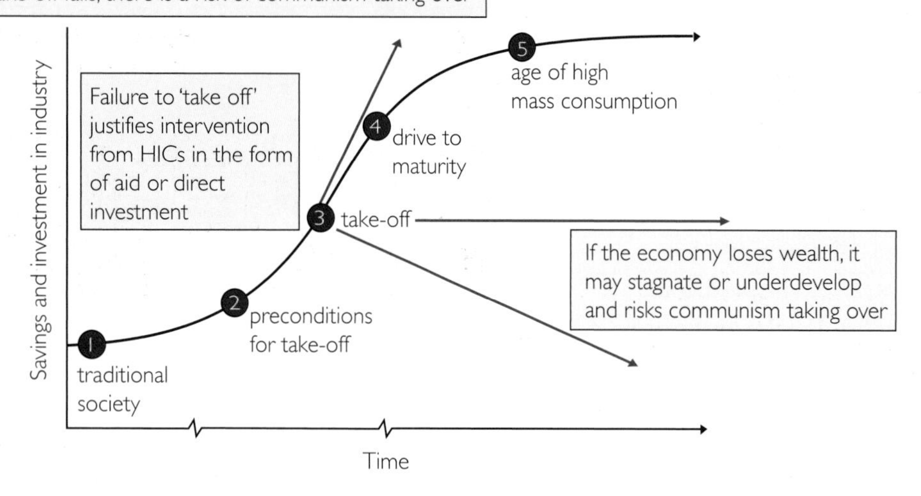

◀ ***Figure 5.8*** *Rostow's model of development – cutting the risk of communism*

Population modernisation

The second half of the twentieth century saw dramatic changes in population in many of the peripheral countries, largely brought about by aid. The aid, in the form of a transfer of western technology (see the text box) reduced infant mortality rates and death rates, and increased life expectancy. But birth rates fell more slowly than death rates.

As a result:

- some countries, such as India, have growing youthful populations which provide a large labour pool
- children are seen as economic assets. The International Labour Organisation estimates that 120 million children are in the workforce in Low Income Countries (LICs).

The rapid growth of population in some peripheral LICs means that they do not follow a straight path on the development continuum.

Transfers of western technology

In the 1960s, HICs (High-Income Countries) started to provide technological help to LICs (Low-Income Countries). This took two main forms:

- Death control policies – these were measures aimed at reducing infant mortality and extending life expectancy e.g. through inoculations, vaccines and sanitation.
- The Green Revolution – a package of mechanised farming techniques and high yielding seeds to enable LICs to feed growing populations.

Globalisation – factors and dimensions

Although Columbus (see page 203) began a pattern of trade and development that shaped the global economy, globalisation as we know it has only really taken off since the 1970s. Manufacturing has now shifted to poorer countries, and globalisation is no longer confined to trade.

As we have seen, globalisation is the process by which people, their cultures, money, goods and information can be transferred between countries with few or no barriers.

Globalisation is affected by a number of factors, as the text box here shows.

Flows of capital, labour and services

- Money (currency, capital) flows electronically around the world. HICs invest in LICs to take advantage of cheaper production costs.
- Countries such as India provide a range of financial and IT services for higher-income countries.
- People move around the world for a variety of reasons, e.g. skilled workers within a company, and unskilled workers migrating to find work.

Global marketing

- Global brands such as Adidas and Coca Cola advertised their products at the 2010 World Cup (see page 200) to a global audience.
- Transnational companies use the same adverts to advertise their products in different parts of the world.

Patterns of distribution and consumption

- Transnational companies (see page 230) dictate where their products are made – generally where labour costs are cheaper in LICs.
- Products are distributed around the world to meet the demands of consumers in HIC.

Globalisation affects every aspect of our lives:

- political globalisation: national boundaries fade away
- cultural globalisation: values and attitudes merge as movements of people and products lead to increasing uniformity and loss of individuality
- economic globalisation: the emergence of transnational companies with more wealth and power than individual countries.

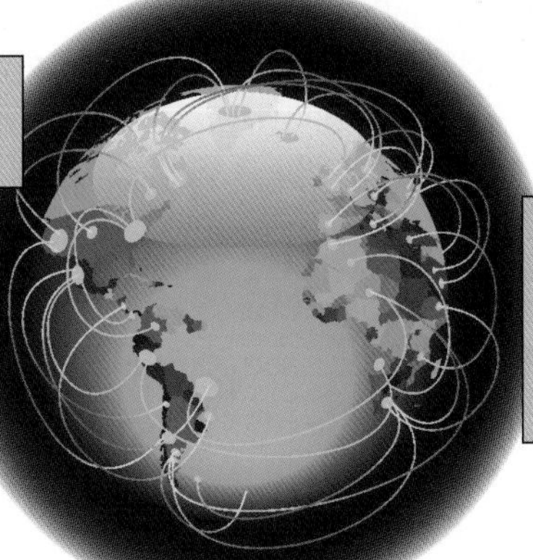

◀ ***Figure 5.9*** *The world becomes increasingly interconnected*

ACTIVITIES

1. Explain these terms: development continuum; core and periphery, modernisation, and dependency theories.
2. **a** Explain the reasons why some countries like Ghana and Guatemala find it hard to progress along the development continuum.
   b How far are these reasons (**i**) the fault of the countries themselves (**ii**) the fault of others?
3. Survey what globalisation means to your class. Find out:
   a Where the class and their parents were born.
   b Where the clothes they wear, the music they listen to and the films they watch come from.
   c Where parents are working and for whom.
4. In pairs make a spider diagram to show what you think the gains and losses are as globalisation progresses for:
   a) you, **b**) your local area, **c**) the UK.
5. Now repeat Activity 4 for Ghana and Guatemala.

Internet research

Research the exports and imports of one HIC and one LIC. (The website of the CIA World Factbook is a good place to start.) How far do they match the trade patterns described in this section? What problems are there with this kind of trade?

Patterns and processes – 1

In this section you will learn about:

- the growth of the Newly Industrialised Countries and the 'Asian Tigers'

Newly Industrialised Countries (NICs)

In order for countries to develop they need to change from what Rostow called a traditional society (see page 206) and begin the process of industrialisation (what Rostow called industrial 'take-off') – i.e. develop manufacturing industries. In the 1960s, some countries in Asia and South America were making this change and became known as the **Newly Industrialised Countries** or NICs. Three groups existed:

- Asian Tigers – Hong Kong, Singapore, South Korea and Taiwan
- South America – Mexico and Brazil
- European countries– Portugal, Greece and former Yugoslavia.

As their manufacturing industries developed quickly, exports and GDP grew rapidly. But how did they do it?

Asian Tigers – three steps to development...?

The aid, loans, investment (see pages 206) and technology that poured into South-East Asia meant that countries there could invest in manufacturing industries to produce goods for export. High rates of population growth and inward migration provided large labour forces. The Asian Tigers took on the world. For three decades their growth was remarkable as the information in the text box opposite shows.

...or fourth step to disaster?

South Korea's economic rise from 1960-90 was dramatic. Its economy grew by an average of 9% a year, with major companies like Daewoo cars, LG and Samsung electronics selling and exporting manufactured goods worldwide. South Korea's economic success was based on taking huge financial loans from the USA, and it appeared that it was moving along the development continuum. However in 1997, relying on loans (which meant piling up massive debts) in order to promote economic growth, proved one step too many. The huge Daewoo conglomerate imploded with $80 billion of debt. This caused a ripple which spread to other countries in South-East Asia and became known as the South-East Asian Financial Crisis.

Figure 5.10 High level manufacturing in an export orientated industry in Taiwan ▲

Disadvantages of the NIC pattern of development

One of the reasons for the NICs' success was their large cheap labour force. But as countries develop, the **comparative advantage** of cheap labour disappears. As people become used to regular wages and rising incomes, they want a better standard of living. Labour costs increase and TNCs start to look elsewhere for cheaper labour.

Indonesia, Thailand, and India all began to expand their own textile industries and undercut the original Tiger economies. To stay competitive, NICs have diversified and moved into less labour-intensive and more skill-intensive industries such as advanced electronics, for example, Apple's iPods are manufactured by Inventec in Taipei, Taiwan (see Figure 5.10).

Step 1
Governments in the 1950s-60s encouraged traditional labour-intensive industries like textiles, clothing, footwear and leather goods which they could produce more cheaply than the old industrial countries. TNCs were attracted by cheap labour and local raw materials.

Step 2
Money earned from exports was invested in new industries which were making items that had previously been imported. These are known as **import substitution industries** and included, iron and steel, shipbuilding, chemicals and fashionable clothes.

Step 3
Money saved by reducing imports was invested into higher-level manufacturing of computers, televisions and cars designed for export. These are known as **export-oriented industries** and earned the Asian Tigers much-needed foreign currency. This currency was used to re-pay loans, invest in expanding industries, and improve productivity, education and skills.

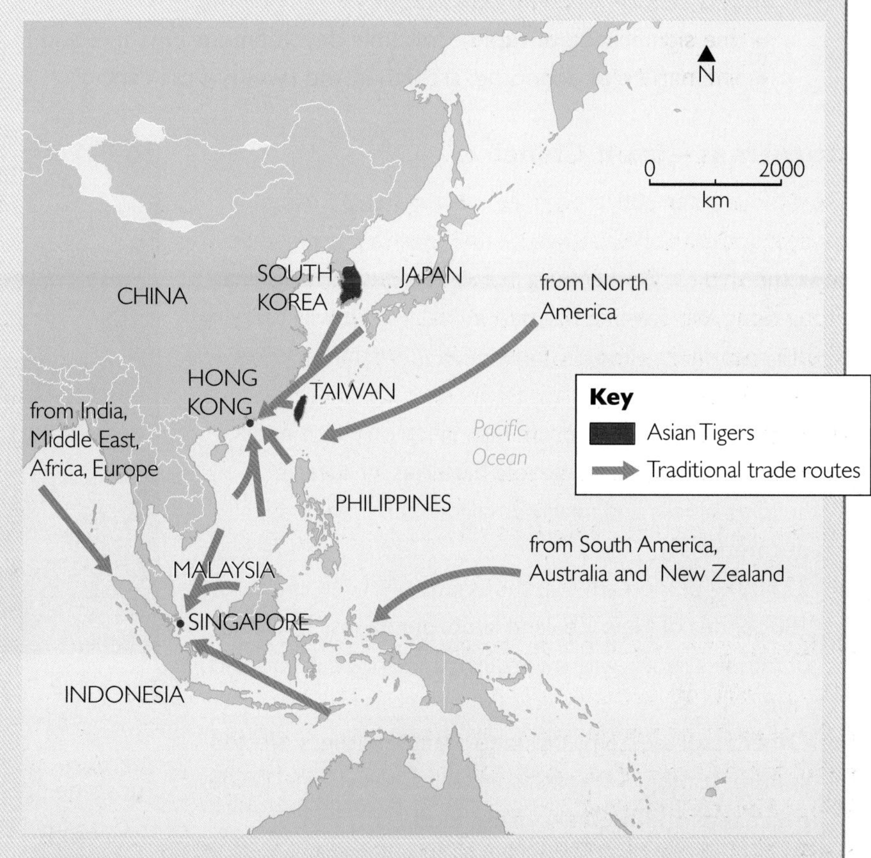

Figure 5.11 The Asian Tigers ▲ *Figure 5.12* The Asian Tigers' development ▼

Why the Asian Tigers developed	Why they might not have developed
• large, well-educated workforce • export-oriented industries • technological transfers from USA and Japan • close to emerging China • located on existing trade routes (Singapore and Hong Kong are major ports) • no or reduced tariffs/duties • aid and investment from USA	• few raw materials • limited local markets (in the 1950s), with low incomes • internal wars in Korea • mass immigration of poor refugees to Hong Kong, Singapore and Taiwan placed pressure on each country's resources

ACTIVITIES

1. Define these terms: NIC, export-oriented industries, import substitution industries, and comparative advantage.
2. Draw a sequence diagram to show how the NICs managed to 'take-off'.
3. Add to your diagram the likely local impacts of the South-East Asian financial crisis in 1997.
4. Draw a table to show the economic advantages for NICs of developing in this way.
5. Why might Ghana and Guatemala (see page 205) find it hard to take off in the same way?

Internet research

Research one of the Asian Tigers and find out a) how well its economy has been doing since 1997, and, b) which sectors of its economy have been growing fastest.

5.4 Patterns and processes – 2

In this section you will learn about:
- the significance of rapid economic development in China and India
- the nature of economic growth in the twenty-first century

Christmas – from China!

On 4 November 2007, the giant container ship *Emma Maersk* docked at Felixstowe, Suffolk after a journey that took the ship to: Gothenburg in Sweden; Yantian in China; Hong Kong and Tanjung Pelepas, in Malaysia. It was carrying 11 000 containers – mostly for the UK Christmas market – holding:

- Crackers, bingo sets, drum kits, electronic toys and gadgets, 40 000 rechargeable batteries, children's building blocks and nearly 2 million Christmas decorations
- 22 280 kg of Vietnam tea, thousands of frozen chickens, 150 tonnes of New Zealand lamb, pumpkins, 10 tonnes of mussels, along with swordfish, tuna, noodles, biscuits, jams
- Other assorted items included potato mashers, slotted spoons, toothpicks, sofas, spectacles, televisions, Disney pyjamas and 138 000 tins of pet food.

China – the world's fastest growing economy

The *Emma Maersk* delivered 45 000 tons of consumer goods from China and is the latest stage in the shift in manufacturing from Europe and the USA to China, a shift which has already taken place in Japan, Taiwan and South-East Asia. In 2006, the port of Felixstowe reported an increase of 16% in Chinese imports. Since 1980, the UK has experienced a 30-fold increase in trade with China, exporting £7.7 billion and importing £22.8 billion worth of goods in 2009. The *Emma Maersk* is a real symbol of globalisation.

China's growth as a manufacturing nation means that European consumers can purchase cheaper goods. Maersk Shipping explained that *Emma Maersk* would be used on its return journey to ship one of the UK's biggest exports back to China – waste plastic – which would be turned into new soft toys and decorations ready for the following Christmas.

Figure 5.13 The Emma Maersk *is a quarter of a mile long (397m), 200ft (61.5m) high and as wide as a motorway!* ▲

But some people are not so happy about this trend, as Caroline Lucas, Green MEP for South East England suggested:

> "The *Emma Maersk* brought the goods that Europe used to make. Whole sectors of global trade are now being dominated by companies operating out of China. The real cost of the goods that the *Emma Maersk* is bringing in should include the environment, the markets destroyed in developing countries and the millions of jobs lost."

Did you know?

China already consumes 50% of the world's concrete, 40% of its coal, 25% of its iron and nickel, and 19% of its aluminium.

How has China's economy grown?

China has undergone massive change. Since the 1970s China's economy has doubled every eight years, while the US economy has doubled just once. By 2010, China had become the world's second-largest economy. China has the largest sustained GDP growth ever seen and is no longer an LIC. Its largest city, Shanghai, has been transformed. Its shiny business towers offer a glimpse of China's future when manufacturing will have moved elsewhere and the 'new' knowledge and information sector take over. How can China's economic growth be explained? And how is its growth different from that of the NICs on pages 208-9? The following national factors help to explain it.

National factors

- From 1949-76 Mao Zedong's People's Republic kept communist China deliberately separate from the rest of the world. The economy was centrally planned, goods produced for the consumption of China's own people, and little individual wealth or overseas business was permitted.
- Following his death in 1976, new rulers developed an 'Open-Door Policy' to investment from overseas from 1986 onwards.
- In the 1990s, China transformed to a more capitalist economy, allowing individuals to accumulate wealth by producing goods and services, without State interference.
- China's massive population and huge natural resource base (Figure 5.15) provide both the workers and raw materials for industrialisation.
- China's increasing need for energy will be helped by the Three Gorges Dam – its 25 generators providing enough energy to supply 11–15% of China's energy.
- Prolonged spending on health and education over 50 years has provided a healthy, literate and skilled workforce.
- The creation of special industrial **export processing zones** has stimulated mass manufacturing on a scale unimagined 25 years ago.

Global factors

In 2001, China joined the World Trade Organisation. TNCs now invest in China to take advantage of low labour costs and the Special Economic Zones. Since 2000, China has been the largest recipient of overseas investment and 53% of its exports are produced by foreign-owned companies or those in partnership with Chinese companies.

60% of the increase in world trade between 2006-07 was a result of China's industrialisation, and in 2007, China overtook the USA, consuming:

- 67 million tonnes of meat (USA 39 million tonnes)
- 258 million tonnes of steel (USA 104 million tonnes).

Before 2050, China will be consuming more oil and paper than the world now produces.

> • **Export processing zones** are a type of **Special Economic Zone** where businesses are free to import raw materials, process and manufacture them, and re-export without paying duties or tariffs – helping to keep costs down.

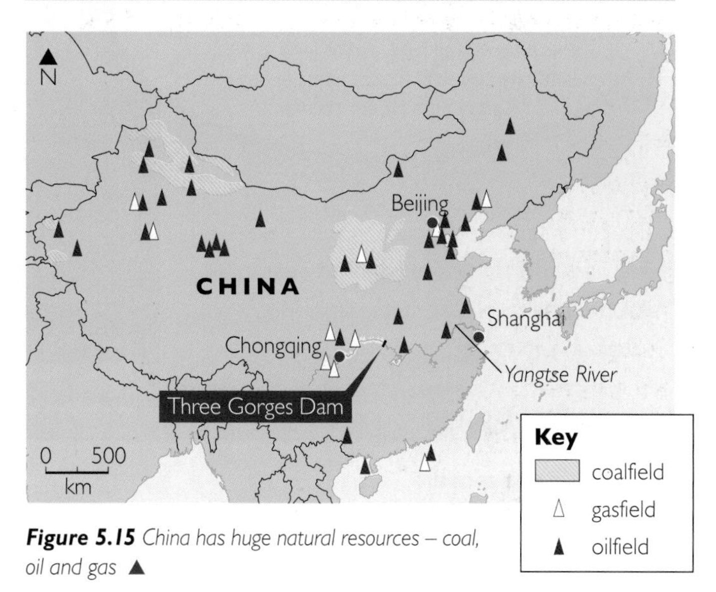

Figure 5.15 *China has huge natural resources – coal, oil and gas* ▲

◀ **Figure 5.14** *Shenzhen was China's first* ***Special Economic Zone****. It offered tax incentives for foreign companies to build new factories there, and is now one of the world's fastest-growing cities*

Consequences of China's growth

Chongqing, on the Yangtze River, is the world's fastest-growing city. At the heart of developments around the Three Gorges Dam, this port is the centre of Sichuan province. 10 million people already live there (up from 4 million in the 1980s) and it is expected to double in size by 2020. The city represents modern China, with giant factories and skyscrapers. 50% of the population are migrant and half a million new workers arrive each year. Since 2000, over US $250 billion has been spent on roads, bridges, dams and power supplies. Living conditions are often crowded and squalid for workers like Yu Lebo (see below). But incomes are 3 times higher than on the farms.

Chongqing is one of the dirtiest cities on earth (China now has 16 in the top 20) and its filthy air causes thousands of premature deaths. The air quality was so bad in 2005 that it did not reach the government's own safety standard 25% of the time.

Voices of Chongqing

Yu Lebo shares a three-room apartment with his wife and three other couples, all of whom are porters or cleaners or odd-job men. There are two double beds in one room, separated by a thin sheet, a third in a tiny room next door and another in the kitchen.

'I used to be a farmer, but I could not afford to raise my two children. So we left them behind with relatives. I earn about 20 yuan (£1.50) for 12 hours work. Most of this, and the money my wife earns as a cleaner, goes on rent and food, but as long as we stay healthy we can send money home to buy clothes and books for our children.'

Education and healthcare – once free – are now the biggest burden on peasants.
Adapted from C4/Guardian Films, March 2006 'Invisible City'

Figure 5.16 *Chongqing, China – the world's fastest growing city* ▲

People on the move

As China's economy has grown the movement of people to the cities has been staggering.

- China has seen the largest rural-urban migration ever recorded (8.5 million people a year).
- Over 140 million Chinese have left the countryside since the 1980s.
- China now has 90 millionaire cities (cities of over 1 million people).
- 45 million people are expected to move by 2012, and most head for the factory towns by the coast.
- The interior of China is being transformed by industrial development.

India and the globalisation of services

If you visited one of India's cities you'd see signs of extreme poverty, such as slums, alongside high levels of wealth – upmarket apartments. Despite the poverty, India, along with China, is one of the world's emerging superpowers. Since the government's economic changes in the 1990s India's economy has grown by an average of 6% a year. So, what has happened?

- The Internet has been at the root of India's rapid economic rise.
- India has a large youthful population, providing a huge labour force (see Figure 5.17).
- India produces 2 million English-speaking graduates a year – so the workforce is well educated.
- TNCs take advantage of India's lower wages to **outsource** work to the country.
- Services now earn India US$25 billion a year.

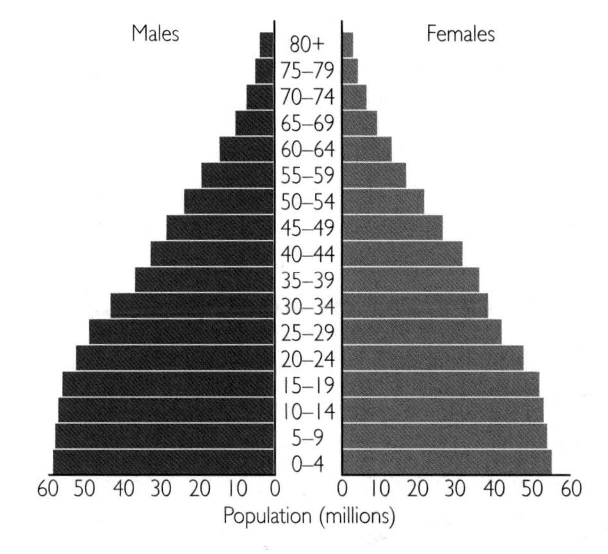

Figure 5.17 *India – population pyramid* ▲

- **Outsourcing** is the employment of people overseas to do jobs previously done by people in an HIC. Outsourcing is usually associated with IT software development, banks and service companies such as call centres. Outsourcing is an example of the **globalisation of services**.

Bangalore

Unlike most Indian cities, which grew as industrial ports, Bangalore is the centre of new technology, banking, finance and the knowledge economy. 40% of India's 1.3 million workers in the IT industry are based in Bangalore. A whole range of outsourced operations are based both there and in other Indian cities, including:

- technical development – e.g. companies such as Deutsche Bank, one of London's biggest investment banks, which has set up teams of software developers in India (IBM alone had 100 000 employees in India in 2010)
- support – people who provide technical support, e.g. for BT Broadband
- call centres – e.g for rail companies in the UK selling tickets or handling refunds.

Motorola, Hewlett-Packard, Cisco Systems and other high-tech giants rely on teams in India to devise software platforms and multimedia features for next-generation devices. Google principal scientist Krishna Bharat is setting up a Bangalore lab to work on core search-engine technology. Indian engineering houses use 3-D computer simulations to tweak designs of everything from car engines and forklifts, to aircraft wings, for such clients as General Motors and Boeing. Financial and market-research experts at outfits like B2K, OfficeTiger, and Iris analyse the published accounts of blue-chip companies for Wall Street.

Figure 5.18 *Adapted from an article in* Business Weekly, *August 2006* ▲

Did you know?

Some British companies are now returning outsourced services such as call centres to the UK.

Electronic City

In the 1990s, Bangalore set up designated areas such as Electronic City to become 'hubs' for high-tech firms. They were attracted by tax breaks (reduced tax levels) and the cheap labour. So successful have these schemes been that today Indian firms have developed which provide a range of technical and support services alongside the operations that started up for Western companies.

Further growth is certain. A new 'Knowledge City' is being built in Bangalore. In 2008 Oracle estimated that India would need 8 million extra workers in out-sourced industries. Life for the people who work in the industries has changed beyond all recognition, as the speech bubble shows. A ripple effect is starting as other cities across India – such as Chennai – are also attracting large companies providing similar services.

"Software professionals are earning money their parents never dreamed of, driving Mercedes and partying all night. I am making more money than my parents could imagine and for an Indian woman that is totally liberating. I don't need to depend on my parents, I don't need a husband. I can choose when to marry, I might not even need to get married"
Devika – (a call centre worker in Mumbai)

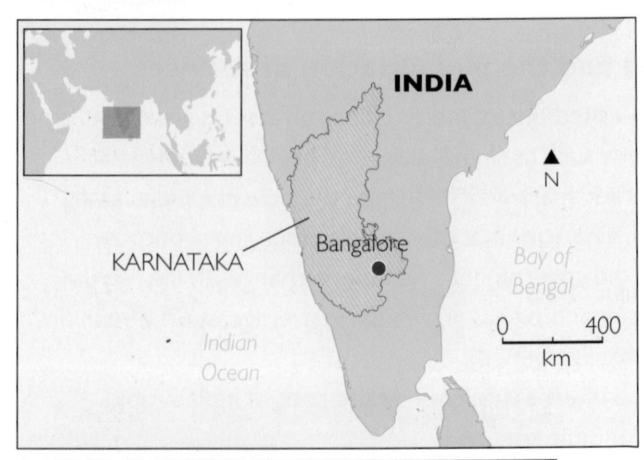

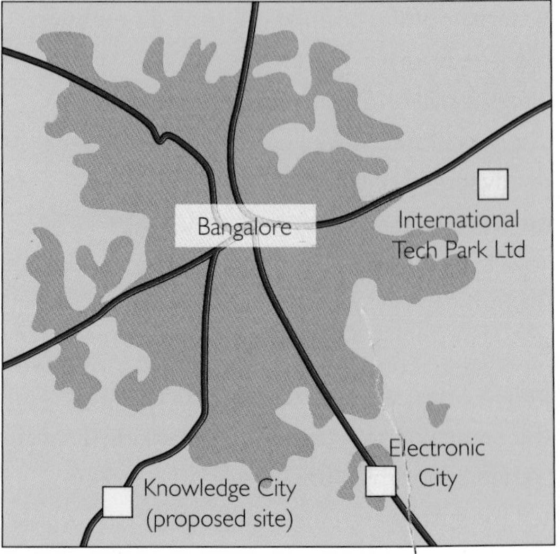

Figure 5.19 *Bangalore's hubs for high-tech firms* ▲

It's not all good news

Bangalore is India's fastest-growing city. It grew from 1.6 million in 1970 to 7.2 million in 2010, and has India's highest-paid workers. But within the city you find evidence of the development continuum. The gap between rich and poor remains – roadside tents, overcrowding and squalor sit next to the bright lights of the city centre. In Mumbai 60% of the population live in poverty – not for them the benefits of globalisation, but the daily reality of life in the slums.

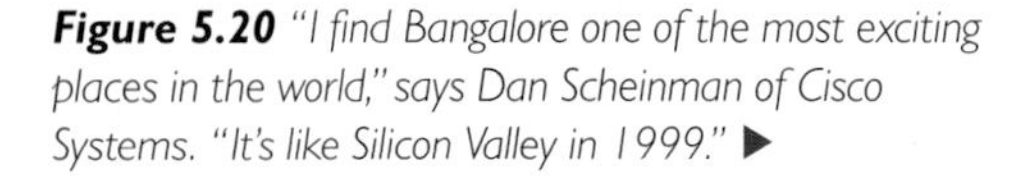

Figure 5.20 *"I find Bangalore one of the most exciting places in the world," says Dan Scheinman of Cisco Systems. "It's like Silicon Valley in 1999."* ▶

Growth in the twenty-first century

Throughout the twentieth century the global economy was managed by the G7 (a group of industrialised countries) – the USA, UK, Japan, Germany, Canada, France and Italy. But global economic power is shifting.

- China has overtaken Japan as the world's second-largest economy (China overtook the UK in 2007).
- The G7's power is threatened by the emerging markets of Brazil, Russia, India and China (the BRICs). These countries already have a bigger share of the world's trade than the USA. Figure 5.21 shows the economic growth forecasts for India and China.
- By 2050 the BRICs will be in the G8 (the world's richest industrial countries). Only the USA and Japan will remain from the current members.

The BRICs' economic growth is based on these facts:

- Brazil and Russia are two of the world's largest suppliers of raw materials.
- Russia controls massive reserves of oil and natural gas which it supplies to Europe.
- 50% of Chinese investment is in Latin America, the Caribbean and Africa.
- Chinese, Indian and Russian companies are acquiring European businesses and opening plants around the world.
- Brazil is devoting increasing amounts of land to food exports and biofuels.
- China and India collaborate – India designs and innovates, and China manufactures.

Constant updates in technology and communications, along with falling costs, mean many transnational companies are already having their products designed in India and built in China. Every year, a million engineers and scientists graduate in these two countries, providing an educated and skilled labour force.

And in the Middle East...

Oil, the dominant energy source for so long, may be reaching a peak of production and the Organisation of Petroleum Exporting Countries **(OPEC)** are looking to a future without oil. Dubai has been switching to service sector activities like the new knowledge economy, ICT and tourism.

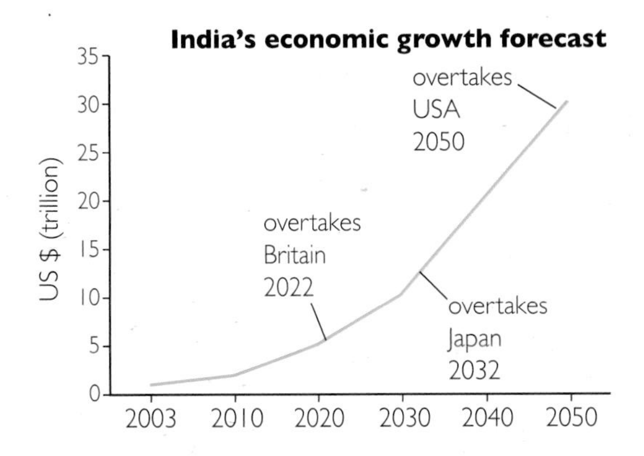

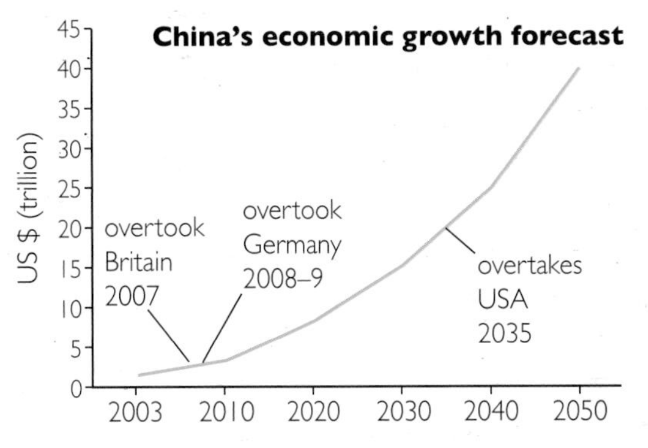

Figure 5.21 *The balance of global power is shifting – who will be the next economic superpower?* ▲

ACTIVITIES

1. Define the following terms: special economic zone, export processing zone, out-sourcing.
2. Explain why most economic growth in China and India has taken place in urban and not rural areas.
3. In what ways are China's and India's growth similar to and different from each other?
4. In what ways is the growth of China and India's economies similar to and different from that of the earlier NICs (see pages 208-9)?
5. In what ways might the BRICs pose threats and opportunities to the old economies?

Internet research

In pairs use Google and YouTube to research social and environmental issues in the Chinese city of Chongqing. Type in 'Chongqing industries', 'Chongqing pollution' and 'Chongqing housing.' Research:

a How and why Chongqing's economy has grown.
b What it is like living there.
c The benefits and problems of living in a rapidly growing economy.

5.5 Countries at very low levels of economic development

In this section you will learn about:

- the issues facing countries with very low levels of development

The world's Least Developed Countries

The development continuum means that some countries are richer than others (the HICs), and some are poorer than others (the LICs). Within the LICs there's a core of Least Developed Countries (LDCs). The United Nations classifies LDCs as countries with:

- the lowest incomes
- limited opportunities for economic development
- extreme vulnerability to external shocks like earthquakes, droughts, floods and shifts in the global economy that lead to increased food prices, reductions in trade and aid.

The core of LDCs are hard to change. The countries that make up this group are the same as they were 50 years ago. The map shows the world's 50 LDCs (as defined by the UN in 2006). Three criteria were used to identify this group of countries:

- GNI per capita – under $900 (Gross National Income is a measure of a country's wealth – see page 221).
- Human Asset Index – this is made up of the percentage of the population undernourished, death rate of children aged 5 years or under, secondary school enrolment rates, and adult literacy rates.
- Economic Vulnerability Index – the extent to which a country would be affected by unplanned 'shocks'. It is based on population size, remoteness, importance of agriculture, the risk of homelessness, dependence on exports.

Figure 5.22 *The world's 50 LDCs* ▼

"No mother wants her baby to die. No one wants to have AIDS. No one wants to lose family members to Malaria. No man wants to feel incapable of feeding his family."
Volunteer worker in Niger

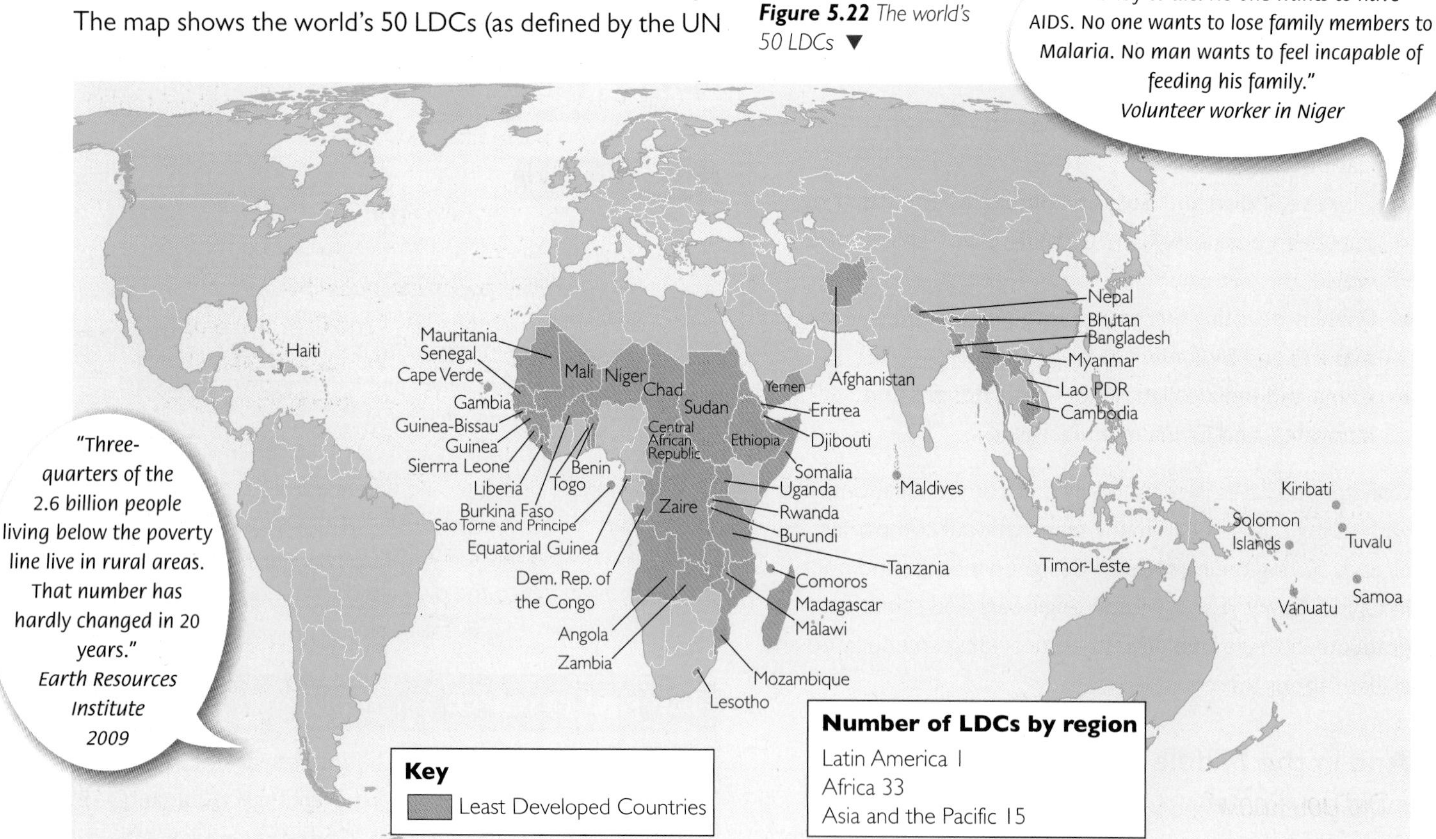

"Three-quarters of the 2.6 billion people living below the poverty line live in rural areas. That number has hardly changed in 20 years."
Earth Resources Institute 2009

Life at the bottom

According to the World Bank, 1.2 billion people live on less than $1.25 a day. For children in the LDCs this means:

- often going to bed hungry
- being unable to read or write
- risking disease because there is no guarantee of safe water
- there is little chance of earning a living by working
- there is every chance of facing a natural disaster and being dependent on aid from overseas.

Issues in the world's LDCs

Quality of life and social problems

The UN measures quality of life using the **Human Development Index** (HDI). This measures life expectancy, knowledge (literacy rate and number of years spent at school) and standard of living (GDP per capita).

In the world's LDCs people may have a poor quality of life because:

- rising food prices bring with them higher rates of malnutrition
- climate change is undermining the ability of rural people to feed themselves
- diseases like malaria are widespread, and preventable diseases like tuberculosis, diarrhoea and measles kill 11 million infants each year
- some suffer high rates of HIV/AIDS prevalence
- 500 000 women die during childbirth each year
- they lack piped water, sewers, electricity, surfaced roads, hospitals and schools.

Quality of life and social problems in the world's LDCs go hand in hand. The table below shows some development indicators for the world's 10 LDCs in 2009. What isn't clear from the table is that girls have an even harder time in the LDCs than boys.

- 100 million primary aged children don't go to school – most of those are girls. This means girls marry young, and start families in their teens.
- The economic value of a girl's work at home is greater than the perceived value of education – so girls drop out of school early.
- Discrimination towards infant girls means that mortality rates for girls are higher than boys.

On top of those problems:

- Population growth is high because fertility rates are high. Yet life expectancy is low and infant mortality is high.
- HIV/AIDs has reduced the number of teachers and education has suffered.

Debt

What makes the situation worse for the LDCs is their extreme debt. Debt repayments take money away from things like healthcare and education that would help real development. How did the LDCs get into debt?

- Oil price rises in the 1970s boosted the earnings of oil-exporting countries.
- These countries invested in Western banks, who then lent the money to LICs who spent it on development projects like dams, power stations and mechanisation of agriculture.
- By the 1980s, global interest rates had more than doubled – increasing the repayments needed to service the loans taken out in the 1970s. The LICs couldn't meet their debt repayments, so unpaid interest was added to the original loan.
- Every year, the amount owed increased and the debt burden got worse.
- To prevent a collapse of the global banking system the International Monetary Fund (IMF) developed **Structural Adjustment Packages (SAPs)**. SAPs involved re-scheduling loans to make them more affordable – in return for cuts which the IMF imposed on Government budgets and spending.
- Governments in LICs such as Uganda cut spending on health and education and the poorest people suffered.

The poorest ten countries 2009 (by HDI rankings)	Life expectancy	Infant mortality per 1000 live births	% Adult literacy	% under weight children	% without safe water	GDP per capita PPP$*
173 Guinea-Bissau	47.5	98.0	64.6	19	43	477
174 Burundi	50.1	63.3	59.3	39	29	341
175 Chad	48.6	97.0	31.8	37	52	1477
176 Congo (DR)	47.6	79.3	67.2	31	54	298
177 Burkina Faso	52.7	82.9	28.7	37	28	1124
178 Mali	48.1	113.6	26.2	33	40	1083
179 Central African Republic	46.7	101.6	48.6	29	34	713
180 Sierra Leone	47.3	80.16	38.1	29	47	679
181 Afghanistan	43.6	151.0	72.0	39	78	1054
182 Niger	50.8	114.0	28.7	44	58	627
21. United Kingdom – for comparison	79.3	4.7	100	Less than 1%	0	35 130

*GDP per capita is a measure of a country's wealth (see page 221)

Figure 5.23 *Development indicators for the world's LDCs, 2009* ▲

CASE STUDY

Uganda

Unlike some parts of Africa, Uganda is a green and fertile country with plenty of resources of copper and cobalt. Uganda should be a wealthy country, so why does it struggle to provide a decent standard of living for its growing population?

- By depending on low-valued primary products such as coffee and tea, Uganda earns little foreign money.
- It receives limited tax revenues from its exports.
- Idi Amin's military regime of the 1970s expelled the wealthy Asian business community and exports collapsed.
- Huge loans were used to buy military weapons.
- Uganda's debt burden increased in the 1980s (see previous page) and reached $19bn in 1992. During the 1990s its annual debt repayments exceeded its export earnings.
- HIV/AIDS has reduced life expectancy. Babies born in 2007 had only a 62% chance of reaching 60.
- 56% of the population are under 18, the fertility rate is 6.3 and birth rate 47.5 per 1000.

The table below shows some of the development indicators for Uganda. Look at these figures and you should begin to see why Uganda is one of the world's LDCs.

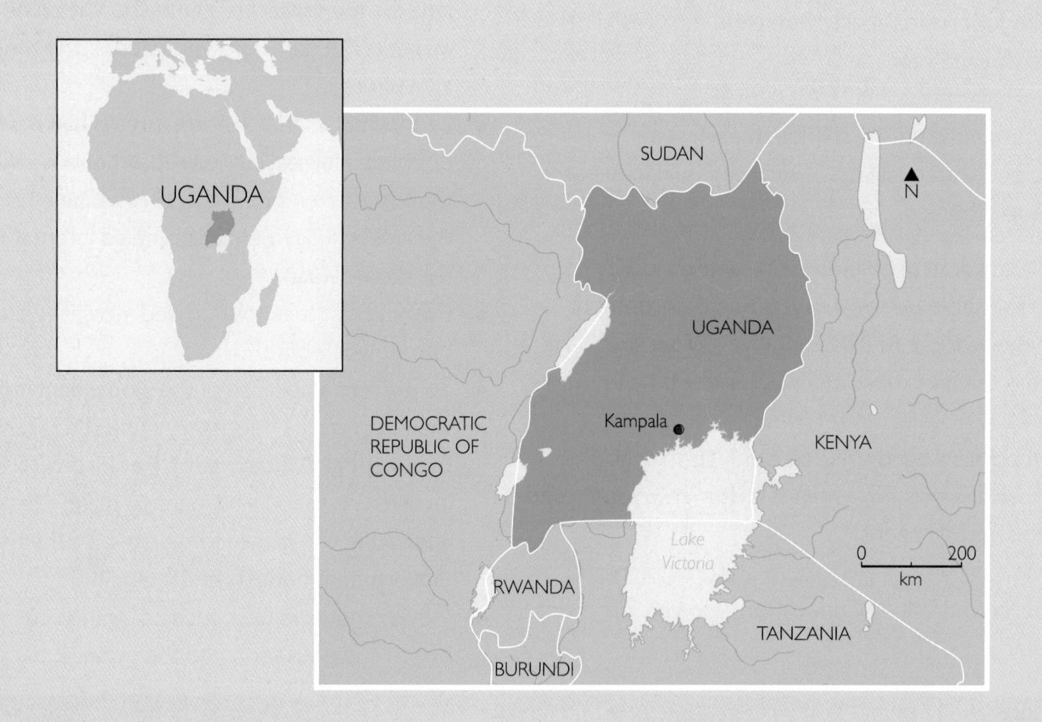

Figure 5.24 *Uganda – a country at a very low level of development* ▲ ▶

Population	31.6million
HDI value	0.514
Life expectancy at birth (years)	51.9
Infant mortality per 1000 live births	85
Adult literacy rate	73.6%
% without electricity	75%
Children underweight for age (under 5yrs)	20%
People not using an improved water source	36%
GDP per capita (PPP US$)	1,059

Did you know?

Since being relieved of most of its debts Uganda now has one of Africa's fastest-growing economies.

Struggling for a brighter future

One of the secondary schools in Ibanda District in southwest Uganda, has the slogan 'Struggle for a brighter future' (see Linda's blog – Figure 5.25). This could be Uganda's own slogan. The cancellation of $1.5bn of its debt under the HIPC initiative (see below) is helping Uganda to make progress:

- free primary education has been introduced – 5 million extra children attend school
- nearly 10% more people have access to clean water
- AIDS awareness and sex education have been made a priority.

In addition:

- Improving roads and building up mobile phone networks are helping to reduce the remoteness of some areas.
- Wildlife parks are beginning to earn much-needed overseas revenue as tourism comes to Uganda.
- Production of bio-fuels is increasing as well as fair trade crops of tea and coffee for export.
- Kampala and other cities are developing service sector jobs.

But Uganda's girls and women still remain the poorest. Women rarely own land and are most likely to work as landless labourers as, and when, needed. They gain respect by having many children, even though maternal and infant mortality rates are high. For them, life is still tough.

..I visited a class of 79 pupils ranging in age from 13 to 19. They had all been to school for 3 years. Boys outnumbered girls by 6:1 as many girls leave school by the time they're 14. The pupils' school uniforms were immaculate, and they had an impressive desire to learn.

Gloria told me that her parents don't see the point of paying for her education as it will be her future husband's family who gain. It's no wonder that traditional practices of early marriage and large families remain the norm.

Figure 5.25 *An extract from Linda's blog after she visited the Ibanda District in April 2010* ▲

The **Highly Indebted Poor Countries** (HIPC) are a group of 38 of the LDCs with the greatest poverty and debt, and they include Uganda. The HIPC initiative began in 1996 to reduce debt burdens. Further debt reductions came in 2005, and as a result Uganda has had $1.5 billion of its debt cancelled. However there were two conditions:

- the government had to demonstrate good financial management and a lack of corruption
- the money saved had to be spent on poverty reduction, education and health care (see Figure 5.26).

Figure 5.26 *HIV/AIDS awareness poster in Uganda* ▶

ACTIVITIES

1. Draw a spider diagram to show the key factors that have kept the LDCs poor, and how these factors keep them poor.
2. In pairs, discuss these factors and rank them according to importance. Justify your ranking.
3. How many of these factors are beyond the control of people living in these countries?
4. Is it fair to say that girls suffer most in countries of low levels of development?
5. How far might the HIPC initiative resolve the problems in Activies1-4?

Global social and economic groupings – 1

In this section you will learn about:
- the North-South divide
- measuring differences
- different ways of grouping countries

NORTH-developed core	SOUTH-developing/ undeveloped periphery
• 25% of global population • 80% of global income • life expectancy now over 75 years • most people well fed • high levels of literacy and educational achievement • 90% of global manufacturing output • 95% of global spending on research and development • 99% ownership of patents • controls global finance and trade	• 75% of global population • 20% of global income • life expectancy of 50 years in 1980, but mostly 65 now • 20% suffer from malnutrition or hunger • fewer than 50% attend school • 10% of global manufacturing output • limited opportunities for research and development • dependent on North for medical supplies • dependent on North for loans and aid for development

Figure 5.27 *The North-South divide* ▲

Global differences

Three worlds …?

The term **third world** originated during the Cold War. The advanced industrialised, free-market economies became known as the **first world** (West) and the communist (also industrialised), became the **second world** (East). However, these groups only accounted for 25% of the world's population, so the remaining 75% became known as the third world.

Differences within the third world group of countries were enormous, but images of poverty and low levels of human well-being dominated the media, and people's perceptions of the Third World have been difficult to shift ever since.

…or two worlds…?

In 1981 a landmark report about global development was published, called The Brandt Report. It acknowledged that human well-being depended on more than just economic growth. It established a wider range of criteria for judging levels of development around the world. The report showed a divided world and referred to the differences as the **Development Gap**. It proposed global reforms and urged the '**developed** North' to share its wealth with the '**undeveloped** or **developing** South'.

…or one world?

If globalisation – together with the loosening of trade barriers and decades of international aid – had been successful, the gaps between richer and poorer countries should have declined. But the text box below shows what has really happened.

Figure 5.28 *The North-South divide and world wealth* ▼

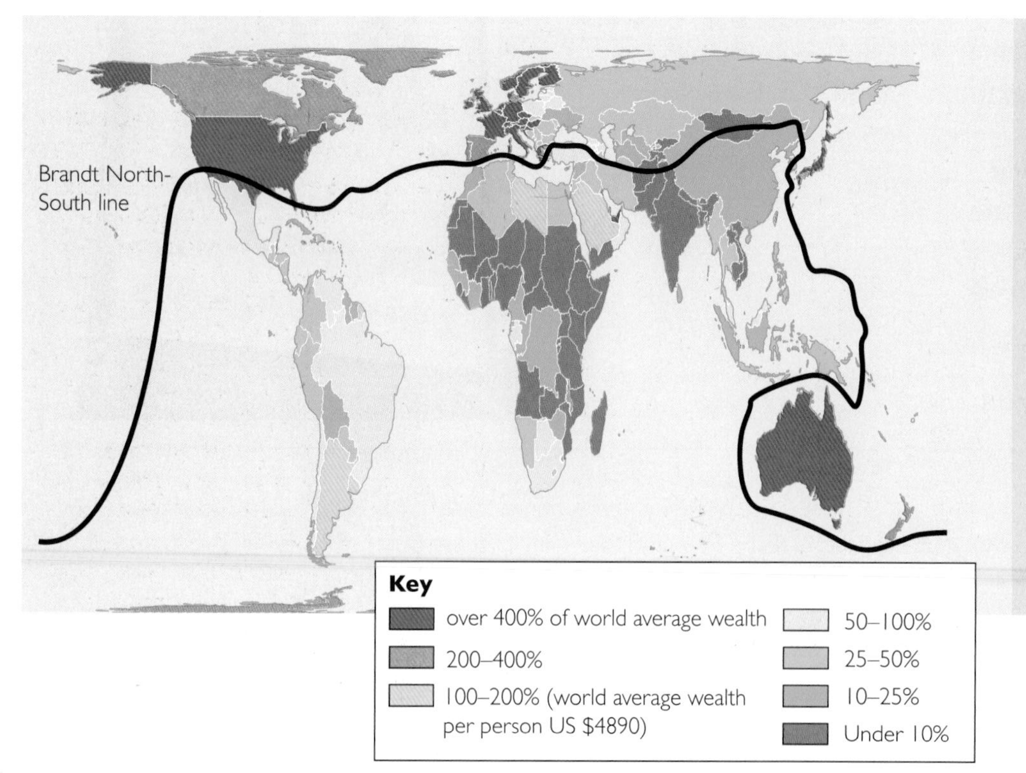

- The poorest 20% of the world's population received 1.5% of global income in 2006 (down from 2.3% in 1998).
- The richest 20% increased from 70% to 85% of global income in the same period.
- Sub-Saharan countries have seen their incomes fall to below 1980s levels.
- The richer countries now consume more than 80% of the world's natural resources.
- The developing world spent $13 on debt repayment for every $1 it received in aid in 1998, and very little has changed since then.

Measuring differences

Measuring development isn't easy. It needs economic and social data. Economic data tend to be based on the formal economy and ignore informal work. But in any case, whether economic or social, all data show the same gap between the world's richest and poorest countries.

Economic Indicators

Economic activity is measured in US dollars using key indicators:

- **GNP (Gross National Product)** is the value of all goods and services earned by a country including companies working abroad. The term **GNI (Gross National Income)** is used more frequently these days. These measures are good indicators of a country's wealth – especially HICs who earn a great deal from their overseas investments.
- **GDP (Gross Domestic Product)** is same as GNP but excludes foreign earnings
- **Per capita** statistics provide an average 'per person' figure
- **Purchasing power parity (ppp)** relates average earnings to prices and what it will buy, because a dollar buys more in some countries than others. This is the spending power within the individual country and reflects the cost of living. China's GDP per capita ppp in 2010 was US$ 7400. When PPP$ values are higher than GDP per capita it means prices in a country are cheaper than average and vice-versa.

Figure 5.29 *Street traders in Uganda are part of the informal economy, where no record is kept of trading or income. So traders like this do not figure in most development data about countries.* ▲

Human Indicators

To balance the emphasis on economic data, the UN uses key indicators about human well-being to produce the HDI (see page 217) as a way of measuring progress.

Each indicator that makes up the HDI is given a score, and the HDI is the average of the four scores.1.000 is the highest score, and 0.000 is the lowest. Countries are ranked according to their overall scores. High scores indicate a better quality of life. The implication is that countries spending money on health care and education have a higher level of development. It also means that people are benefitting from economic growth.

Figure 5.30 *Global HDI in 2006* ▼

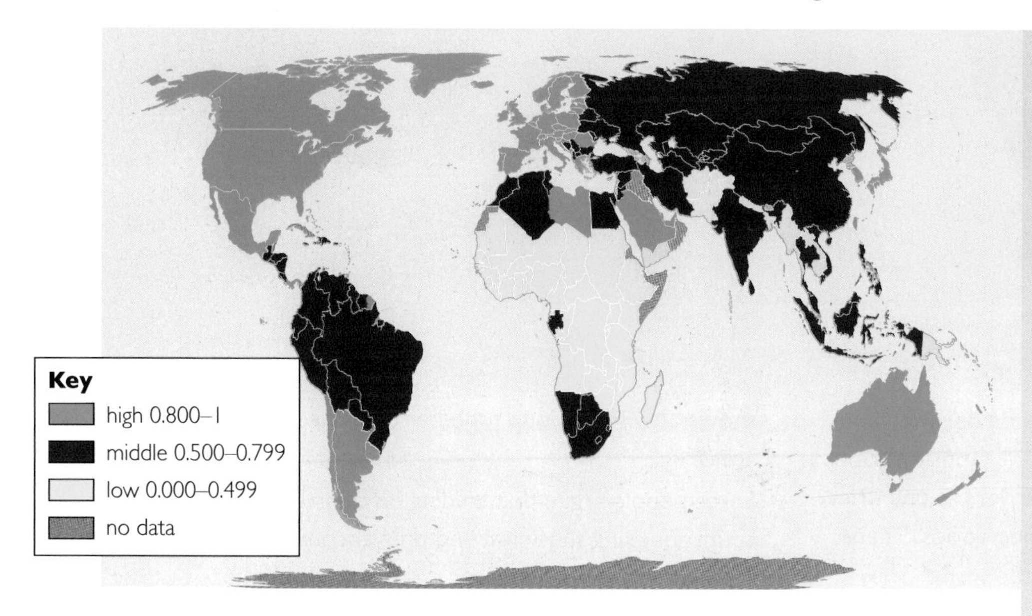

Human development trends—the HDI and beyond

- Since the mid-1970s almost all world regions increased their HDI score.
- East Asia and South Asia have accelerated since 1990.
- Central and Eastern Europe and the Commonwealth of Independent States (CIS), declined in the early 1990s, but have recovered.
- The major exception is Sub-Saharan Africa. Since 1990 it has stagnated, mainly because of the catastrophic effect of HIV/AIDS on life expectancy.
- Eighteen countries have a lower HDI score today than in 1990—most in Sub-Saharan Africa. Today 28 of the 31 low human development countries are in Sub-Saharan Africa.

Changing groupings

Brandt's way of dividing the world into North and South seems too simple now. Rapid development has made the world more complicated:

- Brazil, along with other Latin American countries was already developing fast in the 1970s, so even in 1981 Brandt's classification was out-of-date.
- Singapore, Dubai, Hong Kong have all experienced strong economic development since 1981 and are now highly developed. Other countries in South East Asia (e.g. Thailand, Malaysia) grew rapidly in the 1980s and 90s. China and India are growing fast. These Newly Industrialising Countries (NICs) have surged away from the 'South'.
- Other countries have barely changed – especially across Africa, so the gap between rich and poor has widened substantially.

Income groups

The UN now uses four levels of income: **High**, **Upper Middle**, **Lower Middle** and **Low** to classify the world's countries. There are huge differences within each division; low-income countries include India – one of the world's largest and fastest-growing economies – and Ethiopia, one of the smallest and slowest.

Sub-Saharan Africa contains 26 of the poorest 30 countries in the world. Brandt placed these countries in the South in 1981 and they remain poor today, even further behind the rest of the world.

Figure 5.31 *The UN's divided world based on four levels of income (2006)* ▼

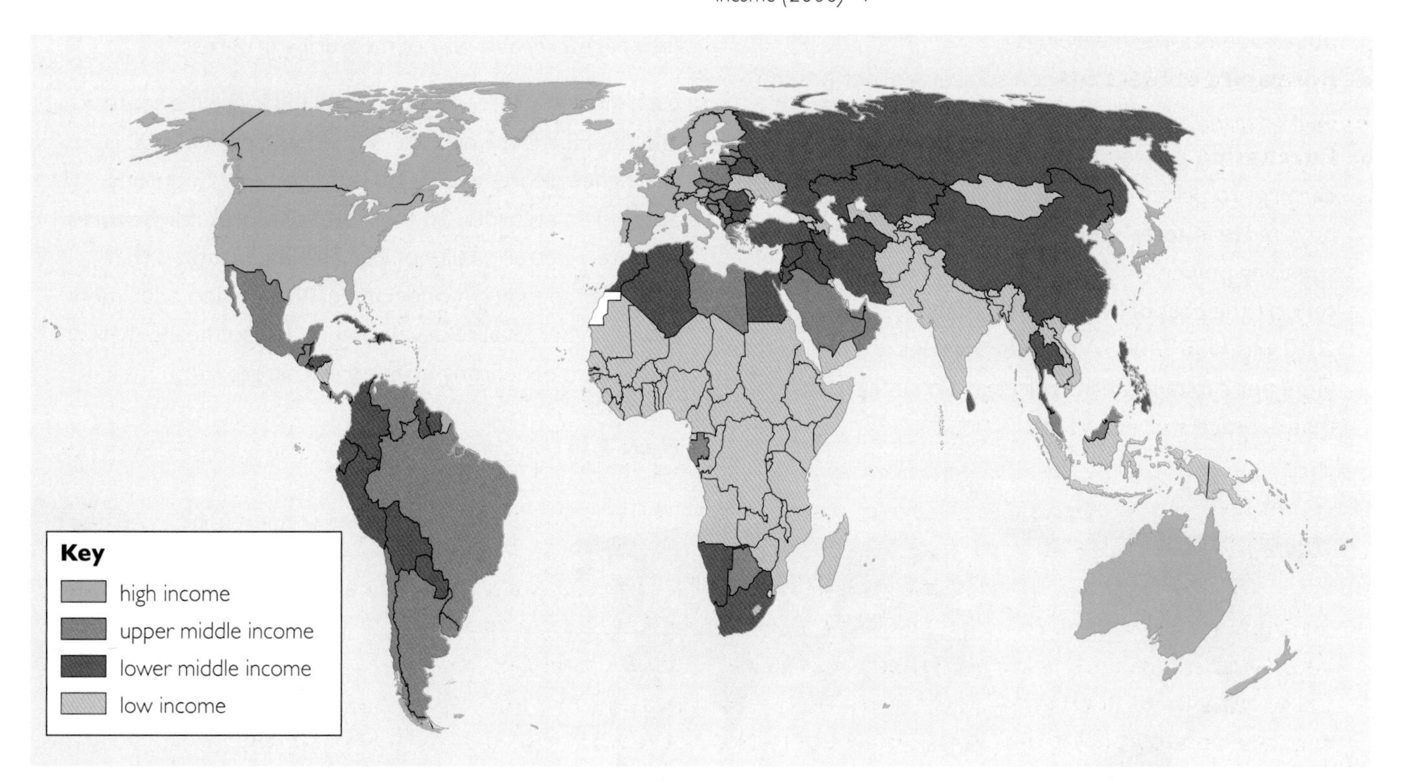

Trading groups

Trading groups, or blocs, encourage trade between a group of countries by removing duties or tariffs on goods for member countries, and creating barriers to countries outside the bloc by placing tariffs on their goods. There are a number of trading blocs around the world. They include the **EU**, **NAFTA**, **ASEAN**, **MERCOSUR**, **The Cairns Group**, **APEC** and **OPEC**. Figure 5.32 opposite shows how they group together countries in geographical areas.

Some people argue that trading blocs protect the member countries using subsidies, and only encourage trade between member countries rather than with those outside the bloc. By promoting some goods and preventing others, they prevent the development of a global economy.

Who runs world trade?

World trade has been dominated by three regions: North America, Europe and East Asia. These regions produce most of the world's goods, control most trade and absorb most investment.

- In 2001 these three regions generated 85% of the world's manufacturing output and 81% of global exports (an increase from 76% and 71% in 1980 respectively).
- East Asia has grown most since 1980, increasing output from 16% to 29%, and trade from 17% to 26%.
- In 2007, China became the world's leading exporter and has held that position ever since.

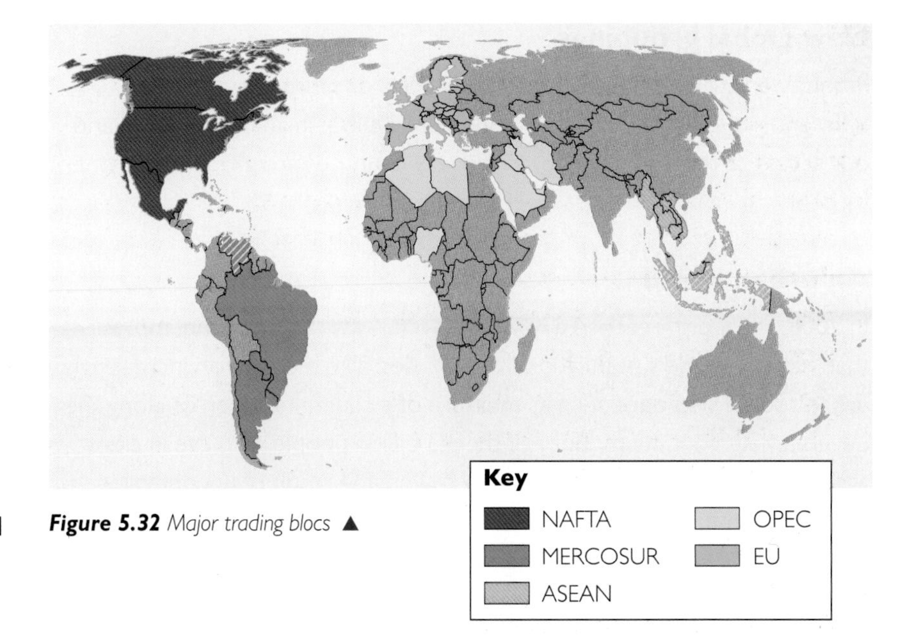

Figure 5.32 *Major trading blocs* ▲

Trading organisations

As well as trade blocs, several other organisations exist to promote trade of different types and from different countries. They attempt to govern and set rules of trade and include:

- **WTO** (World Trade Organisation). Formed in 1993, it aims to cut trade barriers (subsidies, tariffs and quotas) that stop countries trading freely, so that goods can flow more easily. It set conditions under which trading can occur. It has been accused of acting in the interests of wealthy countries, especially the USA, through its fixation with reducing trade barriers.
- **OECD** (Organisation for Economic Cooperation and Development). It is a global 'think tank' for 30 of the world's wealthiest nations.
- **OPEC** (Organisation of Petroleum Exporting Countries). Established to regulate the global oil market, stabilise prices and ensure a fair return for the 11 member states who between them supply 40% of the world's oil.
- **G8, +5** The Group of 8 (G8) consists of Russia, the USA, UK, France, Canada, Germany, Italy and Japan. The G8 represents 65% of the world's trade and meets annually to discuss economic development. In 2005, G8+5 was formed to include China, India, Brazil, Mexico and South Africa in order to create deeper international cooperation and an understanding of climate change and international trade.
- **G20** This actually has 23 members from the developing world, and was formed in 2004 with a focus on agricultural trade.
- **World Bank**. Exists to promote investment globally and provides loans for countries who agree to conditions (like the IMF).
- **IMF** (International Monetary Fund). It forces countries to privatise (or sell off) government assets, which are then bought by large TNCs, and open up trade in return for re-financing debt. Many believe that this has forced poorer countries to sell their assets to wealthy TNCs.

New global groupings

In the twenty-first century global differences, or similarities, have led to different groupings – from where and how we live, in an urban world and our access to the digital world, to the way countries are grouped together by debt.

An urban world

In 2008, the world reached a new milestone – for the first time, more than half the world's population lived in cities. The move to an increasingly urban society was once seen as measure of a country's progress along the development continuum. But half the 3.3 billion people who live in cities are living in poverty. Sprawling shanty towns, the result of uncontrolled growth, have spread across the LICs.

Figure 5.33 *More than half the world's population live in urban areas* ▼

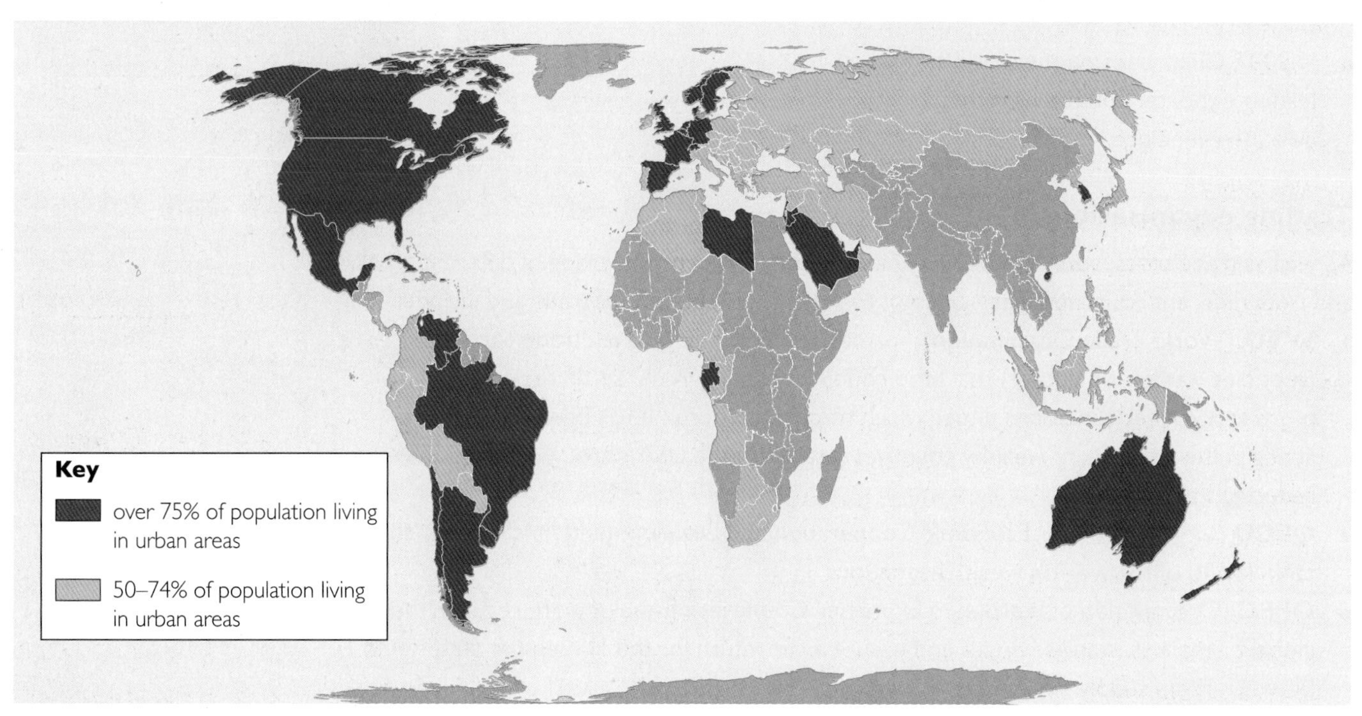

A digital world

Mobile phones, texts, faxes and e-mail now dominate business, and enhance learning, and communication. Rapid movement of data via the Internet brings people even closer together. Slow movement, or a lack of movement keeps them apart, and this creates a **digital divide** between countries.

In Africa many people do not have access to the Internet. Africa only had 3.9% of the world's Internet users in 2009 compared with 42.6% in Asia and 24.1% in Europe. Whilst digital social network sites break down geographical boundaries and group people in new and different ways, a lack of access to the digital world can increase the gap between richer and poorer countries. However, it can also help to protect areas from westernised culture and influence, and preserve the uniqueness of places.

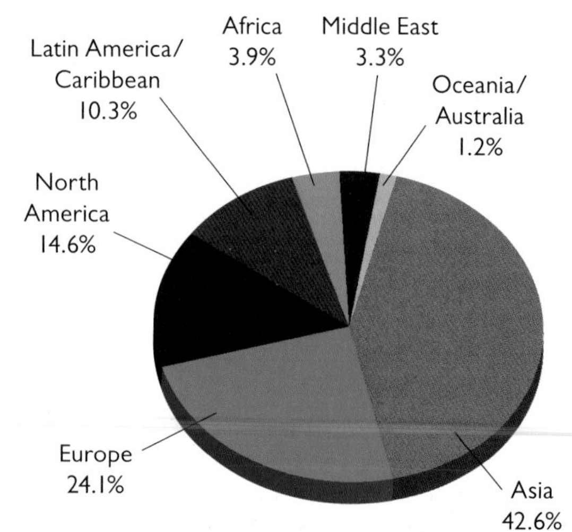

Figure 5.34 *Internet users by region, 2009* ▲

A world of debt

For over 30 years African countries have been trying to recover from debts dating from the 1970s when loans were cheap (see page 217). The HIPC are a group of 38 of the world's LDCs with the greatest poverty and debt, and 29 of them are in Africa. Figure 5.35 shows the cost of servicing their debts in 2003.

But debt has become a bigger issue for the world since 2008, and a financial crisis has divided the world. In 2010 the debt of the UK and Japan was over 400% of each country's GDP. National debt or personal debt limits spending power and restricts economic development. Interest payments on debts mean that money is diverted away from items like healthcare and education.

New economic groupings

The WTO (see page 223) believes in Free Trade, without subsidies or tariffs, and removing these is known as 'liberalising trade'. International agreements are needed to achieve the WTO's aims, but talks have stalled. Countries have started to make their own deals:

- The USA has agreed, or is negotiating, free trade arrangements with 25 countries, including strategic allies like the UAE, Bahrain, South Africa and Thailand.
- China has been building up deals with ASEAN countries to form a new East Asian bloc and is working closely with New Zealand, Australia, Korea and several African countries.
- The EU is establishing new economic partnership agreements with 69 African, Caribbean and Pacific nations (former colonies).

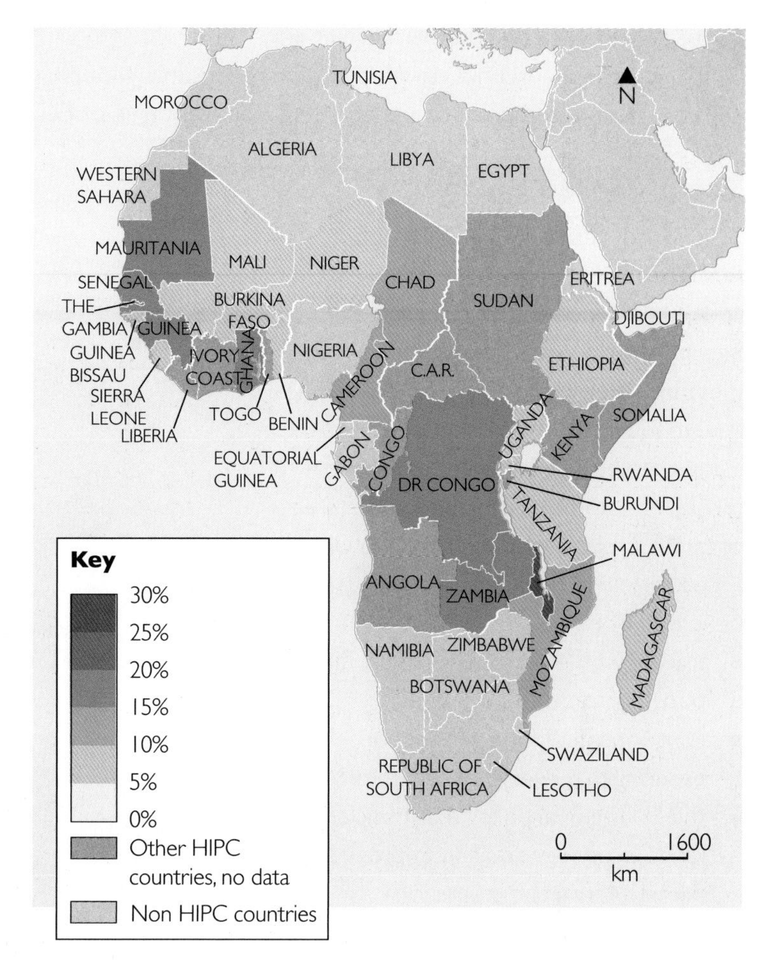

Figure 5.35 *The cost of servicing debts in some African countries in 2003, as a percentage of government revenue* ▲

Internet research

1. In pairs, research one of the trading blocs shown on Figure 5.32, so that your class covers all of them. Design a leaflet explaining the purpose of one of these trading blocs.
2. Compare your findings. Which blocs have similar and different aims?
3. Using examples, suggest why the membership and influence of blocs might change over time.

ACTIVITIES

1 Construct a table as shown

Country	Life expectancy	Infant mortality	GDP per capita ppp/GNI per capita ppp	HDI	North or South today

a Use information from the World Bank and the UN (http://www.worldbank.org/ and http://hdr.undp.org/en/statistics/) to collect key data on 30 countries representing the Low, Middle and High Incomes (10 of each).

b Complete the table and indicate whether they fall North or South of the old Brandt Line.

c Comment on the advantages and disadvantages of using **i**) GDP, **ii**) GNI, **iii**) HDI as indicators of development.

d How appropriate are the terms Third World and North-South Divide today?

Global social and economic groupings – 2

In this section you will learn about:

- how the EU has developed as an economic grouping of nations
- about some of the benefits and disadvantages of EU membership

The European Union

The EU was set up by the Treaty of Rome in 1957 to achieve economic and political cooperation following the Second World War. Originally it had six members (Belgium, France, Germany, Italy, Luxembourg and the Netherlands) and its aim was to bring economic cooperation through a common European market free of tariffs. Denmark, Ireland and the United Kingdom joined in 1973. This trading group benefited from:

- trading with each other without tariffs – boosting trade and wealth
- joint control of food production – ensuring that everyone had enough food
- the EU regional policy that transfers funds to poorer areas to improve infrastructure and create jobs.

By 1987, Greece, Spain and Portugal had joined, and the 'Single European' Treaty was signed with the aim of creating a 'Single Market' allowing the free flow of trade across EU borders.

- This concept was taken a stage further in 1993 with the free movement of goods, services, people and money permitted across the EU.
- Further expansion in 1995 saw Austria, Finland and Sweden become members, and in 2004 and 2007 the community grew to a group of 27 countries, as Figure 5.36 shows.
- Other countries are seeking permission to join, and Croatia, Turkey, Iceland along with other East European nations could become part of the EU by 2020.

A combined Europe is also able to trade more effectively with the rest of the world – a single market with common trading rules, and in many cases a common currency, is more efficient than 27 competing with each other.

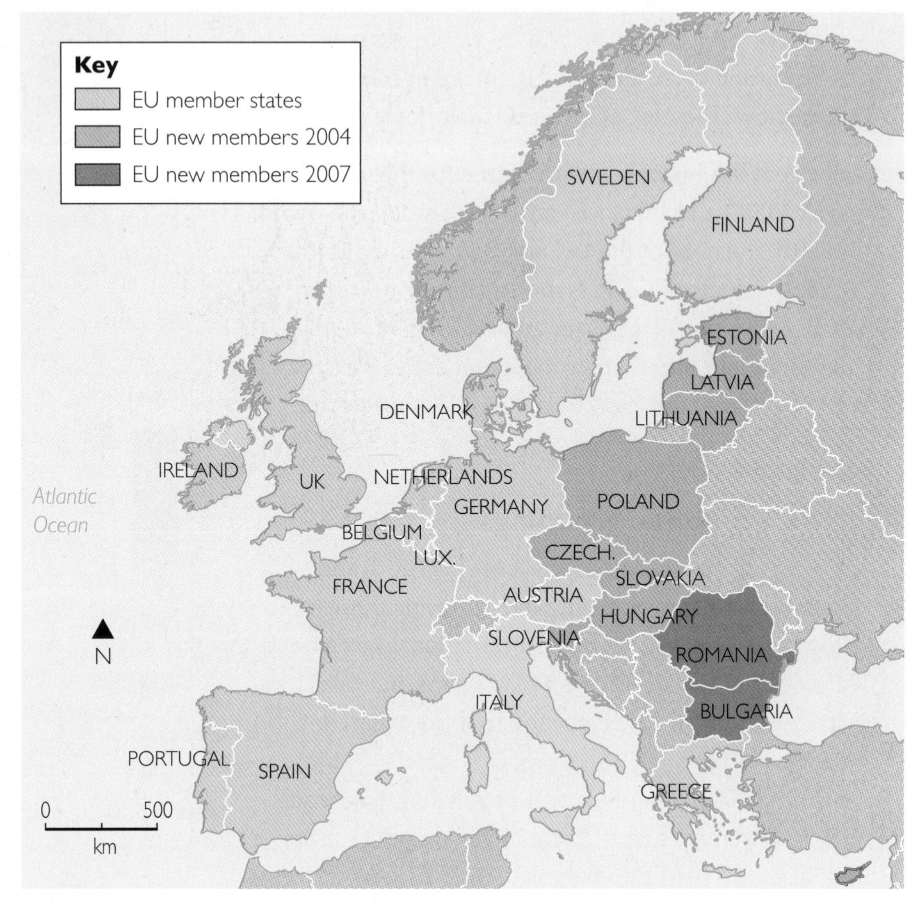

Figure 5.36 *EU member states* ▲

Differences in development

There are large differences in levels of wealth both between, and within, countries of the EU, as the graph opposite shows.

- Urban areas like London, Brussels and Hamburg tend to be the richest.
- Rural areas in the newest members – Bulgaria and Romania tend to be the poorest.

Increasing membership should help the poorer areas. All member states of the EU contribute taxes to a central fund. This money is then re-distributed to areas where economic stagnation or decline has occurred.

Did you know?

If all the countries who are considering EU membership, and who have expressed a wish for closer links joined the EU, then membership would total over 50 countries!

Figure 5.37 *London's financial sector creates wealth for the UK* ▲

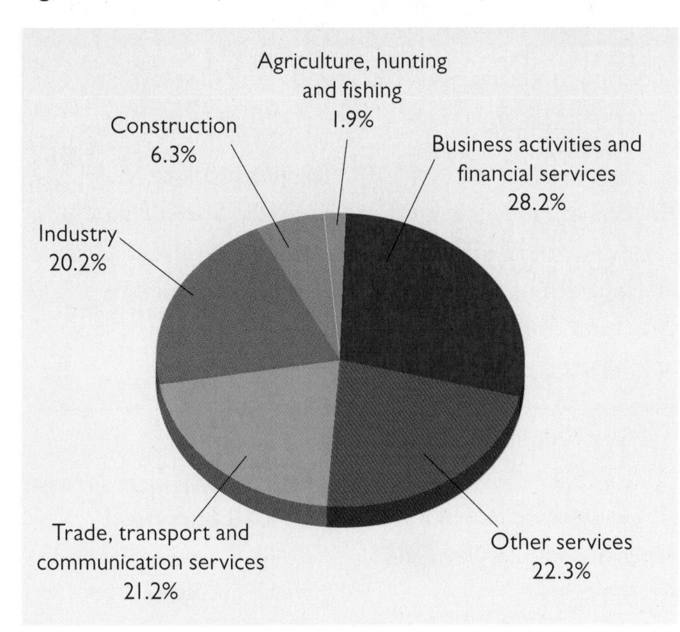

Figure 5.38 *Europe's GDP – from different employment sectors* ▲

Figure 5.39 *A high percentage of people in EU countries such as Bulgaria and Romania still work in farming* ▲

Figure 5.40 *Differences in wealth between EU member states* ▼

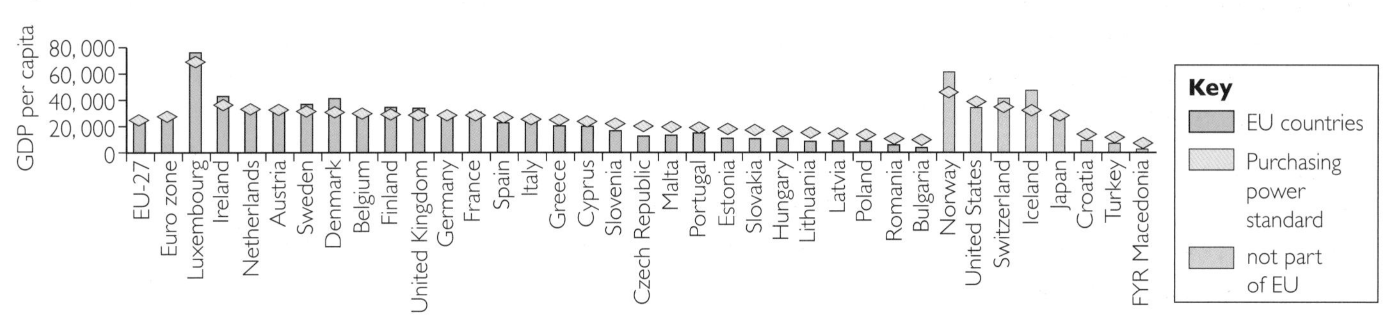

Europe expands

Fireworks and all-night celebrations in Bulgaria and Romania marked the arrival of these countries as part of the EU in January 2007. They added another 30 million people to the EU, but just 1% to its economic output.

Bulgaria and Romania are relatively poor compared with the rest of Europe.

- Their GDP's are just 33% of the EU average.
- Wages are low at €180 in Bulgaria and €300 per month in Romania.

But both countries had been growing their economies since they asked to join the EU in 1993 – by 5.5% and 7% per year respectively. Both see the EU as a source of economic and political security, and they received more than €600 million each in their first year of entry, to help improve infrastructure, employment opportunities and welfare programmes.

Figure 5.41 *Celebrations in Bucharest as Romania joined the EU in 2007* ▲

Bulgaria and Romania – background

Half of Bulgaria and Romania's population live in rural areas and most of these people farm (in the UK just 1.5% of the workforce is in agriculture). The extension of the EU into old Communist countries of Eastern Europe presented these countries with serious difficulties. Farmers that were used to 'collective' organisation now faced the free-market economies of the EU with no guaranteed market for their produce. Full employment under communism, was exchanged for unemployment under capitalism.

Bulgaria used to sell most of its farming produce to Eastern communist countries. When the Soviet Union collapsed in 1991, Bulgaria's economy suffered badly. In preparation for joining the EU, the government re-structured the economy, unemployment fell and TNCs have started to move in.

Consequences of EU membership

2010 saw protests across parts of the Eurozone – the EU countries using the Euro. Being a member of the Eurozone means that countries have to stick to strict economic rules. And as some countries faced recession they found it difficult to keep to the rules.

A number of countries in the EU, including these in southern Europe – Portugal, Italy, Greece and Spain – were faced with massive government debts. The IMF agreed to work with the European Commission and The European Central Bank if programmes of public spending cuts were introduced in these countries. The cuts are known as **austerity measures** and meant reduced wages, higher taxes, and cuts to spending on education, healthcare, welfare and infrastructure projects.

To the people living in these countries the austerity measures imposed on them appeared undemocratic and unfair. That's why they took to the streets to protest. Suddenly, being part of the EU seemed less popular, as the newspaper headlines suggest.

EU to bail out Greece in £104bn deal

A Greek rescue package worth €120bn (£104bn) was put to an emergency meeting of Eurozone finance ministers after the International Monetary Fund (IMF) finalised negotiations in Athens

May 2010

'Hands off our rights! IMF and EU Commission out!'

May 2010

Benefits of EU membership

Three things are central to the EU's activities:

- convergence
- competitiveness
- cooperation.

The intention is that together they will reduce the gap in income between the rich and poor. The EU's regional policy is used to achieve these aims by investing in people. 36% of the EU's budget between 2007-2013 (a sum of €340 billion) will be used to help the poorer regions reach similar levels of development enjoyed across the rest of Europe. Just over half this spending will target the 12 newest EU members from Central and Eastern Europe. Although they represent less than 25% of the EU's population they are the most in need. The remaining spending goes on regions with a GDP below 75% of the EU average, to help develop their economic and human potential. There are three sources of funding:

- the European Regional Development Fund (ERDF) – invests in infrastructure and innovation
- the European Social Fund (ESF) – invests in skills training, job creation and support
- the Cohesion Fund – invests in the development of renewable energy, environmental and transport infrastructure projects. This is reserved for countries whose living standards are below 90% of the EU average.

Together, these funds are designed to keep the EU attractive for investment, enhance accessibility, provide high quality services and maintain the environment. By encouraging innovation and the development of the knowledge economy, the EU hopes to compete with the other major trading groups of the world.

Figure 5.42 *Bulgaria's green economy is set for a huge EU funding boost. Renewable energy is something Bulgaria could not afford before joining the EU* ▲

ACTIVITIES

1. In pairs, draw a table to show the advantages and disadvantages of EU membership for a) the UK, b) recent members from Eastern Europe such as Bulgaria. Highlight two from each column which are most compelling. Feed back your ideas to the class.
2. Now score your advantages and disadvantages between 1 and 5, where 1 is weak and 5 is strong. Add the totals; which scores most?
3. Why should UK political parties such as UKIP protest about EU membership? What would be the implications of **a**) trying to change the EU from within, **b**) leaving the EU?
4. In 750 words, justify how far you support the concept of **a**) The European Regional Development Fund, **b**) The European Social Fund, **c**) The Cohesion Fund.

Internet research

1. In groups, use the Internet to research the economic and social characteristics of the 27 EU members. On an outline map of Europe use a choropleth technique to show regional variations in wealth, standards of living and social well-being. These sites will help:
   - **EUROSTAT** http://epp.eurostat.ec.europa.eu/portal/page/portal/eurostat/home/
   - **European commission** http://ec.europa.eu/agriculture/trade/index_en.htm
   - **EU/UN** http://www.europa-eu-un.org/articles/en/article_8059_en.htm
2. With reference to one wealthy and one poorer country examine the benefits and disadvantages of being a member of the EU.

Aspects of globalisation

In this section you will learn about:

- how TNCs are organised and how they grow
- about the impacts they have
- how TNCs can shape the modern world

TNCs – who are they?

TNCs are transnational companies. They're companies like Wal-Mart, BP and Toyota, who are involved in a range of different activities. We think of TNCs as recent developments, but from the seventeenth century parts of India were run by the East India Company. It controlled trade routes and ruled 20% of the world's population. This was probably one of the world's first TNCs. Today, there are over 60 000 TNCs and the top 200:

- employ just 1% of the global workforce
- but account for 25% of the world's economic activity.

In 2006, six of the top ten TNCs originated in the USA. By 2010 only 2 originated from the USA and 3 originated from China. What hasn't changed is that all of the top ten make more money in a year than some countries produce in GDP.

Rank	Company	Country of origin	Revenue ($ billions)	Profit ($ billions)	Employees	Revenue compared with country GDP
1	Wal-Mart Stores	USA	408.2	14.3	2 100 000	more than Saudi Arabia
2	Royal Dutch Shell	Netherlands	285.1	12.5	101 000	more than Saudi Arabia
3	Exxon Mobil	USA	284.6	19.3	102 700	more than Saudi Arabia
4	BP	UK	246.1	16.6	80 300	more than Finland
5	Toyota Motors	Japan	204.1	2.3	320 590	equal to Hong Kong
6	Japan Post Holdings	Japan	202.1	4.8	229 134	equal to Hong Kong
7	Sinopec	China	187.5	5.7	633 383	equal to Egypt
8	State Grid	China	184.4	-0.3	1 533 800	more than Singapore
9	AXA	France	175.2	5.0	103 432	equal to Singapore
10	CNPC	China	165.4	10.2	1 649 992	equal to Pakistan

Figure 5.43 *The world's top ten TNCs, 2010* ▲

How are they organised?

Take Toyota. In 2010 it was the fifth-largest TNC.

- It has 51 overseas manufacturing companies in 26 countries and regions (these are shown on the map).
- It has Design and Research and Development centres in the USA, Japan, Belgium, the UK, France, Thailand and Australia.
- Its headquarters are in Japan.

Toyota, like many TNCs in general, undertakes much of its production in LICs (its largest manufacturing companies in terms of the number of people employed are based in China and Thailand), to meet the demand for its goods from HICs.

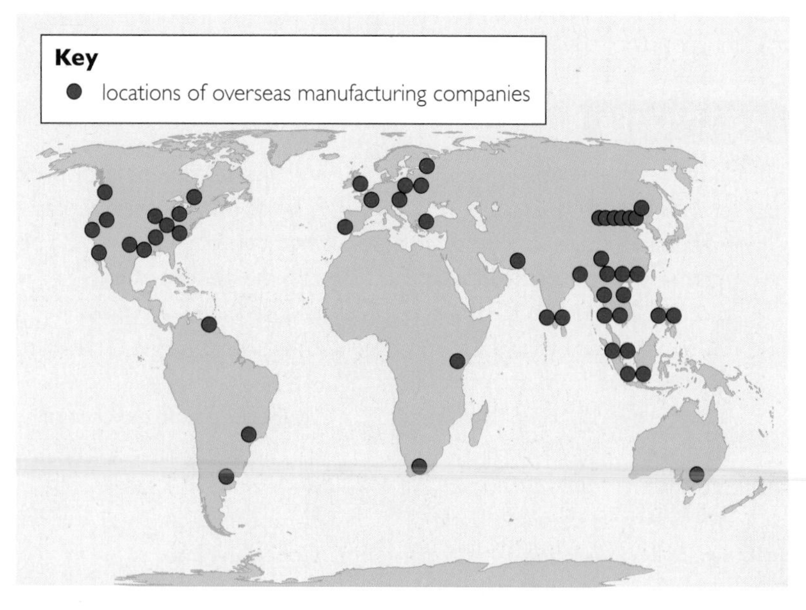

Figure 5.44 *Toyota's overseas manufacturing companies* ▲

How do they grow?

How and why do companies become so large? Three factors are important: motive, means and mobility.

Motives

Under a capitalist system, there is one motive – profit. Therefore, those companies which become dominant do so by controlling costs – which means mostly the price of raw materials and the costs of producing goods and services. This involves mergers and take-overs, which, occur in three ways:

- **Horizontal integration** – by buying up the competition. Ford expanded from mass-market sales to up-market sales by acquiring Jaguar, Volvo and LandRover.
- **Vertical integration** – by controlling and owning every stage of production from exploration and research through to sales, e.g. Exxon owns oil wells, refineries and petrol stations.
- **Economies of scale** – companies expand production (e.g. Apple computers) to increase efficiency and reduce unit production costs.

Means

Banks provide the means for companies to grow by providing finance. HSBC supports businesses throughout the world. Companies invest overseas to boost their market presence or to take advantage of labour and environmental laws, e.g. where environmental regulations are less strict than in the home country. Flows of money around the world connect businesses and countries in complex webs. Since 2008, these flows of money have slowed and many businesses have reduced their levels of investment.

The top five countries investing overseas have for a long time been: the USA (21%), the UK (15%), Germany (7%), France (8%), and Japan (5%). In 2005, half of all global investment came through London's banks.

But a growing trend in the twenty first century has been for TNCs from developing countries to invest overseas. Hong Kong, Singapore, Taiwan, China, South Korea, Malaysia, India and Brazil now account for more than 15% of overseas investment. Examples from 2007 included:

- Indian firms, led by Hindalco and Tata Steel, who bought 34 foreign companies for a combined US $10.7 billion.
- Big Blue sold its personal-computer business to a Chinese multinational, Lenovo.
- Russia overtook India, investing US $11.4 billion abroad.

And China continues to buy a wide range of Japanese household brands.

Mobility

Other factors which have helped companies to grow and move around the world include:

- Faster and cheaper transport – including the use of ever-larger container ships.
- Faster and cheaper communications systems – using fibre optics, satellite and digital technology.
- Production technology – 'just in time' and 'flexible production' systems that can provide cheap and fast turn-around (see next page).
- Global Production Networks – linking source, manufacturing and sales systems.

While globalisation has opened new markets to rich-world companies, it has also given birth to a pack of fast-moving, sharp-toothed new multinationals that is emerging from the poor world.

The Economist, April 2007

CASE STUDY

Disney

These days Disney is a globally recognised brand. It ranks third in the global brand league (after Coca Cola and McDonald's) and competes with AOL/Time Warner for the top slot in the world of media and entertainment. In 2009 the company's income was $36.1bn. It has 130000 employees of its own, as well as 40000 suppliers in 50 countries. From a small animation studio in California in the 1950s, Disney expanded rapidly in the 1980s taking advantage of the communications revolution to become a truly global company.

Disney's social and economic impacts

Disney is a global company which is typical of the **new economy**. The new economy is where companies (or countries) are based more on creativity in terms of things like finance, media and management, and less on the manufacturing of goods. Disney's ideas originate in, and decisions about its products are made in, the USA. The merchandise it sells – the toys, t-shirts, books and so on – are manufactured overseas (generally in LICs) to meet the demand from the HICs.

Disney, and the way the company operates brings with it a range of impacts on the countries where it manufactures its goods.

- Disney demands quick delivery times from its overseas manufacturers. It places orders **just in time** once it knows how successful its merchandise is likely to be. This means that Disney doesn't have to tie up money in large volumes of stock, but it does put pressure on suppliers.
- Workers overseas, e.g. in China, Macau and Vietnam may receive low wages or be paid late, and may use toxic substances banned in the USA. In 2007, toys manufactured in China were recalled from shops because of dangerous levels of toxic lead in their paint.

However, Disney is aware of the poor publicity that can result from using production methods of this type, and increasingly monitors overseas suppliers to ensure higher standards. It cancels contracts with companies who abuse workers.

The Disney Corporation is involved in many global activities, such as:

- 230 linked satellite and cable TV companies
- 6 film/TV production and distribution companies
- 12 publishing companies, 15 magazines and newspapers
- 728 shops worldwide, plus galleries and toy companies
- 5 record labels and music publishing
- 3 theatre production companies
- 6 theme parks and resorts and a cruise line
- sports franchises and teams
- multimedia – producing CDROMs and electronic games, plus Disney Interactive Media Group
- property development and human resources agencies.

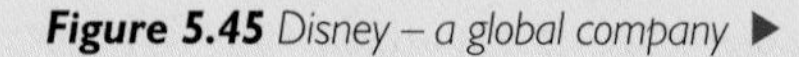

***Figure 5.45** Disney – a global company* ▶

Disney's environmental impacts

The Disney Corporation has attracted a lot of attention about its environmental impacts. In 2006, Disney's company-wide CO_2 emissions were estimated at 566 000 tons, with 48% coming from its two cruise ships. Its image was damaged by the waste materials dumped on the abandoned private Abaco (Treasure) island resort. It also introduced invasive plant species and insects that threaten natural fauna and flora. But, on Earth Day in 2008, Disney launched the Green Standard initiative with strict controls over emissions, waste, water, energy, and ecosystems. The aim is to inspire all business partners, employees, guests and consumers to follow 'environmentally responsible behaviour'. It means big changes in the years ahead for an industry based on overconsumption.

Disney's impacts on cultural globalisation

Disney owns 40 Spanish-speaking radio stations, foreign language television channels, and a Chinese-language radio station in Hong Kong. Several Disney films have targeted specific markets:

- *'Mulan'* marked Disney's decision to enter the Chinese market.
- *'Hunchback of Notre Dame'* was launched to rebrand Disneyland Paris.
- *'The Lion King'* was aimed at African markets, *'Aladdin'* at the Middle East, while *'Rescuers Down Under'* and *'Finding Nemo'* targeted Australia.

Disney aims for global markets, although its characters remain Americanised. Its influence spreads wider than entertainment – urban planners have imitated Disney's ways of managing theme parks and people movement, as Figure 5.46 shows. Is the world in danger of becoming 'Disneyfied'?

Impacts and influences on urban planning

- Shopping malls, like Disneyland, are often on suburban edge-of-town developments.
- Disney-themed fast-food outlets.
- Crowd monitoring using CCTV cameras.
- Resort tourism with everything on site.

Impacts and influences on the media

- Disney TV channels broadcast 24 hours a day in North Africa, the Middle East, Europe, Australia, Malaysia and even Cuba. Disney also owns shares in commercial TV channels in Europe and Brazil.
- Chinese state TV uses Disney's ESPN for sports coverage.

Impacts and influences on governments

- The US government enforces copyright protection for US companies such as Disney.
- The French government paid $2 billion towards Euro Disney, providing 30 000 jobs.

Figure 5.46 *The 'Disneyfication' process – Disney's approach is seeping into our everyday lives.* ▲

ACTIVITIES

1. In pairs, complete a table showing the benefits and problems of the ways in which goods are produced for:
   a Disney
   b its workers
2. On a world map, locate and show using different symbols where:
   a Disney's company decisions are made,
   b goods are produced,
   c products are consumed,
   d governments are influenced.
3. How far is Disney a truly 'global company'?
4. Explain what the 'new economy' means for those in **a**) a wealthy countries and **b**) poorer countries.
5. Research and write a short report on 'The pains and pleasures of working for Disney'.

Trade or aid?

In this section you will learn about:

- whether trade or aid is the best route to social and economic development

Trade versus aid

For many decades the process of development, or managing poverty in LICs has been linked to giving aid. Individuals usually respond to short term crises by donating generously – in 2010 for example, the Haiti earthquake appeal reached millions of dollars within days. This aid helps in emergencies and big fund-raising events like LiveAid, and Comic Relief provide longer relief to ongoing problems.

But governments are less generous when giving aid. In 1981, all wealthy countries agreed to donate 0.7% of their GDP to international aid. By 2008, only five countries had reached this target (Norway, Sweden, Denmark, Luxembourg and the Netherlands). The UK promises to meet this target by 2013.

However, in some instances aid can cause problems for the receiving country. Farmers can't compete with food that has been provided free as food aid, so local markets collapse. Dependence on aid may do more long term harm than good and restrict development.

Trade is generally more effective in the long term in helping countries develop because it creates jobs, and they provide wages that people spend, thereby creating further demand. This process has a positive multiplier effect and helps countries to break out of the vicious cycle of poverty shown in Figure 5.47, and move into a virtuous cycle of development. In order to do this countries need to add value to their raw materials and crops by processing them into finished products.

As Bob Geldof said in 2006 "Aid isn't the answer. Africa must be allowed to trade its way out of poverty. Africa accounts for just 1-2% of world trade. But a 1% increase in trade from Africa would be the equivalent of five times the amount of aid the continent currently receives."

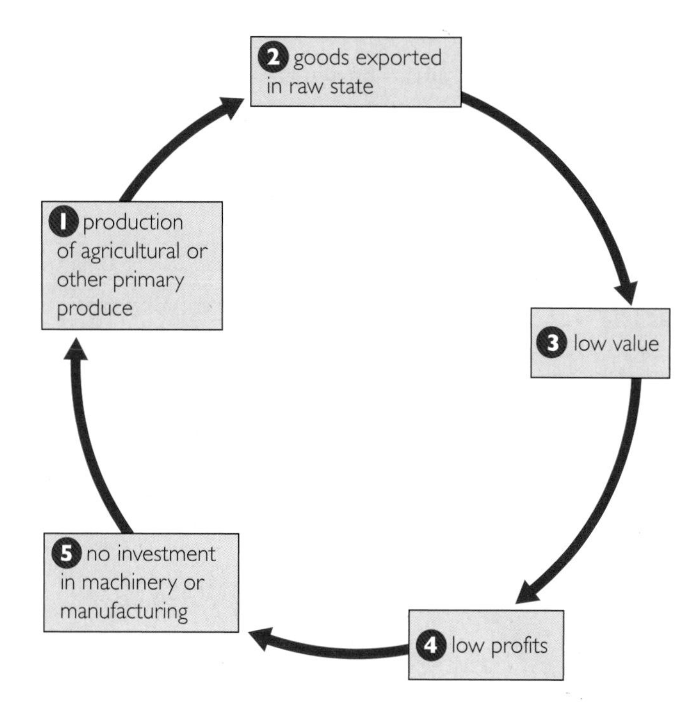

Figure 5.47 *The vicious cycle of poverty* ▲

CASE STUDY

Converting aid into trade – Uganda

After Uganda's independence from Britain in 1962, its economy was strong – it exported commodities including sugar, tea, coffee, cotton and bananas to HICs. During the politically unstable times of Idi Amin in the 1970s foreign investments and export earnings collapsed, leaving a subsistence economy in which farmers mainly grew crops to feed their own families.

After Amin's fall from power, Uganda's economy slowly recovered and growth averaged 6% during the 1990s. Critics argue that this growth has not filtered down to the poor. To counteract this, the current President has emphasised processing agricultural products within Uganda.

The aim is to shift the country away from a primary export, aid-based economy, to one with growing secondary and tertiary sectors. This helps because:

- Manufacturing adds value.
- Fewer major global price fluctuations occur for manufactured goods.
- Manufacturing creates jobs – and this is important because 35% of Ugandans live below the poverty line – mostly due to under-employment or unemployment.

Trade liberalisation

When Uganda accepted loans from the IMF in 1987 to help its economy, it agreed to conditions as part of **structural adjustment** (see page 217). This involved **trade liberalisation**, by reducing import tariffs and export taxes. 82% of Ugandans earn a living through farming, so this had a great impact on them. Many switched from subsistence to cash crops, such as coffee and tea, because at that time prices were high for these crops. However, a crash in the price of coffee in the late 1990s reversed this.

Trade liberalisation didn't lead to more wealth for Uganda's individual farmers. Most Ugandan farms are small, and the country's poor transport infrastructure made it unrealistic for them to export their own produce. As a result middlemen exploited the situation by buying from individual farmers and selling on the produce from several farms at a profit.

Uganda joined the WTO in 1995. The removal of trade barriers should have helped Uganda to export commodities and earn foreign currency. However, when organisations like the EU continue to subsidise their own farmers so they can sell their produce more cheaply, any natural price advantage the Ugandans possess is undermined.

- **Structural adjustment** is a package of cuts in Government spending, trade liberalisation and privatisation, which is imposed on governments seeking help from the IMF.

- **Trade liberalisation** (also known as free trade) means removing barriers such as duties or customs. The theory is that the fewer barriers there are the greater trade will be.

Economic Partnership Agreements

Since 2001, Uganda has been part of a new trade relationship. Economic Partnership Agreements (EPA) are a set of trade rules between the EU, Africa and the Caribbean and Pacific countries. The aim is to promote sustainable development in LICs by bringing them into the global economy.

These agreements mean that Uganda must remove its import duties and subsidies to gain access to EU markets. This should increase Uganda's trade. But, Uganda may suffer because:

- it may not be able to meet demand
- its infrastructure (e.g. banking and transport) is weak
- it is a landlocked country with little port capacity.

It may take time to work, but perhaps global competition will prevent Ugandan industries from ever getting off the ground.

Figure 5.48 *Growers cycle their pineapples to the village market where middlemen buy up the crop* ▼

Does Fair Trade help?

Free trade often means that the factory worker, or grower of a commodity gets the smallest share of the product's value. Fair Trade aims to return a bigger proportion to producers.

Fair Trade is like a cross between aid and trade, as it has its roots in the concept of helping people through **bottom-up development** schemes. In 1990 Sainsbury's supermarket began buying bananas from the Windward Islands, under the Fair Trade scheme. Now all of its bananas are Fair Trade and they are supporting 33 000 producers around the world into new markets. These include peanuts from Malawi, dried pineapples and bananas from Uganda, roses from Kenya and cotton from Tanzania. The Fair Trade system:

- guarantees prices for its producers that are always more than the world market price
- pays a 'social premium'. An extra dollar a case, over and above what is paid to the farmer, goes into a fund to be invested into community projects like crop storage, water supplies, schools, health clinics and road maintenance schemes. Sainsbury's has paid a total Fair Trade premium to banana suppliers of £3.9 million per year since 2007.

- In **bottom-up development** schemes local people have control over projects and improving their lives with the help of technical experts.

Figure 5.49 *Some of the benefits of Fair Trade* ▲

ACTIVITIES

1. In pairs, show the advantages and disadvantages for LICs of encouraging **a**) free trade, **b**) Fair Trade, **c**) economic partnerships
2. Outline the advantages of aid and trade as ways of generating economic growth.
3. In pairs, suggest arguments why all wealthy nations **a**) should donate 0.7% of GDP as aid, **b**) why they should not be forced to donate 0.7%.
4. Make a copy of Figure 5.47, the vicious cycle of poverty diagram, and annotate it (with extra arrows as needed) to show how loans, aid, and trade can help countries move into a virtuous cycle of development.
5. How would you try to persuade **a**) a major supermarket chain, **b**) a local grocer's shop in a village or suburb why they should adopt Fair Trade products. What arguments do you imagine they would use against you?

Internet research

Look up some of the projects that have been undertaken as part of the Fair Trade initiative and identify the strengths, weaknesses, opportunities and threats posed to the local communities, local economies and natural environments. This website will help: http://www.fairtrade.org.uk/ .

5.10 Economic or environmental sustainability?

In this section you will learn about:

- the challenges that face economic and environmental sustainability

Economic sustainability?

The global financial crisis that began in 2007 created problems for HICs. At one point in 2009 world trade was down 90%, millions of jobs had been lost around the world and debt was mounting. In February 2009 President Obama, of the USA, made this speech, raising the development of a **green economy** to the top of America's agenda:

> *"To truly transform our economy, protect our security and save our planet from the ravages of climate change, we need to ultimately make clean, renewable energy the profitable kind of energy".*

He said that America needed to create jobs in the green economy so that it could lead the world after falling behind Japan, Germany and China. He went on to announce that America's economic recovery would benefit from:

- $100bn of investment, making private homes and government buildings more efficient
- $15bn of investment a year developing wind and solar power
- the expansion of public transport
- changing the vehicle manufacturing industry to produce more fuel-efficient cars.

President Obama wanted to drive the USA's economic sustainability by providing jobs, incomes and social well-being.

A sustainable future?

Could developing the green economy be the way forward? In 1992, 178 countries met at the Earth Summit held in Rio de Janeiro. They produced a plan for Sustainable Development in the 21st century and it became known as Agenda 21. Future economic progress combined with greater care for the natural environment was at the heart of Agenda 21. But economic progress and environmental sustainability don't always go hand in hand, and lead to arguments between HICs and LICs, as the comments below from the International Debate Education Association show.

Figure 5.50 *Industrial pollution in Wuhan, China, caused fish to die in their thousands* ▼

The environmental impacts of globalisation

China's economic growth

Can sustainable economic growth be achieved without damaging the natural environment? China's recent rapid growth has created serious environmental consequences. How many of the following are really sustainable, economically or environmentally?

- Subsistence farmers now grow cash crops and have to buy the food they used to grow.
- 20% of China's population lives on less than US$1 a day.
- 70% of China's rivers and lakes are polluted.
- 100 cities suffer from extreme water shortages.
- 360 million people do not have access to safe drinking water.
- Industrial and urban effluent are turning the Yangtze into the world's largest cesspit. The Three Gorges Dam prevents natural flushing of the river.
- Water quality in 207 of the Yangtze's tributaries is not fit for spraying on farmland, yet 1 in 12 of the world's population depend on this river for drinking water.
- Tap water in Chongqing contains 80 of the 101 substances forbidden under Chinese law.
- The Huang He River (one of China's biggest) dried up because of over-extraction.
- 85% of trees along the Yangtze have been cleared, causing erosion and dust storms.
- 30% of China suffers from acid rain caused by emissions from coal-fired power stations.
- China overtook the USA as the largest producer of CO_2 emissions in 2007 – over 6 billion tonnes.

Figure 5.51 *China is trying to clean up its act. In Guangzhou illegal motorbikes are confiscated by the police in an effort to curb air pollution, and are then destroyed at this recycling plant.* ▶

Deforestation, debt and trade

Deforestation, economic growth, trade and debt – they are all linked. Central America once had about 500 000 km^2 of rainforest cover, but by the late 1980s this had fallen to an estimated 90 000 km^2. Over 30% of Honduras' rainforest has been lost since 1960, with more than 800 km^2 being lost every year for ranches, banana plantations, small farms and fuelwood. The expansion of fruit plantations is partly due to the country's need to earn foreign money to repay debt. Exports of hardwood also add to the money earned to repay debt. 20% of Honduras' export earnings is spent on debt repayment every year.

Deforestation and the lack of forest cover means that:

- ecosystems are destroyed
- there is more rapid runoff of rainfall
- there is increased soil erosion and the risk of flooding, and landslides.

"There is no chance whatsoever of saving the rainforest unless the debt crisis is resolved. It's entirely impossible to expect a country which is desperately struggling to meet interest payments, to divert funds to their long term needs, and the conservation of natural resources."
Barbara Bramble (International Director, National Wildlife Federation)

Deforestation and climate change

In the 1980s the plight of the rainforests regularly made headline news. It was recognised that local destruction of rainforests had global effects. Loss of biodiversity became a matter of global concern.

'Debt for Nature Swaps' were one way out of tackling the debt and deforestation issue. They were designed to free up resources in debtor countries for much-needed conservation activities. Debt swaps involve purchasing and converting the debt into local currency, and using the proceeds to finance local conservation projects. Figure 5.52 shows some examples of Debt for Nature Swaps.

Did you know?

Tropical rainforest originally covered 15 million km². About half of that has gone, and it is likely that by the end of this century they will be reduced to 10-25% of their original extent.

Figure 5.52 *Debt for Nature Swaps* ▼

Date	Size of debt swap	Countries involved	Conservation activities
2010	$21 million	USA, Brazil	Protection of the Atlantic Rainforest, Caatinga (dry tropical forest) and Cerrado (grassland) ecosystems, plus development of sustainable livelihoods
2008	$20 million	France, Madagascar	Funding of the Madagascar Foundation for Protected Areas and Biodiversity
2008 (and 2002)	$40 million	USA, Peru	Protection of 27.5 million acres of Peru's biologically diverse and endangered rainforests
2007	$26 million	USA, Costa Rica	Tropical rainforest conservation activities
2006	$24 million	USA, Guatemala	Local conservation activities
2004	$10 million	USA, Colombia	11 million acres of tropical rainforest protected
2002	$25 million	France, Cameroon	Protection of part of the world's second-largest tropical rainforest

As the world turns to green economics, there are now new concerns:

- The EU intends to source 10% of its fuel from African palm oil plantations by 2020. But, many now see biofuels as more damaging than the fossil fuels they replace.
- Forests have been cleared to fund development projects.

Deforestation has major impacts on climate change:

- Burning forests accounts for 25% of global carbon emissions (the next-highest source after burning fossil fuels).
- Deforestation means the loss of a major carbon sink.
- One day's burning of a rainforest is equivalent to 8 million people flying across the Atlantic.
- Indonesia and Brazil became the 3rd and 4th-largest emitters of greenhouse gases in 2007 through burning forests.
- The annual global rate of destruction equals the area of England, Scotland and Wales, releasing 2 billion tonnes of CO_2.

CASE STUDY

Reducing deforestation in Africa

By the side of many roads in Malawi, charcoal and fuelwood are for sale – one of few opportunities for poor households to earn an income. But this is an environmentally unsustainable practice. Damage to, or felling of, trees, causes soil erosion, which in turn leads to food insecurity as fertile soil is lost.

However, since 2005, a national tree-planting scheme has helped Malawi's forest reserves. Until the trees are large enough to harvest, cleaner fossil fuels such as LPG offer an alternative fuel which takes pressure off the forests, and reduces greenhouse gas emissions.

In Kenya and Zimbabwe traditional building bricks are made and baked by burning fuelwood, leading to deforestation. But now, low-cost, low-energy building blocks are being made from local soil mixed with cement. Water is added and the mix placed in a block press. The bricks are then dried in the sun. The process uses little water and produces no waste.

Other environmental impacts of economic growth

Our economic growth is destroying entire ecosystems. At current rates of economic growth we will need two planets' worth of natural resources by 2050. The Earth is not able to regenerate supplies at the rate they are being used. This raises some serious issues:

- As more countries develop, there will be more pressure on resources.
- Emerging economies will demand more of the world's resources.
- Competition for resources will push prices up and undermine development.
- Scarcity of resources creates insecurity.
- How can resources be managed and distributed equitably?

Environmental degradation at a local level

At a local level people make decisions which have an environmental consequence. The decisions made by indigenous people are often a matter of survival. Unaware of decisions made at a global level about commodity prices, debt repayments and development projects, they may appear to be guilty of environmental destruction when it is not their fault at all.

Direct causes of environmental degradation	Indirect causes of environmental degradation
• Population growth • Agricultural expansion • Logging for domestic purposes – fuel wood, building • Clearance for roads and infrastructure • Accidental fire	• Poverty – driving people to 'sell' forest products • Debt repayment and Structural Adjustment Policies • Foreign demand for hardwoods – furniture, buildings, toilet seats! • Foreign demand for commercial crops – soybeans, cattle ranches, plantations for tropical fruits and biofuels

Figure 5.53 *Local causes of environmental degradation* ▲

ACTIVITIES

1. Define the term debt-for-nature-swaps.
2. Why is the green economy seen as being economically and environmentally sustainable?
3. **a** In pairs, draw spider diagrams to show the environmental impacts of globalisation.
   b Identify which impacts are deliberate, accidental, destructive, or neutral.
4. Take each direct cause of environmental degradation in Figure 5.53 above.
   a Draw a web diagram to show how each is linked – e.g. how population growth can lead to logging for domestic purposes.
   b Add explanatory notes to each link that you make
5. Now repeat question 4 but for indirect causes of environmental degradation in Figure 5.53.
6. Why can debt be associated with environmental destruction? Are there ways of avoiding this? Discuss your ideas in pairs.
7. In 1000 words, discuss the idea that 'economic growth is only possible with considerable environmental costs'.

Internet research

1. In class, divide into pairs and draw lots – take one direct and one indirect cause of environmental degradation shown in Figure 5.53.
   a Research one country with significant areas of rainforest (e.g. Peru, Brazil, Indonesia, Cameroon) and how far each of the two causes has contributed to environmental degradation there.
   b Present your findings and assess how far other countries that the class has researched are affected by these same causes.
2. Select one major environmental issue: **Deforestation, desertification, soil erosion, water pollution, air pollution, biodiversity loss.**
   Research the extent of the problem in two countries at different levels of development.
3. Write a 500 word report entitled: "The environmental costs of globalisation are too high."

Sustainable tourism – myth or reality?

In this section you will learn about:

- whether sustainable tourism can be a reality

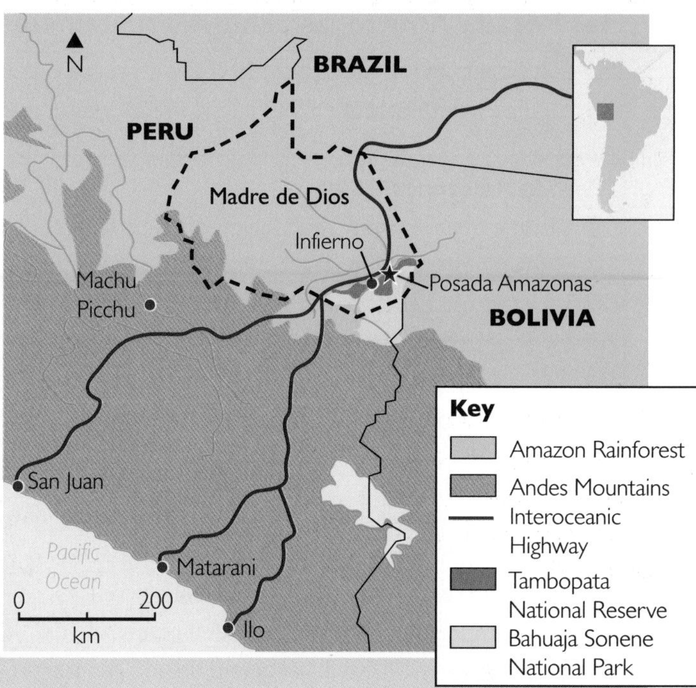

Figure 5.54 *The location of Posada Amazonas* ▲

CASE STUDY

Save the rainforest!

The *Posada Amazonas* eco-lodge is located in the Amazon rainforest of south-eastern Peru – much of it protected by the Peruvian government. The Bahuaja Sonene National Park was established in 1990, to protect 1 million hectares of rainforest from any development. Tambopata National Reserve of 275 000 hectares (also established in 1990) is located next to the National Park. Some limited, small-scale sustainable development is permitted within the Reserve, and Posada Amazonas was opened in 1998 on its border – within land owned by the local indigenous community.

Posada Amazonas is a 30-room eco lodge owned by the local Ese'eja community of Infierno, and managed in partnership with Rainforest Expeditions, a local Peruvian company. The ecotourism project offers tourists an insight into the forest and also boosts the local economy.

Sustainable tourism, when successful, can create new forms of income for local communities, while also encouraging care of the environment. Instead of cutting down trees for timber, the forest can be marketed as a resource for adventure and education. Keeping the number of tourists down to just 40-50 arriving by small boat limits their environmental impact. In this way, tourism can begin to become sustainable, and valuable landscapes can be marketed many times over without being destroyed.

However, by effectively selling their identity, culture, way of life and nature it is clear that the indigenous people have to alter their values. At Posada Amazonas, there is currently a profit-sharing agreement – with the Infierno tribal community receiving 60% for providing knowledge, labour, culture and access to 10 000 hectares of tribal lands, and Rainforest Expeditions receiving 40% for managing the operations and staff. The Infierno community will take over total responsibility for Posada Amazonas in 2016.

Ecotourism offers an alternative route to development without destroying the very things visitors go to see. In Infierno, television and radio had already given the community a view of the outside world and the Posada Amazonas project does not set out to deny anyone the chance of an improved life – levels of literacy, healthcare and nutrition have all improved as a result of the benefits brought by this project.

Figure 5.55 *Posada Amazonas eco-lodge* ▼

Posada Amazonas eco lodge:

- was constructed from local, natural materials
- avoids the use of non-renewable materials and uses recycled material where possible
- fits in with its surroundings
- offers educational, conservation and research facilities.

Has Posada Amazonas achieved sustainable tourism?

The text boxes on this page provide you with some resources to help you to judge whether tourism at Posada Amazonas is sustainable.

Posada Amazonas, 2000

As income from ecotourism conflicts with earning a living from more traditional activities, it is not surprising that 4 years after the project was established the normal subsistence activities seem to be subsiding. Equally, modern tools are being used to cut trees and cultivate the land, and motorboats replace canoes.

Handicrafts are sold as souvenirs and the forest children can now see the economic value of local fauna and flora. Tourism places a value on natural species and, as more tourists arrive so their value increases – along with the significance of protecting them. Ecotourism adds to sustainability and the indigenous community are stakeholders in this.

Local farmers gain as their market now extends to both tourists and those locals who no longer have time to farm, although there is a need to designate land for domestic production. A proposed new road designed to boost development was rejected by the community because they feared the impacts on wildlife – the very attractions that tourists come to see.

Defining Ecotourism

Ecotourism is 'responsible travel to natural areas that conserves the environment and improves the well-being of local people'. So ecotourism activities should:

- minimise impact
- build environmental and cultural awareness and respect
- provide positive experiences for both visitors and hosts
- provide direct financial benefits for conservation
- provide financial benefits and empowerment for local people
- raise sensitivity to host countries' political, environmental, and social climate
- support international human rights and labour agreements.

Posada Amazonas, 2008

Rainforest Expeditions shares the management of Posada with the elected 'control committee' from Infierno. Eduardo Nycander, one of the founding members of Rainforest Expeditions listed the following benefits of the Posada Amazonas to the Infierno community and the area by 2008:

- Profits of $130 000 in 2007, plus $140 000 in wages – because most of the staff are from the local community.
- The provision of training programmes in readiness for full handover of control to the community in 2016.
- Improved literacy, healthcare and nutrition levels in Infierno.
- Reduced levels of hunting in the rainforest because of the income received from tourists and the value of wildlife for the success of the project.
- Keeping the rainforest unspoilt and undegraded.
- Benefits to conservation and social development as the profits are retained locally.

Nycander believes that, by protecting his own interests he is helping conservation and making money at the same time. But he warns of potential problems ahead. The success of the 70+ eco lodge projects across Peru's Madre de Dios region is leading to improvements in local infrastructure.

The road that was previously rejected has become a reality. Road crews are completing an upgrade of the old dirt tracks and by 2010 the last 700 km will have been paved to form the Interoceanic Highway to Brazil. (The Highway was opened in January 2011.) This upgraded link between Peru's Pacific coast and Brazil will reduce journey times from 3 days to 1 and open the area up to more visitors. It will run just 15 km away from Posada Amazonas.

Traditionally, new roads through the rainforest tend to lead to development and rainforest destruction up to 50 km deep on either side of the road – through deforestation for logging, mining and agriculture. Nycander hopes that the creation and promotion of an ecotourism corridor alongside the new road will lead to the preservation of 150 000 hectares of forest which would otherwise be under threat.

Tourism and its impacts

This page provides you with some information on the impacts tourism has. Given these impacts, can tourism be sustainable?

Changing destinations – changing impacts

The development of the tourist industry inevitably changes the character of a tourist destination. Benidorm's high-rise hotels, which represented its success in the 1960-80s, later appeared tacky and down-market. Butler's Life Cycle model (Figure 5.56) explains how Benidorm, once successful, reached saturation point, and a lack of investment resulted in its deterioration. Meanwhile other destinations were catching up and improving on Benidorm as an experience, so its decline became inevitable.

Any tourist destination is at risk from this pattern of development. Sustainable tourism may be a way of safeguarding the natural environment, but if it is not economically sustainable everything could fail.

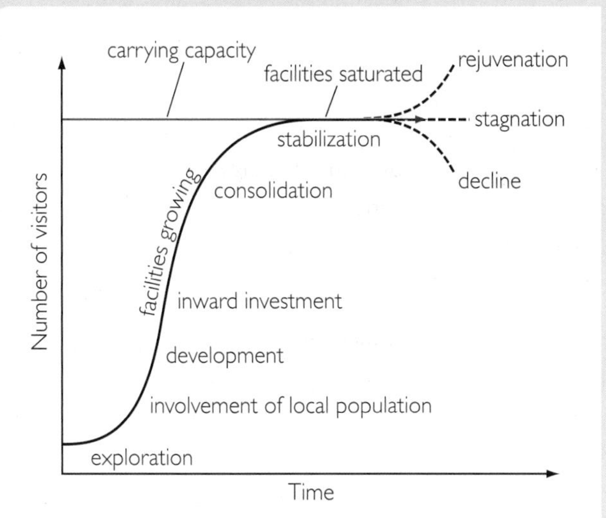

- The exploration and early growth phase involves small numbers of visitors who enjoy real contact with local people. Few changes are experienced and the local community has full control. The visitor is exploring and opening up a new territory.
- The involvement phase sees benefits from investment, and from this point on, a growth in numbers inevitably draws in non-local interests and significant change to local characteristics.
- Further growth brings the arrival of mass tourism.

Figure 5.56 *Butler's tourist destination life cycle* ▲

Pushing the boundaries

Increasing leisure time, higher disposable incomes, greater personal mobility and cheaper travel mean that more people have more time to travel further for pleasure. In effect, the boundaries created by travel costs and journey times have been broken and tourists now reach areas that were once considered remote and inaccessible. The rural-urban fringe and coastal areas little more than an hour from home used to be on the boundary. Those peripheral zones are now being leapfrogged as tourists demand new experiences further away. The furthest distance which tourists travel is known as the **pleasure periphery**.

Wearing out our welcome

Tourism brings with it a range of other impacts:

- Visitors discard rubbish in environmentally sensitive areas.
- The development of attractions such as golf courses, uses up land and destroys habitats.
- Tourists can have detrimental effects on wildlife and vegetation.
- Driving to, and within, sensitive areas creates air and noise pollution, burns fuel, and disturbs the environment.
- Air travel contributes to air and noise pollution.
- Hotels and theme parks can create more sewage and rubbish than areas can handle.
- Cruise ships are criticised for illegally dumping rubbish, fuel and human waste.

ACTIVITIES

1. Define these terms: ecotourism, pleasure periphery, and describe Butler's tourism life cycle.
2. Make a list of the key aspects of the Posada Amazonas project that suggest **a**) it could be sustainable, **b**) it might not be sustainable.
3. From your own knowledge, explain why such schemes sometimes fail.
4. In pairs, discuss whether tourist destinations can ever be totally sustainable.
5. Write 1000 words to discuss 'Sustainable tourism – myth or reality?'

bottom-up development Where decisions about development are made by local communities

colonialism The rule of one weaker power by a stronger power, including take-over of its government

commodity Raw material items, usually either farm crops or minerals

core and periphery A theory that shows how different economic development between regions leads to a prosperous 'core' region and a poorer 'periphery'

demographic Relating to population

dependency theory A theory by economist André Frank, which shows how economic development of core regions occurs at the expense of the peripheries. The core depends on raw materials from the periphery areas, while the periphery depends on the core as a market for its goods

development Socio-economic change which aims to improve wealth and standards of living

development continuum The span of levels of economic development, from poorest to wealthiest countries

development gap The differences between poorer countries of the developing world (or LICs), and wealthier developed countries (or HICs)

export processing zones A type of Special Economic Zone (see separate entry) where businesses are free to import raw materials, process, and manufacture them, and re-export without paying duties or tariffs

export-oriented industries Those industries established largely for the purpose of increasing exports

Fair Trade A system of trade whereby producers are paid fair prices to give them a reasonable standard of living

G8 The group of the eight largest economies in the world (Russia, the USA, UK, France, Canada, Germany, Italy, and Japan)

GDP (and GDP per capita) Gross Domestic Product (and per person). Usually given in US$, this is a figure which sums the total value of all goods and services produced in a country in a year. It may either be expressed for a whole country, or per capita (per person) by dividing the total by the population

globalisation The way in which people, cultures, money, goods, and information 'move' between countries with few, or no barriers

green economy An economy based on sustainable development, e.g. low carbon use

Green Revolution A package of mechanised farming techniques and high yielding seeds to enable LICs to feed their growing populations

Human Development Index (HDI) A UN index with a range between 0 (worst) and 1 (best), which measures life expectancy, knowledge (literacy and number of years spent at school), and standard of living (GDP per capita)

IMF The International Monetary Fund, an organisation which funds development in poorer countries, using bank deposits from wealthier countries

just-in-time A process used by transnational companies to reduce stock, so that goods are produced just in time before sale, rather than being held in warehouses

modernisation theory A theory stating that capitalism generates economic development in ways that will benefit all; it dates from the 1950s-70s when US investment was used in South-East Asia to prevent the expansion of Russian communism

new economy Also known as the 'knowledge economy', which is based on creativity and specialised expertise in finance, media, and management, rather than manufacturing goods

out-sourcing The employment of people overseas to do jobs previously done by people in a HIC. Usually associated with IT software development, banks, and service companies, e.g. call centres

purchasing power parity (PPP) The relationship between average earnings and prices, and what it will buy, because a dollar buys more in some countries than others. It expresses the spending power within individual countries and reflects the cost of living

Special Economic Zone Areas in which governments offer tax incentives for foreign companies to build new factories there. Usually found in NICs

tariffs Also known as 'duties'. Charges imposed on the import of goods from certain countries

top-down development Where decisions about development are made by organisations, e.g. government or large companies, and imposed on a population

trade liberalisation Also known as free trade – means removing barriers such as duties or customs. The theory is that the fewer barriers there are to the flow of goods, the greater trade will be

World Bank An organisation set up after World War II to promote investment globally. It provides loans for countries who agree to conditions; like the IMF

WTO (World Trade Organisation) A group of (in 2011) 143 nations agreeing to trade with each other without the use of tariffs or duties. It deals with the rules of global trade, with the aim of easing trade and getting rid of anything hindering it

Section B

1 (a) Study Figure 5.6 (page 204). Briefly explain how the patterns shown have led to a 'development gap'. *(7 marks)*

(b) Explain the impacts of relying upon primary products for developing countries. *(8 marks)*

(c) Explain why many developing countries are now industrialising rapidly. *(10 marks)*

2 (a) Study Figure 5.22 (page 216). Comment on the location of the world's Least Developed Countries. *(7 marks)*

(b) Explain why the countries in Figure 5.22 find it hard to develop. *(8 marks)*

(c) Explain the effects of either debt reduction or changes to global trade upon the world's Least Developed Countries. *(10 marks)*

Section C

1 Referring to examples, assess the impacts of rapid industrialisation on the Newly Industrialising Countries. *(40 marks)*

2 Explain how far globalisation has brought benefits or problems to places you have studied. *(40 marks)*

EXAMINER'S TIPS

(a) Think about the value of the different goods and services that 'flow' between different parts of the world.

(b) Think about over-reliance, and the reasons for and impacts of price changes in commodity markets.

(c) Consider different reasons, e.g. growth of manufacturing resulting from cheaper labour, relief from debt, changing trade links. Use examples.

(a) 'Comment on' means 'describe', but also providing reasons. Think about debt.

(b) Consider the effects of prolonged debt and how global trade works against the poorest countries.

(c) Consider the effects of debt reduction on countries in this chapter, e.g. Uganda, or of trade patterns and how these can bring benefits and problems to the Least Developed Countries.

(1) Try planning what you will say first. Think about different impacts of rapid industrialisation, e.g. economic, social, and environmental. Remember that to hit the top levels you must assess – i.e. weigh up which impacts are greatest and least, and why.

(2) Consider a range of places here (e.g. Uganda, China) to maximise the range of benefits or problems. Because the question asks 'how far', you must assess – i.e. weigh up which benefits or problems are greatest and least, and why – to hit the top levels.

6 Contemporary conflicts and challenges

Soldiers in action

Look at the photo. What's happening?
Where might the photo have been taken?
How might the local people be feeling?
Where in the world are people fighting today?
Why is there conflict in these places?

Introduction

What can Geography have to do with conflict? Everything!

For example, take water. Throughout history, people have come into conflict over this vital resource – because it is unevenly spread. Today, it's one of the issues at the heart of Israel's relations with its neighbours, and it impacts California's relations with neighbouring states in the USA. Conflict over water is likely to increase as the world's population grows and consumption goes up. And now think of oil.

There are many types and scales of conflict. In the UK, debate began in 2011 about a proposed high-speed rail link between London and Birmingham, which passes through some of southern England's most scenic – and middle class – areas.

Beyond specific conflicts, global poverty continues to be one of the major challenges facing the world.

In this chapter you will learn how Geography can explain and help us understand conflicts. And you will find out about the issue of global poverty.

Books, music, and films

Books to read

Afghanistan – A Cultural and Political History by Thomas Barfield
Injustice by Daniel Dorlin
Open Shutters Iraq by Eugenie Dolberg

Music to listen to

'American Idiot' (album) by Green Day
'If White America Told the Truth for One Day Its World Would Fall Apart' by Manic Street Preachers
'We'll Live and Die in These Towns' by The Enemy
'England's Dreaming' by Cornershop

Films to see

United 93
9/11 (a documentary, 2002, CBS Television)
The Hurt Locker
Green Zone
Water First – reaching the Millennium Development Goals

About the specification

'Contemporary conflicts and challenges' is one of three Human Geography options in Unit 3 Contemporary Geographical Issues – you have to study at least one.

This is what you have to study:

The geographical basis of conflict

- The nature and origin of conflict: identity (nationalism, regionalism, localism), ethnicity, culture, resources including territory, and ideology.
- The patterns of conflict: national, regional, and local.
- The expression of conflict: non-violent, political activity, debate, terrorism, insurrection, and war.
- The resolution of conflict.

Conflict over the use of a local resource

- Identification of the local resource, e.g. land, buildings, space.
- The reason for the conflict, and the attitudes of different groups of people to the conflict.
- The processes which operate to resolve the conflict.
- Recognition that some people benefit, whereas others may lose, when the outcome is decided.

The geographical impact of international conflicts

- The social, economic, and environmental issues associated with major international conflicts within the last 30 years.
- One or more case studies, e.g. the conflicts in Gaza and the West Bank in the Middle East, Afghanistan, the Darfur region of Sudan.

The challenge of multicultural societies in the UK

- Multicultural societies: reasons for their development.
- The geographical distribution of cultural groupings.
- Issues related to multicultural societies.

Separatism within and/or across national boundaries

- The nature of separatism.
- Reasons for separatism.
- Consequences of separatism.

The challenge of global poverty

- The global distribution of poverty.
- The causes of poverty.
- Addressing poverty on a global scale, including the work by international agencies such as the UN.
- The issue: 'No development without security, and no security without development.'

6.1 Why do geographers study conflicts and challenges?

In this section you will learn about:

- why geographers study conflict
- how conflict can arise and the scale at which it operates
- how conflict varies, from non-violent protest through to war

Qatar, October 2010. Talks are taking place to try to end a civil war in Sudan's western Darfur region which has lasted for seven years (Figure 6.1). During this time, the United Nations estimates that 300 000 people have been killed and 2.7 million people forced to leave their homes. The Sudanese government disagrees; it claims 10 000 have died and accuses the UN of exaggerating problems in Darfur.

Progress in the peace talks between the government and rebel groups varies – and so too does the likelihood for peace.

Meanwhile, in southern Cornwall, a local enquiry is leading to further debate about a proposed property development known as *The Beach* (Figure 6.2). Some local people want the development to go ahead, believing it will improve the area and bring high-quality, year-round tourism to an area where incomes are low and seasonal tourism has led to high winter unemployment. But other residents are concerned that the development will deny them access to a popular beach which, while privately owned, has traditionally been used by local residents. If the development goes ahead, residents fear they will no longer be welcome.

Figure 6.1 *Rebel groups in Darfur, in western Sudan* ▲

Figure 6.2 *An artist's impression of the proposed property development 'The Beach'* ▼

What can these two scenarios possibly have in common? And what do they have to do with geography?

- They are each about conflict and challenges in different **places**.
- They are each about **space**. Whose space? What kind of space? Which of the groups involved in the dispute should control the space?
- Though at very different scales, they each involve people trying to **resolve disputes**. Depending upon which resolution succeeds, managing the differences between groups will have implications for different people.

The process is, therefore, the same – conflict has brought people together to fight for a cause in which different groups take opposing views.

Contemporary conflicts and challenges

Conflict can be defined as a '*disagreement caused by the actual or perceived opposition of needs, values and interests between people*'. It does not necessarily mean fighting; it can simply be an expression of different opinions.

Conflict usually involves a clash of ideals and/or actions which can vary on a scale – or **continuum** – between non-violent (such as political activity and debate) to direct action and violence (for example, terrorism or war).

- Politically, conflict refers to a state of hostility between two or more groups of people.
- Geographically, conflict can result from opposing views about ways in which spaces or resources are used and by whom.

Geographers are especially interested in **scales** of conflict, for example, whether the conflict is international, national, regional or local in nature.

What causes conflict?

Most conflicts are complex, but their causes can be classified fairly easily regardless of whether the conflict is about water resources, boundary disputes or religion.

Different people and groups have different outlooks, from local pressure groups fighting a decision by a local council to major political parties seeking to win an election. Once those outlooks are categorised, conflict becomes easier to understand. In general, outlooks vary according to a person or group's:

- shared beliefs and characteristics
- geographic scale of influence.

Shared beliefs and characteristics

Groups differ from one another according to their:

- **Identity** – having a common a set of attitudes or experiences
- **Ideology** – having different ideals and beliefs than others
- **Ethnicity** – having a separate identity from a majority culture, for example, through religion or language
- **Culture** – sharing a language, arts, values and behaviour
- **Wealth** – having more or less wealth than others.

Groups also vary according to the ways in which they are financed, from those that rely entirely on donations to those that are financed by membership fees or grants.

Geographic scale of influence

Groups may also be categorised by their focus on a particular geographic area:

- **Localism** – where issues or values bring local people together, such as against government closures of local services
- **Regionalism** – where, for example, one region tries to take control of functions from a national government
- **Nationalism** – where a particular country asserts its dominance over other people or nations
- **Territorialism** – where a geographic area is defined for a particular purpose, such as for exploration or for finding natural resources such as oil.

Figure 6.3 *Greenpeace – a pressure group fighting local issues about emissions from this power station in Kent, or an ideological group concerned about global climate change?* ▼

Many conflicts have multiple causes, for example, in Northern Ireland where religious beliefs and differences compound what has historically been a conflict about inequalities of wealth. Rarely do conflicts have a single cause.

Look at Figure 6.4 – it lists some of the common geographical causes of conflict, such as the access to water or the right to cross borders, that occur throughout the world.

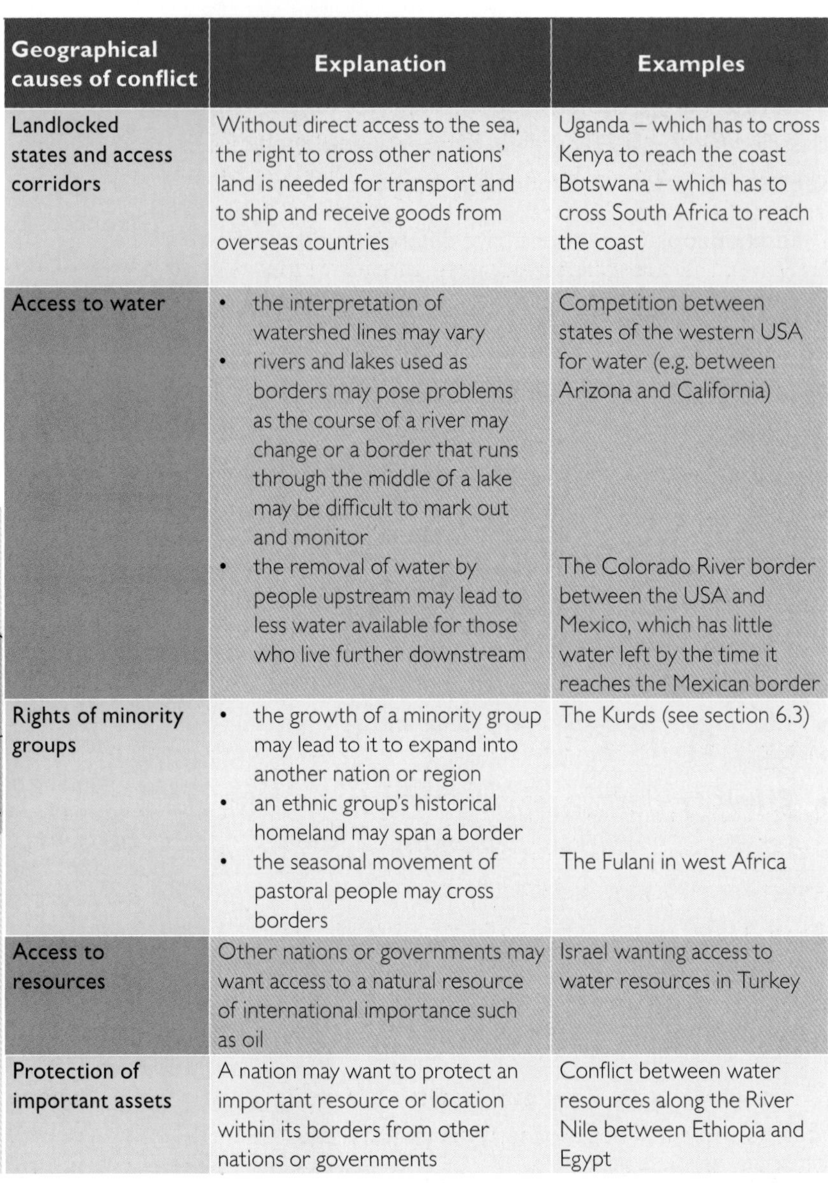

Geographical causes of conflict	Explanation	Examples
Landlocked states and access corridors	Without direct access to the sea, the right to cross other nations' land is needed for transport and to ship and receive goods from overseas countries	Uganda – which has to cross Kenya to reach the coast Botswana – which has to cross South Africa to reach the coast
Access to water	• the interpretation of watershed lines may vary • rivers and lakes used as borders may pose problems as the course of a river may change or a border that runs through the middle of a lake may be difficult to mark out and monitor • the removal of water by people upstream may lead to less water available for those who live further downstream	Competition between states of the western USA for water (e.g. between Arizona and California) The Colorado River border between the USA and Mexico, which has little water left by the time it reaches the Mexican border
Rights of minority groups	• the growth of a minority group may lead to it to expand into another nation or region • an ethnic group's historical homeland may span a border • the seasonal movement of pastoral people may cross borders	The Kurds (see section 6.3) The Fulani in west Africa
Access to resources	Other nations or governments may want access to a natural resource of international importance such as oil	Israel wanting access to water resources in Turkey
Protection of important assets	A nation may want to protect an important resource or location within its borders from other nations or governments	Conflict between water resources along the River Nile between Ethiopia and Egypt

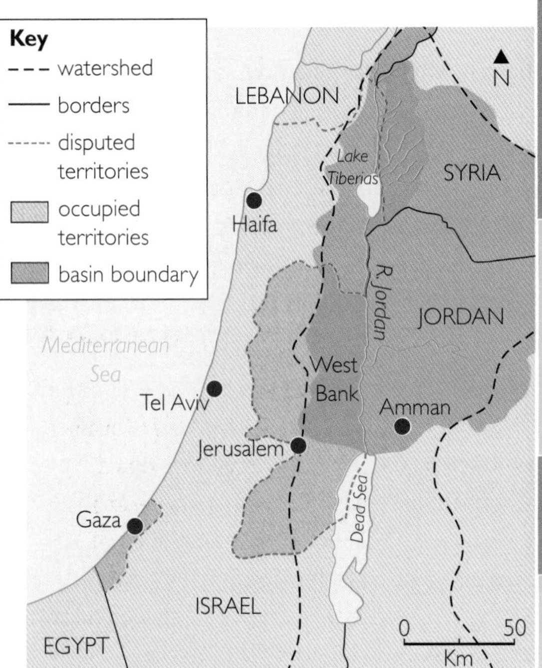

Figure 6.4 *Geography and conflict. Note on the map the number of different political states sharing the same water resources – a situation that has led to decades of conflict.* ▲

It is important to remember that depending on how it is handled, conflict can also be positive. For example, it can be an important driver for change by making people aware of issues or situations, such as how resources are being managed or pollution monitored.

What forms can conflict take?

Regardless of the scale at which they take place, conflicts have common features and can develop in similar ways. It helps to see conflict as:

- varying between peaceful and violent
- a progression between two extremes: peaceful debate and war.

Figure 6.5 *A student protest against tuition fees in London during November 2010. Is this a protest or an uprising?* ▼

Figure 6.6 shows this progression as a continuum, one that applies at different scales. For example, at the local level problems with the opening of London's Heathrow Airport Terminal Five (see page 254) ended at the 'Action' stage with a judicial inquiry. Internationally, the conflict in Afghanistan (discussed in detail in section 6.4) eventually reached the 'Armed conflict' stage – a war.

At every stage, progression along the continuum towards war is likely only if previous efforts fail to draw people together. Equally, at every stage conflict escalation is also possible.

Figure 6.6 *A continuum model of conflict showing a scale of escalation between conflict avoidance and armed conflict. The vertical lines show points at which the nature of conflict can change.* ▼

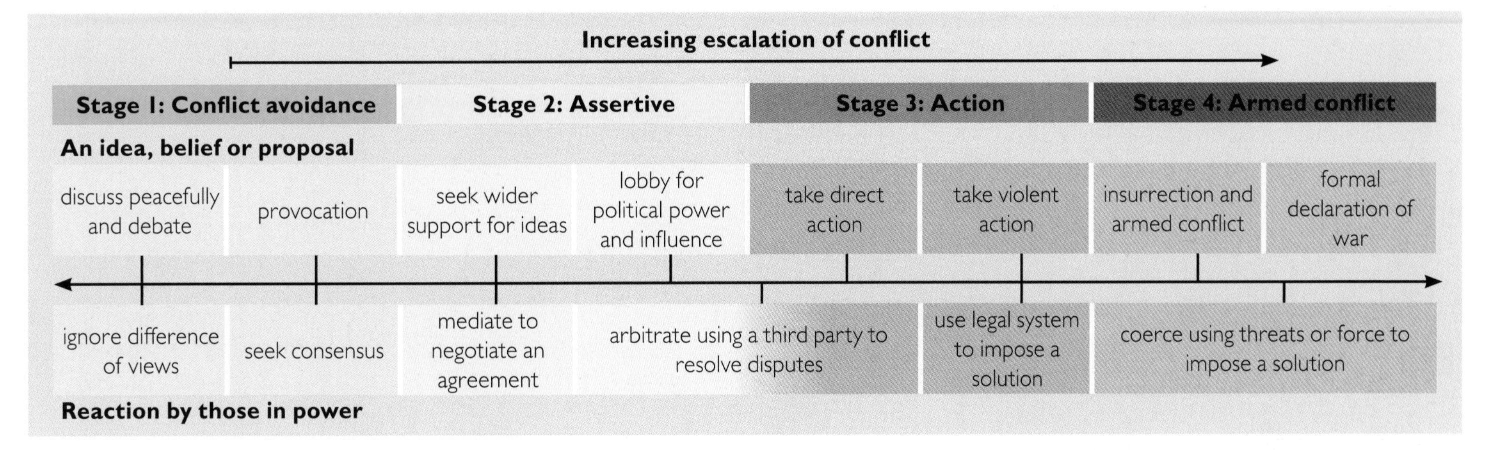

Stage 1: Conflict avoidance

Many conflicts begin as an issue or disagreement, perhaps over the use of a piece of land or access to a valuable resource such as water or oil. They can occur at every scale, from a dispute about the location of a fence between neighbours to a disagreement over water resources between nations.

What begins as a discussion and conflict avoidance may well end there without provocation or threat. Equally, it may become a refusal to accept the other's point of view. Discussion and examination of the issues usually causes views to be expressed peacefully with many conflicts forever remaining at the conflict avoidance stage.

Stage 2: Assertive

This stage in a conflict occurs when a group chooses to put forward its case more strongly and to gain support from a wider public. For example, at a local level, a residents' group might oppose a housing development with the view of preventing it by:

- writing to newspapers
- persuading MPs to support their viewpoint
- taking action such as organising demonstrations.

Nationally at this stage government leaders are lobbied by individuals or companies promoting a particular viewpoint. Councillors and MPs receive hundreds of letters each year about disputes, each trying to persuade them about 'correct' opinions and seeking their help or support. This stage may eventually require arbitration by a third party such as a local council.

Some groups go further by seeking political power, locally or nationally. Each group (or political party) develops a programme to define its beliefs (known as a manifesto) and set out what it would do in office. Much of national government is about working at this stage.

- Domestically, governments draft and develop legislation which is enacted by a parliament or national assembly to help resolve conflict and reflect majority beliefs.
- Overseas, ambassadors and government ministries try to ensure that peace and good relations are maintained.

Stage 3: Action

Sometimes, peaceful actions and lobbying can fail to bring about change. Intense disagreement can continue and lead to people taking direct action or even violence.

- Within a country, violent uprisings can occur against civil authority, established government, or the rules of that government – this is known as an insurrection.
- Governments may respond with force, such as using armed troops, or in more hidden ways as was done by the secret police in the former USSR (KGB) who relied on the use of fear to maintain control.
- At an international scale there may be pressure to send troops. Domestic conflict at this stage is often financed with financial or military support from a sympathetic overseas government.

Stage 4: Armed conflict

This is the most extreme form of conflict where armed, hostile combat develops within or between nations and where, typically, armed forces are used to defend the interests of governments. As more nations become involved, the prospect of global conflict emerges – world war has occurred on two occasions with every continent being involved to some extent. Although the last world war ended in 1945, as Figure 6.7 shows, there is still considerable armed conflict between different nations today.

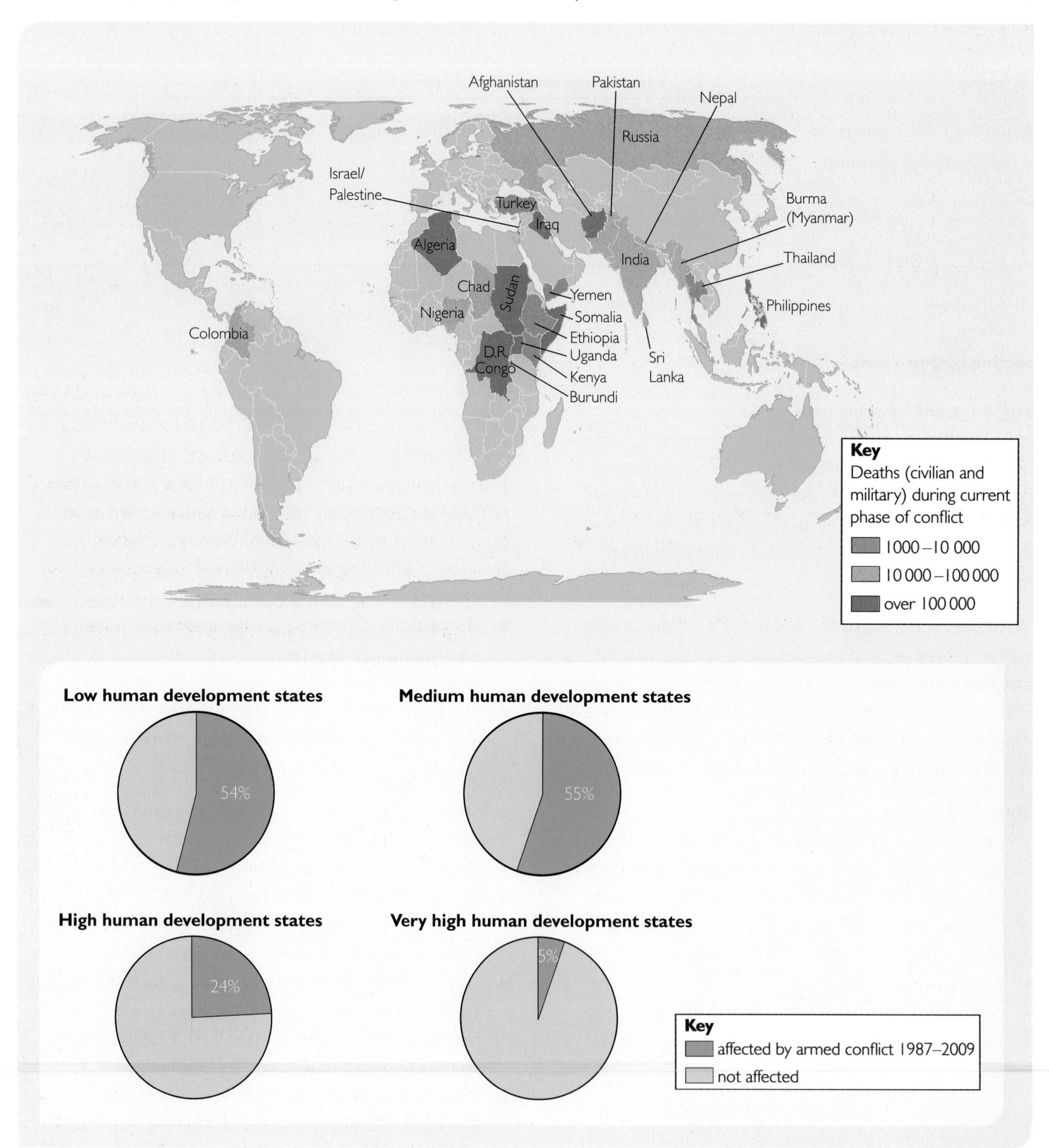

Figure 6.7 *Armed conflicts in the first decade of the 21st century* ▲

Resolving conflict

In the same way that conflict can be seen as a continuum, a range of actions can also be taken to resolve conflict – Figure 6.8 below shows some of these actions.

Figure 6.8 *A continuum of conflict-resolution actions. Note that the actions move along the continuum as the seriousness of the conflict increases.* ▼

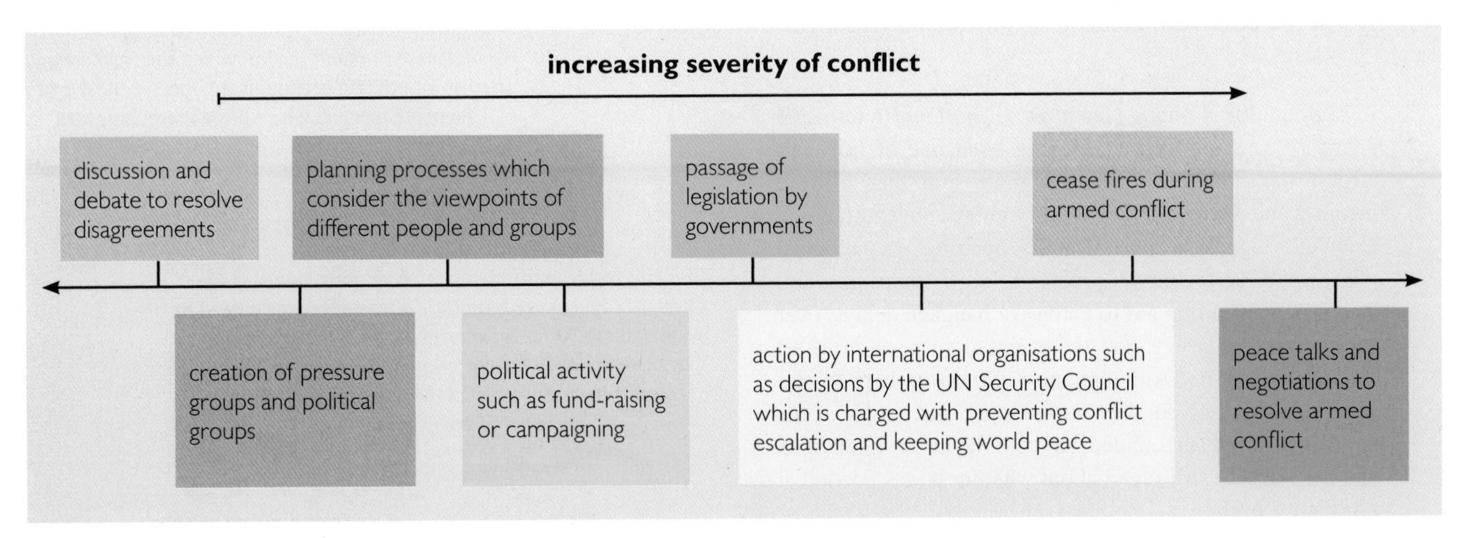

Maintaining peace: the United Nations

Maintaining peace has its costs, but is less than the cost of war. The most important global institution in creating and maintaining peace is the United Nations (UN). It was designed to help manage inter-state wars, though now deals mainly with intra-state wars. Responsibility for the maintenance of peace within the UN lies with the Security Council.

The UN aims to bring the world's nations together to work for peace based on justice, human rights and the well-being of all people. Its basic premise is that global powers can work together to maintain global peace. However, UN peacekeeping has involved other activities, e.g. monitoring or running elections, and delivering humanitarian relief supplies. It has three main areas of work – conflict prevention and resolution, sustainable development, and human rights. A number of UN agencies exist, such as the Food and Agriculture Organisation (FAO) and the World Health Organisation (WHO).

ACTIVITIES

1 Complete the table for each of these groups (you may need to do additional research):

Scottish nationalists, Greenpeace, local protest groups (e.g. about a hospital closure or housing development), campaigners against London Heathrow Airport's third runway, Al Q'aeda, Israeli Jews, and two other groups of your choice.

Group	Common beliefs and characteristics (identity, ideology, ethnicity, culture, wealth)	Scale of geographic influence (local, regional, national, global, territorial)	Actions taken to achieve their ends

2 Form small groups of 3 – 4 and brainstorm the ways in which people are trying to resolve conflict. Use examples from all geographic scales (local, regional, national and international). Draw your answers in a spider diagram.

Internet research

In pairs, research **a)** the aims and work of the United Nations (visit un.org) and **b)** conflicts in which the UN is involved at the moment (visit the BBC News website – bbc.co.uk/news – and use the search facility). How far does its involvement in current conflicts match its intended aims?

Internet research

In pairs, create a short PowerPoint presentation exploring one of the groups listed in Activity 1. Include information about its beliefs, origins, aims, how it's financed, its influence, actions taken to achieve its aims, and whether it has been successful.

6.2 Heathrow Terminal 5 – a local conflict

In this section you will learn about:

- the reasons supporting the building of a new terminal (Terminal 5) at Heathrow Airport
- how different people and groups reacted to the proposal and why
- whether the building of Terminal 5 was a good decision

In March 2008, London Heathrow Airport's fifth terminal – known as T5 – opened. Having been one of Europe's largest building sites, its size, scale and design were meant to be a 21st century quality entry into the UK via London's biggest airport. But its opening was plagued with troubles: technical problems, staff shortages and lack of staff training led to extensive baggage delays, over 30 flight cancellations and the eventual suspension of all customer check-ins. Although most of these early problems were quickly resolved, they served to remind people of the conflict that had surrounded T5 from the onset. Even at its opening, debate still raged about whether it was needed at all (Figure 6.9).

It will be interesting to see how the government proposes to get these extra passengers into the new terminal. I use the motorways around Heathrow and they're inadequate to cope with the current traffic volumes. At the moment, there are six lanes where the M4 meets the M25, how many more do they propose to add to cope with the extra volume?

The approval of T5 means more air traffic, more pollution, more noise, more overcrowding in areas that are already overwhelmed by all of these. As for the economic side of the argument, as there are already four terminals here, couldn't an airport in another part of the country be allowed to benefit from expansion?

I think it is really a case of expand or die. The decision to expand should help ensure that Heathrow keeps its status as a major hub for air traffic, which brings tens of thousands of jobs to the area, and has other indirect benefits, such as making the UK a more attractive place to locate regional headquarters for companies that are expanding into Europe.

Figure 6.9 *Some differing views on the proposal for Terminal Five.* Adapted from BBC News online, 2001 ▶

Background to the debate

Heathrow is the world's busiest international airport. Importantly for London, it carries more profitable Business and First Class passengers than any other airport – in fact, 40% of Heathrow's passengers are wealthier professional and managerial income earners.

In the late 1980s, the airport's owners, British Airports Authority (BAA) believed that new terminals would be needed to meet increasing air traffic. It gained approval for Terminal 4, which opened in 1994, but even then realised that this would be nowhere near sufficient to meet projected growth. It therefore proposed T5, which would be for the sole use of British Airways (BA).

The debate about T5 started in the late 1980s. A design was chosen in 1989 and from then, proposals were made, revised and submitted for approval. The attitudes of different groups of people became central to the conflict and showed that – as with most proposals – once a decision is made, some people benefit whereas others lose.

Figure 6.10 *As the largest terminal at Heathrow Airport, Terminal 5 can handle nearly half of all passenger traffic at the airport* ▲

Terminal 5

- Cost £4.3 billion
- Took 15 years from planning to completion
- Took 6 years to build
- Employed 6000 workers during construction
- Opened on time and on budget
- Has as much capacity as Terminals 1 – 4 combined

Supporters of T5 included:

- BAA, the owners
- British Airways
- The London Chamber of Commerce and Industry
- Major business interests
- Trade unions
- Many local residents, some of whom worked at the airport.

Opponents of T5 included:

- Many local residents and community groups
- 13 London borough councils
- The west London branch of Friends of the Earth (FoE)
- HACAN (The Heathrow Association for the Control of Aircraft Noise) with over 7000 members.

Figure 6.11 Demonstrators against the proposal for T5 ▲

How Heathrow airport functions

Heathrow has two parallel main runways, running west-to-east, which allow take-offs and landings to occur at the same time (Figure 6.12). The decision as to how the runways are used (whether for take-off or for landing) is in part determined by local weather. Most winds at Heathrow blow from south/south-west to west – planes therefore take off into the wind to provide 'lift' on one runway and land into the wind on the other (Figure 6.13). There is also a north-south runway which is used only on occasions when strong southerly or northerly winds make cross-winds a hazard on the two main runways.

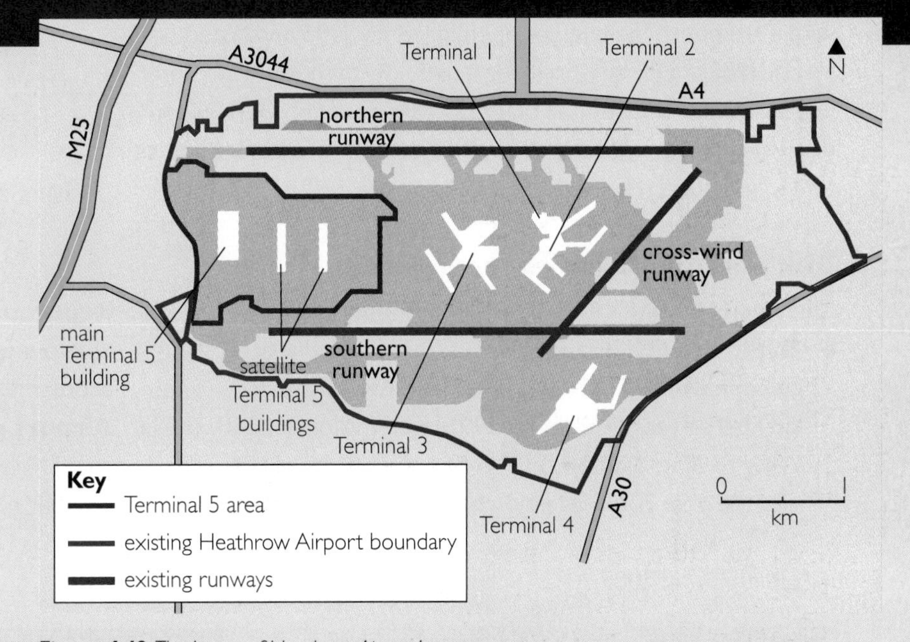

Figure 6.12 The layout of Heathrow Airport's runways ▲

Generally, night-time flying is not allowed after 10 pm or before 6 am, though there are some legally permitted exceptions. The Heathrow day starts just before 5 am when overnight flights arrive from North America, Australasia and Asia. Because millions of people live underneath the approach to Heathrow, landings and take-offs alternate between the two runways in two periods, from 7 am to 3 pm and from 3 pm to 11 pm.

These periods alternate weekly, so for example, the northern runway might be used for landings between 7 am and 3 pm one week and then between 3 pm and 11 pm the following week. During each eight-hour period when planes are flying overhead, they pass every 90 seconds, so that noise is continuous every single day of the year. According to HACAN, even before T5, Heathrow imposed more noise on more people than any other international airport in the world.

Figure 6.13 A wind rose for Heathrow airport showing wind direction over a 10-year period. Wind rose diagrams show the proportion of winds coming from a particular direction. The longer the 'stem' of the rose, the greater the proportion of winds from that direction. ▼

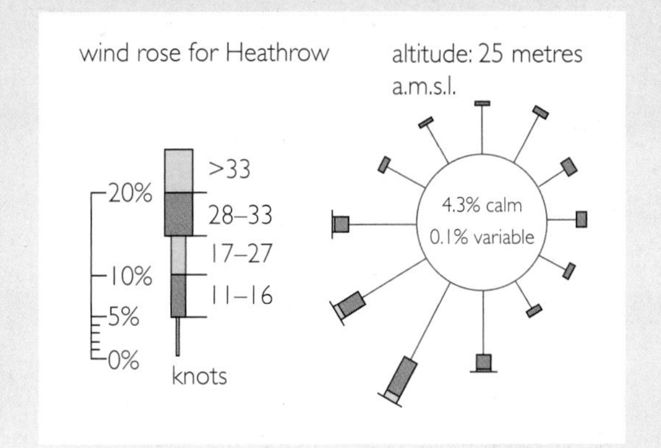

The Heathrow T5 Enquiry

All applications for planning permission go first to local authorities. However, where the application is of national importance the government Planning Inspectorate may order a public enquiry. During an enquiry, evidence is heard in public with individuals and interest groups given opportunities to speak, a process that can make an enquiry both lengthy and costly. The conclusions are published and the Inspectorate makes recommendations which governments may or may not then accept.

In 1993, BAA submitted its planning application for T5 to Hounslow Borough Council together with 35 linked applications, including a spur road to T5 from the M25 and extensions to both the Heathrow Express rail and Piccadilly underground lines. Because of its national significance, a public enquiry was called within a month. The enquiry opened in May 1995 and lasted until March 1999 – a record-breaking length – with more than 700 witnesses heard. Over half of its 313 days' duration were spent discussing three topics: the economic case for T5, surface access to T5, and noise (Figure 6.14).

Number and topic	Number of enquiry days	% of total
1. Economic/aviation case	123	23.4
2. Development pressures and regional impact	37	7.0
3. Land use policy	35	6.7
4. Surface access	117	22.3
5. Noise	73	13.9
6. Air quality	22	4.2
7. Public safety	13	2.5
8. Fuel supply	0	0
9. Construction impact	27	5.1
10. Associated applications (e.g. extension of Heathrow Express)	14	2.7
11. Reducing the impact of T5 (e.g. on local rivers and communities)	64	12.2

Figure 6.14 *Discussion topics at the T5 Public Enquiry* ▲

The arguments for T5

The arguments given to the enquiry in support of T5 were:

- **Rapid growth in air travel** – T5 would be needed to cope with a 70% increase in Heathrow passenger numbers from 40 million in 1991 to 65 million in 2000, rising to 80 million by 2013. In fact, the effects of 9/11 and the post-2008 recession have reduced growth in traffic, with 65.9 million passengers flying into or out of Heathrow in 2009.
- **Reputation** – Heathrow's reputation had weakened, with overcrowding, long security queues, baggage problems and poor cleanliness being major complaints.
- **Airport capacity** – Demand for air travel in south-east England was forecast to double between 2000 and 2020 (Figure 6.15). Without T5, south-east England would run out of airport capacity.

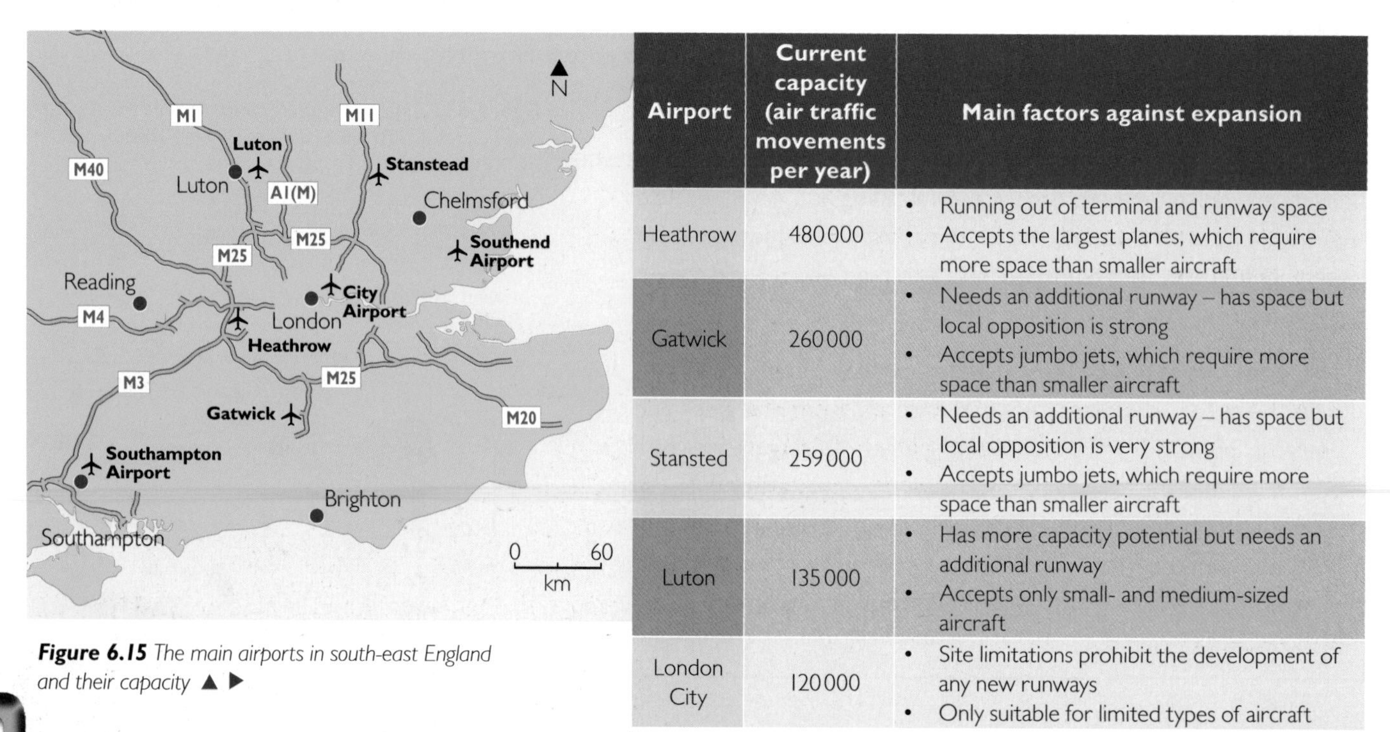

Airport	Current capacity (air traffic movements per year)	Main factors against expansion
Heathrow	480 000	• Running out of terminal and runway space • Accepts the largest planes, which require more space than smaller aircraft
Gatwick	260 000	• Needs an additional runway – has space but local opposition is strong • Accepts jumbo jets, which require more space than smaller aircraft
Stansted	259 000	• Needs an additional runway – has space but local opposition is very strong • Accepts jumbo jets, which require more space than smaller aircraft
Luton	135 000	• Has more capacity potential but needs an additional runway • Accepts only small- and medium-sized aircraft
London City	120 000	• Site limitations prohibit the development of any new runways • Only suitable for limited types of aircraft

Figure 6.15 *The main airports in south-east England and their capacity* ▲ ▶

- **Competition with Europe** – The position of Heathrow as Europe's number one airport is being challenged by other European airports, such as Paris Charles de Gaulle (CDG), Amsterdam Schiphol and Frankfurt, all of which plan to expand capacity (Figure 6.16). Without expanding its own capacity, Heathrow would fight to keep its place.

European hub airports	Number of runways	Number of passengers per year (2008)	Average passengers per flight	Air traffic movements per hour (and targeted)
London Heathrow	2	67.7 million	146	80
Frankfurt/ Main	3	51.9 million	113	80 (120)
Paris CDG	4	53.5 million	113	120
Amsterdam Schiphol	5	44.1 million	112	90 (120)

Figure 6.16 *Heathrow's competitors* ▲

- **Its importance as a global 'hub'** – Several business routes from Heathrow, such as to Chennai and Bangalore (India) and Los Angeles and Seattle (USA), rely on connecting traffic for 50% of their customers. They bring value to the airport and the UK economy by making routes financially feasible that would not be based on local demand alone.
- **Its local economic value** – Locally, Heathrow employs 77 000 people. T5 would either protect or create 16 500 jobs as well as provide an extra 6000 construction jobs. Heathrow and the businesses it supports already contribute £3 billion to the local economy.
- **The wider significance** – Although the enquiry noted that Heathrow's contribution to the UK's gross domestic product (GDP) is probably underestimated (as there are no data to measure its influence in the UK's and London's economy), the economic importance of Heathrow is linked to its capacity. Restricting Heathrow to 60 million passengers each year could affect the attraction of London and the UK for international investors. T5 would contribute substantially to the UK economy, and to the success of London as a financial centre and its ability to attract further investment. London has to compete to remain an attractive city in which to do business and a suitable airport is essential to the city's infrastructure.
- **Tourism** – Tourism, which is worth £10 billion a year to the UK economy, relies on Heathrow. The enquiry heard that premium high-spending overseas visitors would spend an estimated 10 million fewer nights in Britain without T5, costing London's hotels and tourist industry an estimated £1.5 billion (Figure 6.17).

Did you know?

Heathrow's most popular passenger destinations are, in order:

1. New York
2. Dubai
3. Dublin
4. Amsterdam
5. Hong Kong

The arguments against T5

Two broad arguments formed the case against T5 – economic and environmental.

1 Economic

- Aviation's contribution to the economy is overstated. It is only the 26th biggest industry in Britain, half the size of the IT industry and a tenth the size of banking and finance.
- The aviation industry is heavily subsidised. Airlines pay no VAT on aviation fuel, a tax concession worth £7 billion a year in the 1990s and estimated to be worth £16 billion by 2020. Airlines pay nothing towards the noise and pollution they cause as there is no environmental levy.

Figure 6.17 *Overseas visitors to London bring tourism revenue* ▼

HACAN claim that the aviation industry runs up a deficit on tourism, worth over £13bn in 2009, because money spent by Britons flying from the UK exceeds what visitors to Britain spend here. In 2010, UK residents made 55 million visits overseas by air while overseas residents made 30 million visits to the UK. Opponents of this view say that many visitors to the UK bring investment – 38% of Heathrow's incoming passengers are on business while the majority of UK trips overseas are for tourism.

- FoE and HACAN argue that transfer passengers at Heathrow (those who do not stay in the UK but just transfer from one international flight to another) do not directly benefit the UK economy. They bring benefit only to BAA, whose shops they visit, and British Airways, whose flights they use.

The issue of noise at Heathrow

Noise is measured by the LEQ (**L**oudness **EQ**uivalent) and is given in decibels (dB). LEQs around Heathrow are mapped using the 57dB contour and are above the World Health Organisation (WHO) limits which prefers 55dB as the maximum and 50dB as the target levels.

Noise affects the immediate airport environment, but most significantly those who live underneath the take-off and landing 'pathways' (Figure 6.18). Landing aircraft approach on a gentler incline than those taking off so those who live underneath the landing pathway suffer the most noise.

However, the issue of noise is not a simple one:

- Noise readings are added together using a formula for each plane type. When it was operating, one Concorde was considered the equivalent of 120 Boeing 757s. HACAN argued that for most people, four hours of non-stop 757s at a rate of one every 90 seconds is far worse than one loud Concorde with 3 hours 58 minutes relief afterwards.
- Noise readings do not reflect the actual noise experienced when a plane passes overhead (Figure 6.19). Readings are averaged to include quiet times of the day and of the year.
- No noise measurements are taken between 6.00-7.00 am. – one of the busiest periods at Heathrow!

***Figure 6.18** Aircraft landing over houses in Hounslow, west London. Hounslow Borough Council was against T5 in spite of the fact that many in the borough work at the airport.* ▲

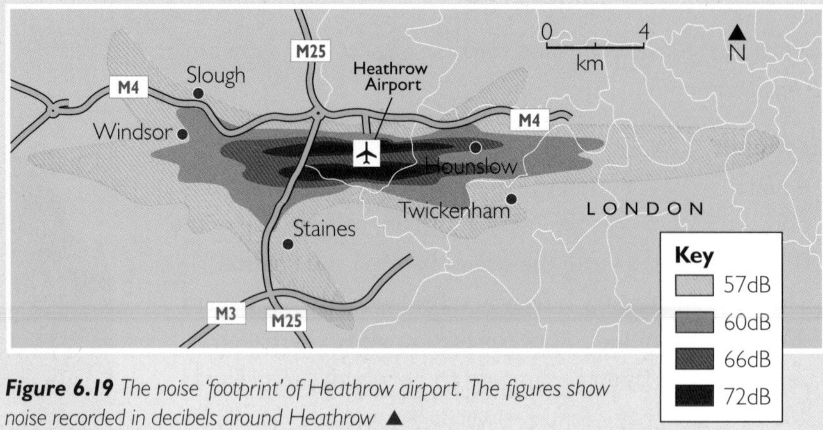

***Figure 6.19** The noise 'footprint' of Heathrow airport. The figures show noise recorded in decibels around Heathrow* ▲

Figure 6.20 A view of the landscape around Heathrow airport ▲

2 Environmental

- **Aircraft noise** – Many individuals and groups, including both FoE and HACAN, were concerned about aircraft noise. To meet these concerns, BAA accepted a limit on noise at 1994 levels (as newer aircraft have quieter engines) and limited the increase in the number of night flights.
- **The green belt** – Although T5 was built on a disused sewage works within the airport perimeter, FoE claimed that it would be the largest structure ever built on London's green belt as well as result in the loss of hundreds of acres of wildlife habitat. The enquiry report noted that the site represented 'inappropriate development' if calculations of land lost to the M25 Spur Road were included. There were different views of this; some felt that the area to the south and east of Heathrow was already urbanised (Figure 6.20) with little green space, making farmland to the west and north more precious. Others regarded T5 as making little difference.
- **Pollution** – T5 would result in more air pollutants because it would allow for more flights and the use of larger planes. Globally, aviation is responsible for just 3% of greenhouse emissions but in the UK this rises to between 6–13%.

  Emissions thin the ozone layer because aircraft release them in sensitive regions of the upper atmosphere. The EU sets an annual 'carbon allowance' (the amount of carbon emissions a country can emit) which member states are not permitted to exceed. In recent years it has tried to force countries to reduce emissions. A study in 2005 found that if current aviation growth continues, the industry could use up the UK's 'carbon allowance' by 2050.
- **Rail instead** – Opponents to T5 claimed that high-speed rail via Heathrow should replace short-haul planes (for example, to Manchester) as city centre-to-city centre it is at least as fast. People from the West Midlands spend over £400 million a year travelling to Heathrow because many long-haul flights leave from there.
- **Air pollution on the M25** – With increasing airline passengers, the number of people arriving and leaving by car and taxi would increase from 19.6 million per year in 1991 to 39.9 million by 2016, increasing flows on the M25 by 7-10%. Friends of the Earth calculated that T5 would lead to 326 million km of extra driving every year at an economic cost of £60 million.

Was it the right decision?

The go-ahead to build T5 was given in November 2001. The Inspector's report stated that *"the decision on T5 is a balance between its benefits to the economy and the travelling public against its impacts on local residents. I have come to the conclusion that benefits would substantially outweigh impacts ....."*.

Two broad conditions were imposed:

- Flight numbers were limited to 480 000 per annum and, from 2016, the area affected by noise levels to over 57 decibels should be limited to an area no bigger than 145 sq km
- To encourage the use of public transport to the airport, fewer car parking spaces were created than planned and a proposal to widen the M4 was rejected.

In addition, British Airways themselves specified environmental qualities for the new terminal.

How correct were predictions for air traffic? The airline market is fickle and depends upon consumer demand. Predictions in the transport industry as a whole have proved wrong in the past. How did T5 fare?

- BAA predicted that the number of flights at Heathrow would not exceed 453 000 a year until 2013. That number was passed in 2000.
- Predictions were that T5 would create a 60% increase in passengers at Heathrow, but only a 10% increase in flights because of larger aircraft such as super-jumbos with 600-800 passengers. But there are no super-jumbos of this size! In 2010 the largest aircraft was a Singapore Airlines Airbus A380 which can hold 525 people.
- Forecasts about air traffic have proved wrong, consistently over-estimating increases in passengers per aircraft and under-estimating numbers of flights. Airline deregulation is leading to smaller aircraft and more frequent services.

The environmental credentials of T5

Water

- Groundwater and rainwater are reused, cutting demand for water by 70%
- Taps and showers include water-saving devices such as flow restrictors

Waste and resources

- 97% of construction waste was recycled
- 50 000 plants, shrubs and trees were planted as part of T5's landscaping
- All furnishings use timber from sustainable sources; all carpets are PVC-free and recyclable
- All lounges are equipped with recycling facilities

Energy

- 85% of T5's heat is supplied from waste at airport heat and power stations
- Light bulbs use less power
- The T5 building is glazed on all sides, reducing the need for artificial lighting (Figure 6.21)
- Automatic lighting responds to light levels, using energy only when needed

Figure 6.21 *The glazed facade of T5* ▼

Looking forward – a third runway?

Today, the debate is about a proposal for a third runway at Heathrow. The fear of those who opposed T5 in the 1990s was that it would simply be the beginning of an expansion programme which would continue with a call for a third runway. So it has proved. This issue is one which is likely to persist; it is interesting to see how the viewpoints for and against it are remarkably similar to those used for T5 (Figure 6.22)!

Did you know?

A new MAGLEV train in Shanghai travels the 30 km from the airport into the city in just over 7 minutes! A similar train network in the UK could link Heathrow with major UK cities and remove the need for most domestic flights!

Figure 6.22 *The arguments for and against a third runway at Heathrow Airport* ▼

Arguments for	Arguments against
1 More capacity needed • Air travel is predicted to double by 2050; Heathrow cannot cope without an additional runway **2 Boosting the economy** • BAA says that a third runway will be worth £7bn a year to the economy • Thousands of jobs will be created • Lack of expansion threatens London's position as global financial capital **3 Pollution is overplayed** • By 2020, new technology will mean quieter, less polluting planes **4 There is no alternative** • Building a new airport in the Thames Estuary is too costly	**1 Need to reduce emissions** • Heathrow generates 50% of UK aviation emissions **2 Health risks** • Heathrow already breaches European Union regulations on levels of nitrous oxide emissions **3 Economic case overstated** • It will cost £12 billion with a return of £17 billion over 70 years – not an especially large return • Many passengers brought in would only be in transit **4 Local impact** • Noise – a third runway could mean an extra 200 000 flights a year over London • Transport infrastructure could not cope • To make way for the runway the local village of Sipson would be demolished and green belt land lost **5 There are alternatives** • Create a new airport in the Thames Estuary, away from the city

ACTIVITIES

1 a Using material in this section, summarise the views for and against T5. You can use hindsight – i.e. data which have emerged since T5 opened – plus further research.

b In pairs, score the strength of each viewpoint between 1 (views to which you give a low consideration) and 5 (high). Add the total scores for each side. Which argument is stronger?

2 How far should economic considerations for T5 outweigh environmental factors? Prepare a 500-word statement.

3 In class, discuss whether:

a issues of national importance such as the expansion of airports should be decided nationally rather than by local councils

b public enquiries waste time and slow down the development process.

Internet research

Research the debate about a third runway at Heathrow. The following websites may help you:

- BAA (baa.com)
- the various airlines, such as British Airways (ba.com)
- HACAN (hacan.org.uk), and
- news sites, such as BBC (bbc.co.uk/news).

Internet research

Research the debate about a high-speed rail link ('High Speed 2') between London, Birmingham and northern England. The Department for Transport website (dft.gov.uk) is a good place to start.

6.3 Breaking away – the trend towards separatism

In this section you will learn about:

- the meaning of 'separatism' and why some political groups fight for it
- the consequences of separatist movements, using the Kurds as a case study

24 January 2011. A bomb has exploded at Moscow's largest airport, Domodedovo (Figure 6.23). Suspicions focus on separatists from the Dagestan Republic in the Caucasus region of Russia. In March 2010, the same group was most likely responsible for another explosion on Moscow's underground that killed 40 people.

Elsewhere in Europe, there have been threats about a resumption of violence by a splinter group of the Irish Republican Army (IRA). Against this trend, the Basque organisation, ETA, has agreed to stop its campaign of violence against the Spanish government.

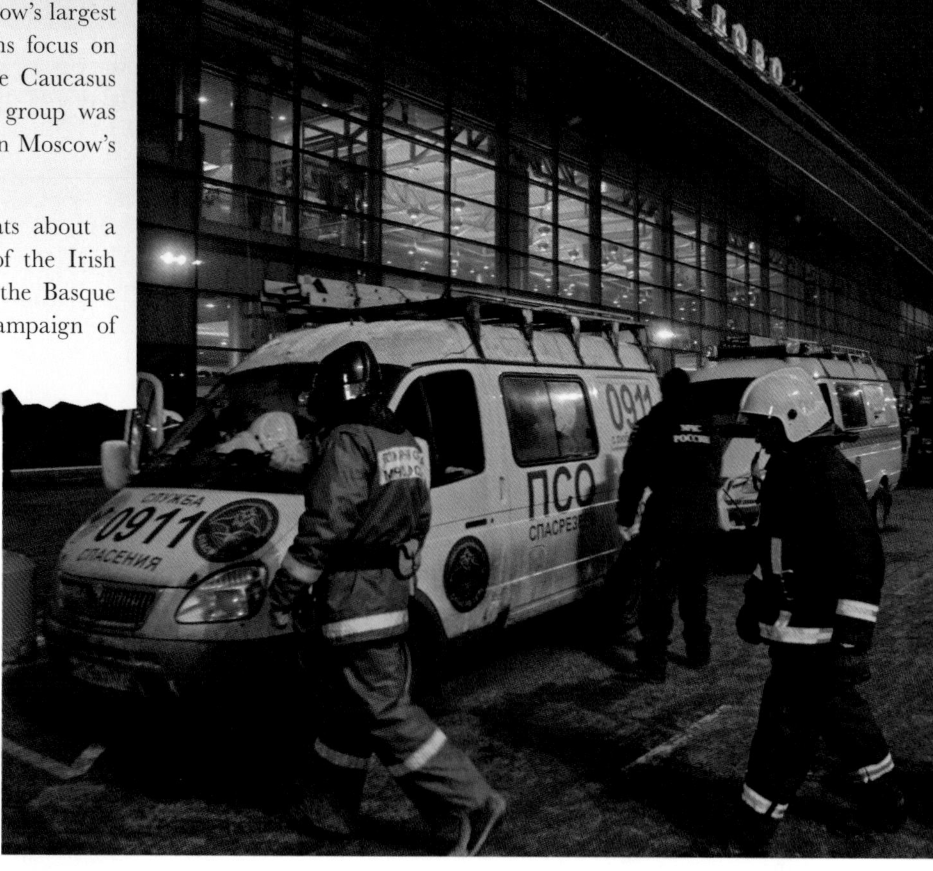

Figure 6.23 *Emergency services respond to a bomb at Domodedovo Airport, January 2011. The bomb was almost certainly the responsibility of minorities in Russia campaigning for greater rights.* ▲

What do these events have in common? They all involve would-be breakaway groups – or separatists – who would prefer a future where they are responsible for their own government, a concept known as **autonomy**. The original aim of the IRA was for a united Ireland, bringing together Northern Ireland with the Irish Republic. Similarly, ETA has campaigned for a breakaway region of Spain, the Basque country, to be governed as an independent state.

What is separatism?

Separatism is the splitting of people or groups from the larger majority on the basis of cultural, ethnic, religious or political differences. It usually means that people of a region or a particular minority want more political control and influence.

These people or groups usually have a different language, culture or religion from the majority. Often, they are on the edge of an area geographically, such as Scotland's Scottish Nationalists who want to be independent from the United Kingdom, and feel remote from or unrecognised by a centralised government. Separatism is usually associated with demands for autonomy and often with demands for full independence and the creation of a new nation state.

- **Autonomy**: the right of self-government and self-determination.

Nations and separatism

To understand separatism fully, it helps to understand what binds people together – a desire for a new nation. A **nation** is a cultural community with five main dimensions:

1 Psychological
    - The idea of a nation must exist in people's minds; if they do not feel part of a community, then it doesn't exist.
2 Cultural
    - Members of a nationality must understand each other through culture, common language or customs.
3 Historical
    - A nation's history is critical – it tells every inhabitant that the nation existed before them and will continue after they die.
    - A nation is greater than the individual, a concept which explains why people say they would 'die for their country'.
4 Territory
    - Nations are usually attached to a specific territory. In a territory:
        - a nation may fight battles for survival
        - sacred lands may exist, such as the sacred lands of aboriginal peoples in Australia
        - injustices are felt if territory is lost, for example in Germany after World War II when the country was divided into two halves.
5 Political
    - People want autonomy to make their own decisions, not always fully independent of the majority but with some freedom. This can occur with nations within larger entities, for example, Scotland within the UK or Tibet seeking to be recognised as separate from China.

Two factors are critical in influencing people to want to separate from the majority: the need for territory and the need for political autonomy. Groups who are psychologically, culturally and historically bound together may fight for separatism if denied territory or political autonomy.

Figure 6.24 *American school children pledging allegiance to the American flag* ▲

Nations and nationalism

For a new nation born out of separatism, a belief in its value contributes a sense of identity for its population – a concept known as **nationalism**. Nationalism creates unity between individuals across social boundaries. For most, nations are the political unit to which they show their allegiance. For example, school children in the United States, regardless of wealth, ethnicity or creed, pledge allegiance to the US flag every morning as a symbol of their national identity (Figure 6.24). This act reflects a positive side of nationalism that helps a country's culture develop, survive and strengthen, even when its peoples are very diverse.

However, nationalism can have a negative side too. Sometimes, a majority population can exclude members of its society through violence, ethnic cleansing, racism and **xenophobia**. A clear example of this was in Hitler's Germany in which Jewish, gay or disabled people were among those excluded and marked as different. Nationalism has also formed part of the past policies of ethnic cleansing in eastern European countries such as Serbia.

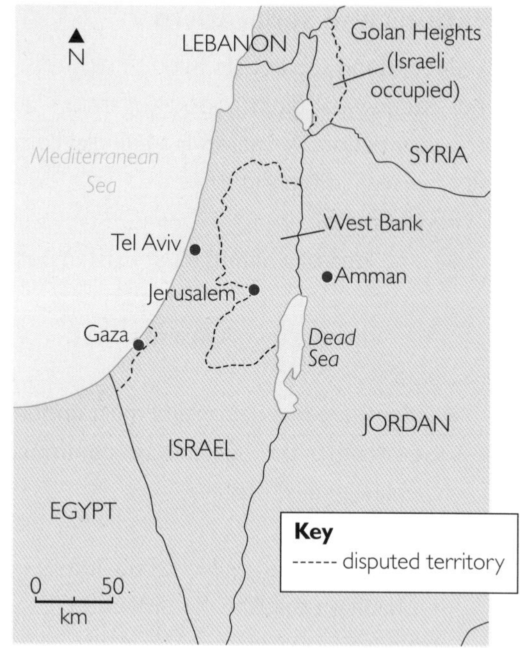

Figure 6.25 *Israel, which became the separate territorial homeland of Jews in 1948* ▲

- **Xenophobia**: an unreasonable hatred or fear of foreigners or strangers.

Ethnicity and separatism

Ethnicity is important in the formation of separatist groups. According to German sociologist Max Weber (1864-1920), an ethnic group is *'a human group that entertains a belief in their common descent because of similarities of physical type, or of customs, or because of memories of colonisation and migration'.* Members see themselves and their community as part of a group with similar beliefs and experiences.

Ethnicity can be fixed, particularly where it is based on religion and a belief in common descent from a deity, but it is more frequently the result of common identity. Such identity can result from oppression or suffering – the Jewish faith, for instance, is not only bonded by religion and the desire for a separate country (Figure 6.25), but is reinforced by the treatment and suffering of Jews under Hitler.

The strength of ethnicity and sense of belonging can vary. At its strongest, it can involve political claims and fights for independent territories as in the case of the Scottish and Welsh Nationalist Parties, the Quebec sovereignty movement in Canada, the Irish conflict, Palestinian conflict, and ETA's Basque struggle in Spain. In milder forms, it involves the revival of cultural languages and customs. For example, the Cornish Nationalist Party, which has members on the Cornwall County Council, promotes greater autonomy and the support for Cornish culture (such as re-establishing the Cornish language) rather than independent separation (Figure 6.26).

Figure 6.26 *A welcome to Cornwall in both English and Cornish* ▼

Between these two, the re-establishment of minority cultures or property issues contributes to the formation of ethnic pressure groups such as aboriginal groups in Australia or American Indian and Inuit communities in the USA and Canada.

Some nations are **diasporas**, meaning that they do not possess a territory. A diaspora (from Greek διασπορα 'scattering' or 'dispersion') is the movement or migration of people who share a national and/or ethnic identity away from their perceived homeland (see the case study on the Kurds on the next page).

CASE STUDY

The Kurds

The Kurds are the world's largest stateless population. There are about thirty million people worldwide linked by ethnicity and language who hail from Kurdistan, a region spanning eastern Turkey, northern Iraq, western Iran, and parts of Syria and Armenia (Figure 6.27).

In all of these countries, the Kurds are a minority group, generally with fewer rights than the majority. They often suffer discrimination and are regarded as second-class citizens. In Arabic idiom, anyone shabbily dressed is said to be 'dressed like a Kurd'.

In Turkey, Kurds are typically less educated and earn lower incomes than the majority population. They struggle to achieve equal rights or incorporate Kurdish language into schools. A 2005 United Nations report noted that Kurds in Iran suffered 'disproportionate inadequacy of services such as water and electricity'.

The origins of Kurdistan

Kurds have occupied the same area for several thousand years. They form one of the Middle East's largest 'nations' but have never had their own state territory. A state of Kurdistan was planned in the post-First World War settlement of territories of the former Ottoman Empire but the idea collapsed when Turkish nationalists won British support in preserving some of the former Empire, helped by knowledge of the substantial oil reserves located there.

Since then, the Kurds have endured minority status within Turkey, Iran, Iraq and Syria. As European influence in the region extended to win contracts for weapons and oil, Kurdish interests were largely ignored. Now, all four countries in which the Kurds live are against them having a separate state.

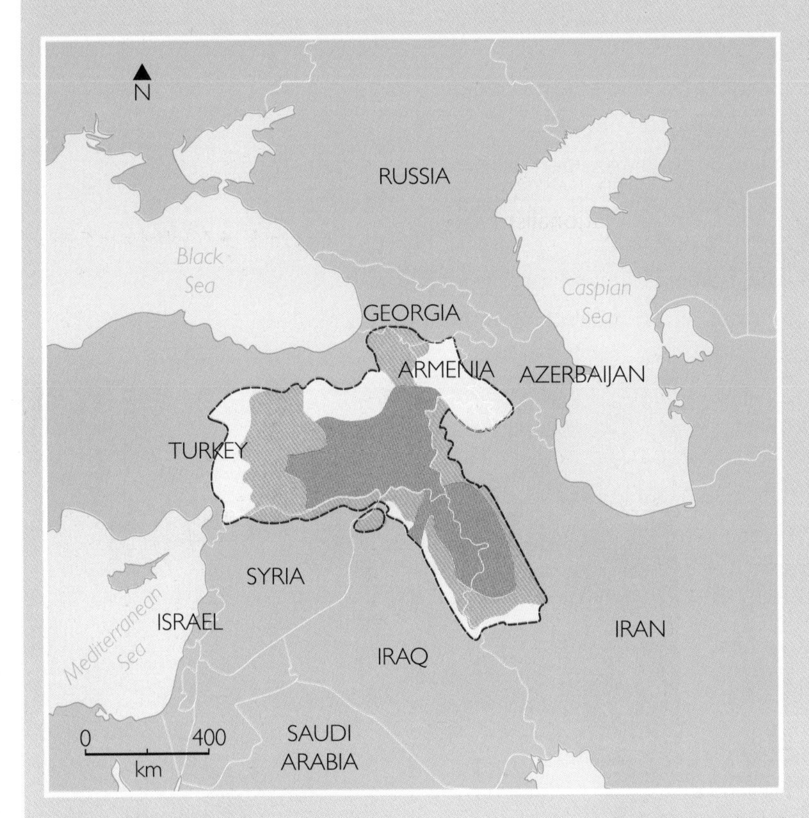

- Kurds comprise 20% of the population in Turkey, 15-20% in Iraq, 10% in Syria, 8% in Iran, 1.3% in Armenia, and a small number in Azerbaijan
- Kurds form the largest ethnic group in Turkey, Iraq, Syria and Armenia
- About half of all Kurds live in Turkey, about 18% in each of Iran and Iraq, and about 5% in Syria
- The Kurdish diaspora is large; there are about 35 000 in the UK and over 400 000 in Germany.

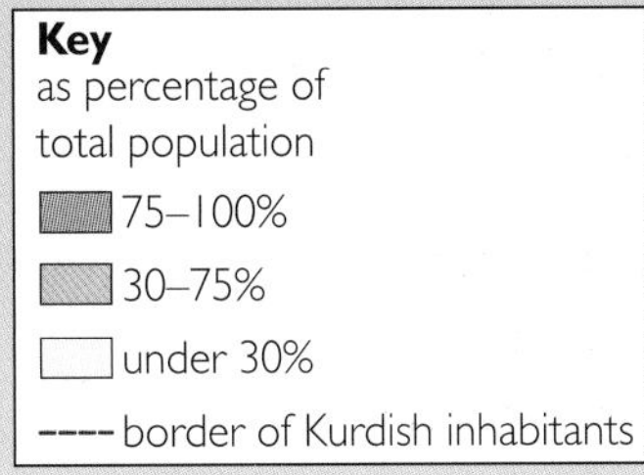

Figure 6.27 *This map shows the territory occupied by the Kurds. This area, which is about 2.4 times the size of the UK, would form the basis of negotiations for a separate Kurdistan..* ▲

Historically, none of the four countries in which most Kurds live openly acknowledged the existence of Kurds. This prejudice is based on deep-seated resistance towards Kurdish independence. Specifically, these countries argue that:

- a new Kurdistan would disrupt their national security and balance of power
- if one country supported an independent Kurdistan it would encourage Kurds elsewhere to seek the same as an autonomous or independent Kurdish nation would unsettle countries with Kurdish minorities throughout the region
- they have no intention of giving up territory to create another rival country, least of all if valuable oil deposits are located there (Figure 6.28).

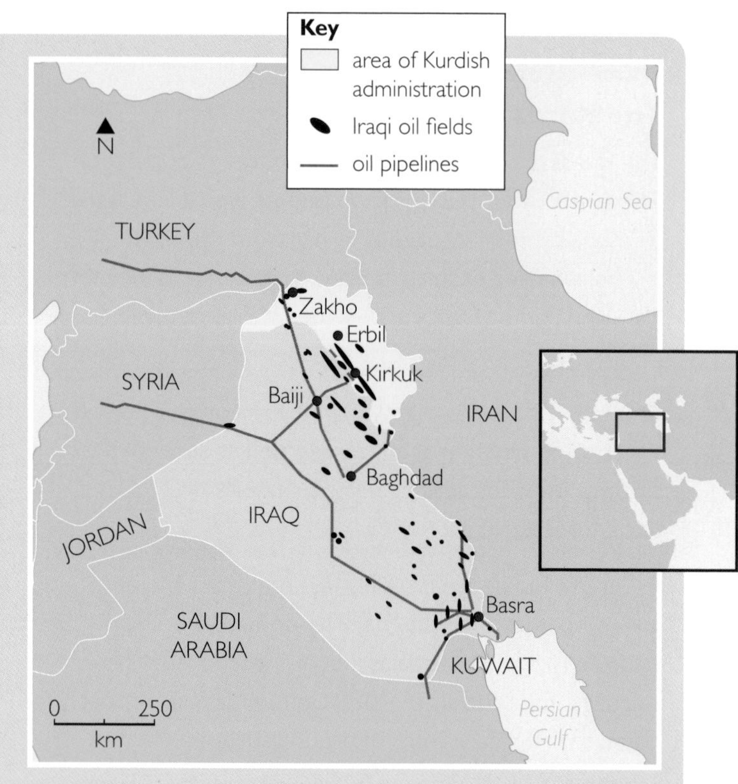

Figure 6.28 *Valuable oil reserves in northern Iraq would form part of Kurd territory if a separate state were ever granted* ▲

Outside the region, the USA is also against an independent Kurdistan. On one hand, it does not want to destabilise northern Iraq where the Kurdish area is the most stable. On the other hand it supports Turkey's efforts to maintain stability by trying to reduce the threat of Kurdish militancy. As far as the USA is concerned, the Kurds live along a major geopolitical fault-line. Kurdistan marks a frontier between:

- weak but rich states of the Gulf and their expansionist neighbours
- countries that the USA regards as enemies (Iran) and those it considers allies (Turkey). The USA fears that a break-up of Iraq or Turkey would strengthen Iran, which it regards as the greatest threat to stability in the region.
- a tinderbox of anti-US feeling to the south and east, namely Iran and Iraq, as well as to the north, for example, the Caucasus with countries and separatist movements in areas such as Dagestan and Chechnya.

The Kurds and Turkey

Turkey's rejection of an autonomous Kurdistan has led to a struggle for recognition by the Kurds. In 1984, a political party called the Kurdistan Workers' party (PKK) launched an armed struggle against the Turkish government in an effort to win civil rights for Kurds and the establishment of an independent Kurdish state within Turkey with its own government.

The PKK's efforts have included violence (Figure 6.29). In the 1980s, it used guerrilla tactics – kidnapping foreign tourists, suicide bombings and attacks on Turkish diplomatic offices in Europe – to further its aim. It has also launched attacks on Turkish security forces (an attack in 2010 killed ten Turkish soldiers) and Turkish sites at home and abroad.

Figure 6.29 *PKK demonstrators clash with Turkish police in Diyarbakir* ▲

In response, the Turkish government banned all political parties seeking rights for Turkish Kurds and prohibited teaching and broadcasting in Kurdish. In an attempt to deprive the PKK of financial support, between 1992 and 1998 the Turkish army destroyed over 3000 Kurd villages and productive land – actions that drove many Kurds to support the PKK.

The PKK is well funded. The Kurdish diaspora is generous; in recent years, London's 25 000 Kurds have collected over £500 000 annually and German Kurds about five times that amount. Some of its income is derived from drugs – the PKK is said to have its own laboratories and to control routes into Europe. Interpol (an international police organisation) claims that much of the heroin from South-East Asia is processed in Turkey and then sold in Europe (it is estimated that 80% of the heroin in Europe comes from Turkey).

Since 2000, the number of PKK members has declined; current estimates suggest that it has between 3000 and 5000 fighters. Recently, it has softened its original aims and now seeks political autonomy within Turkey rather than full independence. It has also recently received support from the EU which, in its desire for Turkey to join the EU, wants a peaceful and stable Turkey.

Yet, the separatist campaign continues. Kurdistan Freedom Falcons, a branch of the PKK, claimed responsibility for bombs injuring UK visitors in 2006. In an effort to undermine Turkey's tourism industry, a major source of income for the country, it warned that overseas visitors should not come to Turkey (Figure 6.30).

Figure 6.30 *A poster discouraging tourism to Turkey* ▲

The Kurds and Iran

There about 6 million Kurds in Iran who face discrimination and repression, including arrest and hanging, by the Iranian government (Figure 6.31).

- The ruling Shia clerics mistrust Iran's Sunni Kurds. Kurdish rebels fought against the Revolutionary Guard in the 1980s during the Iranian revolution while other Kurds fought alongside the Iraqis during the Iran-Iraq war.
- During both conflicts, when government troops finally secured Kurdish areas, the Iranian government massacred people suspected of helping the militants.
- In 2005, the Iranian government used 100 000 troops and helicopter gunships along the Iran-Iraq border to dampen pro-Kurdish demonstrations.

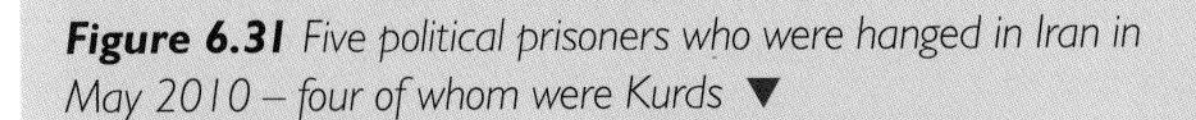

Figure 6.31 *Five political prisoners who were hanged in Iran in May 2010 – four of whom were Kurds* ▼

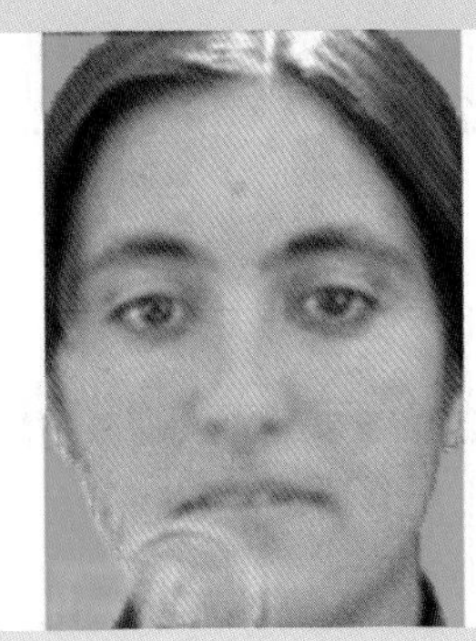

The Kurds and Iraq

The most vicious anti-Kurd action has come from Iraq under president Saddam Hussein.

- During the 1980-88 Iran-Iraq war, Kurdish guerrillas, helped by Iran, opposed Saddam's rule.
- At the end of the Iran-Iraq war in 1988, Kurdish villages were bombed with chemical weapons (such as poison gas) and over 180 000 Kurds were executed (Figure 6.32).
- After the 1991 Gulf War, Saddam tried to 'cleanse' Kirkuk province, which produces 70% of Iraq's oil, by deporting a further 250 000 Kurds.
- Kurds were forced to leave their homes and families, and stripped of their identities, property, documents and food ration cards.

Since the first Gulf War in 1991, Iraq's Kurds have enjoyed autonomy, forming a **de facto** state protected by US and British forces. Kurdish nationalists persuaded Kurdish forces to rebel against Saddam Hussein but they were crushed and 1.5 million Kurds fled to Iran and Turkey. When Saddam Hussein's government collapsed in 2003, the Kurds celebrated the end of genocide and ethnic cleansing against them.

The Kurds and Syria

The Syrian government has tried to reduce the influence of its one million Kurds. The Alawite Muslim minority ruling Syria regards the Sunni Kurds as allies of the repressed Sunni Arab majority and as a threat. For example:

- it is illegal to teach or publish in Kurdish
- a 1961 census stripped over 100 000 Kurds of their citizenship
- in 1962, government land reforms gave Kurdish land to Arab settlers.

- **De facto**: to exercise power or serve a function without being legally recognised.

Figure 6.32 *Anti-Kurd attacks by Saddam Hussein's government left this community destroyed* ▼

Where next?

By 2011, over 30 000 people had died in the conflict for Kurdish rights. In 2007, Turkey positioned 100 000 troops along the border with Iraq to combat PKK fighters within northern Iraq. Fighting continues and Turkish troops occasionally cross the Iraqi border to target PKK camps there.

- **Repatriation**: to send someone back to their country of birth, citizenship or origin.

However, Turkey's application for membership of the EU has led to pressure on the Turkish government to reform its attitude towards the Kurds. It now recognises the right of people within Turkey to use other languages and dialects, Kurds may publish and broadcast in Kurdish, and the Turkish government is **repatriating** hundreds of thousands of Kurds who were evicted from villages for supporting the PKK during the 1990s (Figure 6.33).

Figure 6.33 *The Kurdish language daily newspaper, the Azadiya Welat, now published in Turkey* ▼

ACTIVITIES

1. Distinguish between the following terms: nation, nationalism, nationality, ethnicity, separatism, autonomy.
2. Working in pairs, write 250 words outlining a case for an independent Kurdish state.
3. Draw a copy of Figure 6.27. Annotate it with reasons why the countries on the map are opposed to an independent Kurdish nation.
4. Which of the countries shown on the map do you think has **a**) the strongest, **b**) the weakest case for opposing an independent Kurdish nation? Explain your answer.
5. **a** Draw a table similar to the one below, listing the positive and negative consequences of the formation of Kurdish separatist movements in Turkey, Iran, Iraq and Syria.

Country	Positive consequences	Negative consequences

   b Have the negative consequences of the struggle for separatism been worth it? Write a 500-word argument to support your opinion.
6. As a class, debate the question: *'Are Kurdish separatists terrorists or freedom fighters?'*

6.4 The origins and impact of war in Afghanistan

In this section you will learn about:

- the geographical impact of the international conflict in Afghanistan

Afghanistan, early 2011. A soldier patrols the streets of a small town in Helmand province. It is dangerous territory for British forces; they are there to assume control against the Taliban whose own forces resist both the Americans and the British. Road travel is most dangerous, with roadside bombs causing 1500 deaths among NATO forces since 2003. The soldier is professional; he knows some of those who have died, but believes he is there to do a job. Rumour has it that US forces will take over soon, which might allow him to go home.

Meanwhile, 6000 km away in the Wiltshire town of Royal Wootton Bassett, local people line the streets in silence as another procession takes a body of a soldier killed in Afghanistan from nearby RAF Lyneham through the town on its way to John Radcliffe Hospital in Oxford for the coroner (Figure 6.34). British newspapers are beginning to mirror public opinion, asking why British forces are involved in a conflict that only some fully understand. How, they ask, can Osama bin Laden, the self-confessed architect of the 9/11 attacks on the World Trade Towers in New York in 2001, still be at large – apparently in the caves of Afghanistan? When conflict is complex and a long way from home, it is sometimes difficult to convince people that fighting is worthwhile.

- **Geopolitics** is the study of the ways in which political discussions and processes affect the way space and resources are used. It is the relationship between geography, economics and politics.

Understanding the conflict in Afghanistan

For over 30 years, Afghanistan has been subject to civil war, though the roots of conflict go back much further. **Geopolitics** lies at the centre of conflict in Afghanistan. The western Himalayas in Afghanistan form a huge natural barrier between countries with expansionist pasts (who have tried to bring Afghanistan under their control) and diverse belief systems. These create challenges not only for domestic defence in Afghanistan itself, but also for global superpowers (e.g. the USA) who see the region as a potential tinderbox.

Figure 6.34 *Royal Wootton Basset in Wiltshire, with crowds lining the main street to pay their respects to another soldier killed in Afghanistan* ▼

Its geopolitical location

The geopolitical location of Afghanistan, a deeply religious Muslim state, is highly significant.

- As a landlocked state (Figure 6.35), it shares around 5450 km of boundaries with six countries: China, Iran (936 km), Pakistan (2430 km), Tajikistan (1206 km), Turkmenistan (744 km), and Uzbekistan (137 km).
- Between 1917 and the collapse of USSR in 1991, over 2000 km of its borders were with the former communist USSR – an atheist and politically expansionist state.
- With a population that is 99% Muslim, Afghanistan ought to be a natural ally of other Islamic countries. However, the spread of fundamentalism and terrorism by groups such as Al Q'aeda, means that Afghanistan's future is subject to the scrutiny of world powers and their allies, including international groups such as NATO.

Its geopolitical neighbours

International conflict is not new to Afghanistan. Because of its geographical position, political stability in Afghanistan is of concern to four regions, all of which are potentially unstable (Figure 6.35).

Figure 6.35 *The geopolitical location of Afghanistan* ▼

The **central Asian landmass** to its north, consisting of Russia, and the 'new' states of, for example, Uzbekistan, Turkmenistan and Tajikistan, which until 1991 were part of the USSR. Russia fears internal Islamic uprisings in Chechnya and the Islamic states of central Asia, such as Tajikistan, as these states are only as stable as the governments that rule them.

China to the north-east is probably less concerned about Afghanistan itself, but very concerned about uprisings that could destabilise any part of Asia.

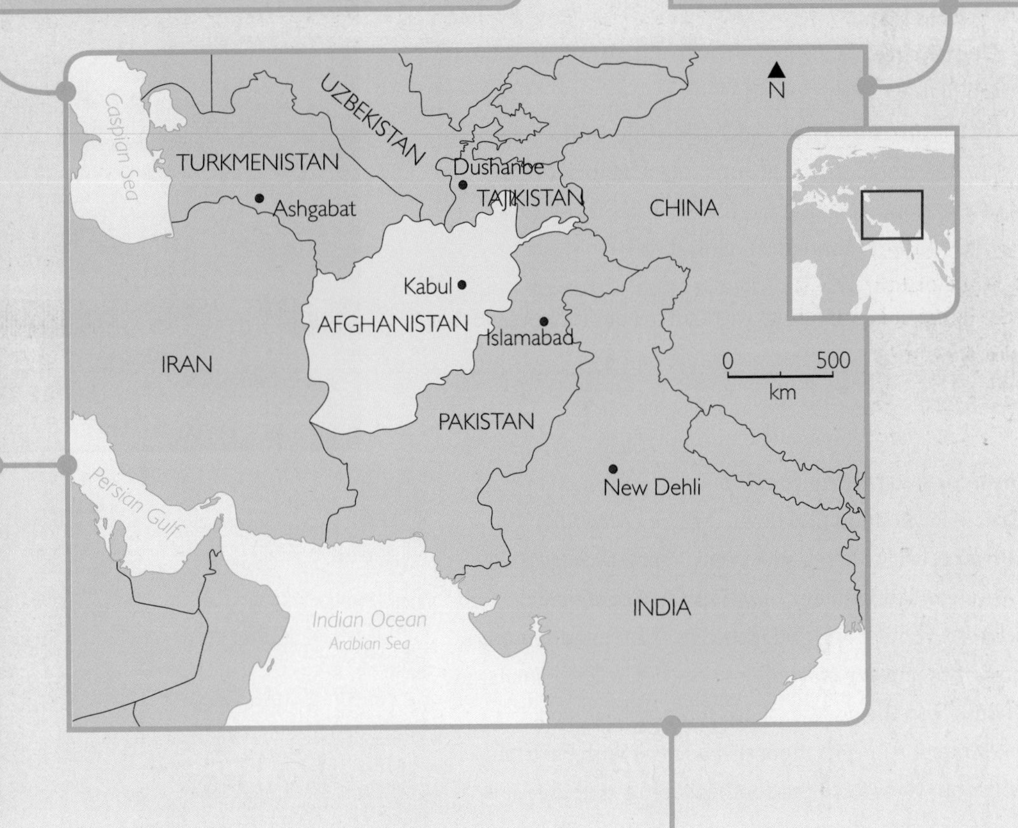

Iran to its west, and beyond that the oil-rich Middle East and Israel. During the 1970s, Iran had its own internal struggles; under its Shah (or King), it resisted the rise of Islamic power until the Shah was overthrown in 1979. An Islamic spiritual leader, Ayatollah Ruholla Khomeini, returned to Iran in 1979 from exile in France to assume leadership. Iran has since been ruled by a traditional Islamist government. Western governments fear Iran's expansionism into nuclear weapons, its hostility towards the west generally and to Israel specifically, and its disregard for human rights.

The **Indian sub-continent** to its south. India has perceived Afghanistan as a buffer to both the aim of the USSR to spread communism (since the Russian Revolution in 1917) and the growth of Islamic fundamentalism in the late 20th century. Like its neighbour Pakistan, it fears destabilisation from Islamic extremists as in the 2008 hotel shootings in Mumbai which were said to be the work of Al Q'aeda. A stable Afghanistan could help to prevent further attacks.

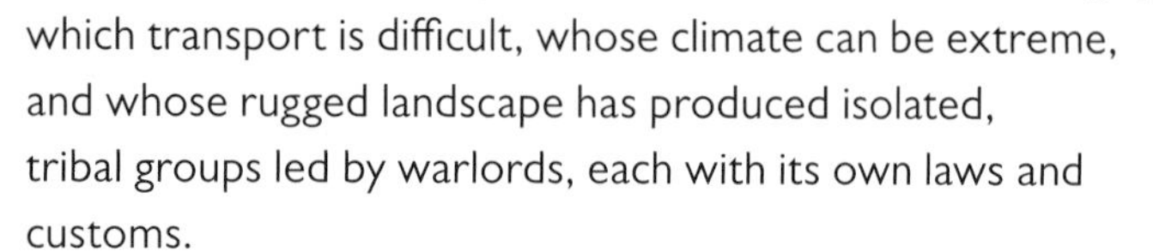

Afghanistan is, therefore, at a crossroads between contrasting political, ethnic, and religious factions in Central Asia, superpowers (China and Russia), and Islamic fundamentalism (Iran). Tactically, it is almost impossible for Afghanistan to defend all of its land borders, not just because of their length, but because of the mountainous landscape (Figure 6.36). Control of its mountain passes has been critical. It is a largely mountainous country across which transport is difficult, whose climate can be extreme, and whose rugged landscape has produced isolated, tribal groups led by warlords, each with its own laws and customs.

Figure 6.36 *Mountain landscapes of central Afghanistan* ▼

Factfile – Afghanistan

Population factfile

Population: 29.1 million (2010)

Age structure: 0-14 years – 43.6%: 65 years and over – 2.4%

Population growth rate: 2.5% (2010)

- ***Birth rate***: 38.1 births per 1000 – 19th highest in the world
- ***Death rate***: 17.7 deaths per 1000 population – 4th highest in the world
- ***Net migration rate***: +4.2 per 1000 population
- Mostly ethnic, tribal groups, e.g. Pashtun (42% of the population), Tajik (27%), and Uzbek (9%)
- 99% Muslim (Sunni 80%, Shia 19%)
- Mainly rural; only 24% live in cities or towns

Figure 6.37 *Population statistics for Afghanistan* ▲

Figure 6.38 *The physical geography of Afghanistan* ▼

Physical geography factfile

Size: 652 230 sq km (2.6 times the size of the UK)

Climate: arid to semi-arid. Precipitation averages 250-300 mm, mostly winter snow. Typical of continental interiors, winters are cold (down to -20°C) and summers can be hot and dry (daytime can be 45°C). The Himalayan Mountains in the northeast are sub-arctic with cold, dry winters. Along its mountain borders with Pakistan, monsoon winds from the southeast bring tropical rains from July-September.

Landscape: mostly of rugged mountains, with lower plains in the north and southwest. Nowhere is lower than 258 metres above sea level, and its highest point (Mt Noshak – part of the Hindu Kush mountains) is nearly 7500 metres. The Hindu Kush mountain range prevents north-south movement, and makes northern parts of the country very remote.

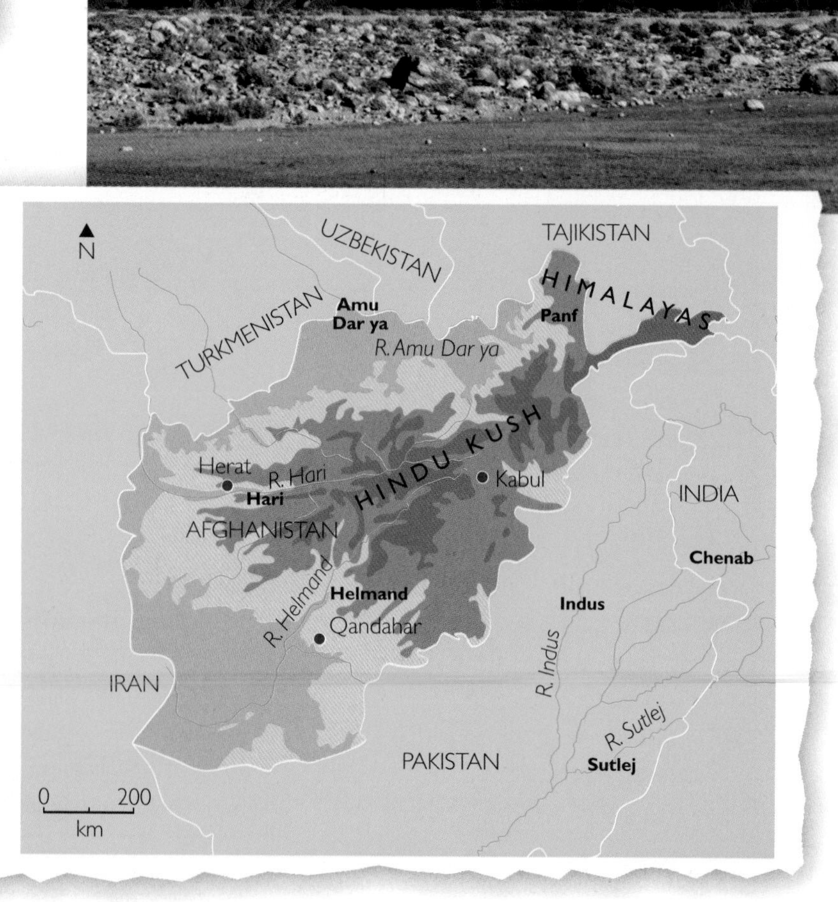

Welfare factfile

Afghanistan has amongst the world's worst health indicators (Figure 6.40). Approximately 20% of all children die before their fifth birthday. It also has one of the world's worst maternal mortality rates – about 2% of all births end in the mother's death. Most women are never seen by health professionals during pregnancy or childbirth, especially in rural areas.

- ***Infant mortality rate***: 151.5 deaths per 1000 live births (2010) – second worst in the world
- ***Life expectancy at birth***: total 44.4 years (similar for male and female) (2010)
- ***Fertility rate***: 5.5 children born/woman (2010) – 13th highest in the world
- ***Literacy total population***: 28.1% (male: 43.1%, female: 12.6%) (2000)
- ***School life expectancy***: 8 years (male – 11 years, female – 5 years) (2004)

Figure 6.39 *Welfare statistics for Afghanistan* ▲

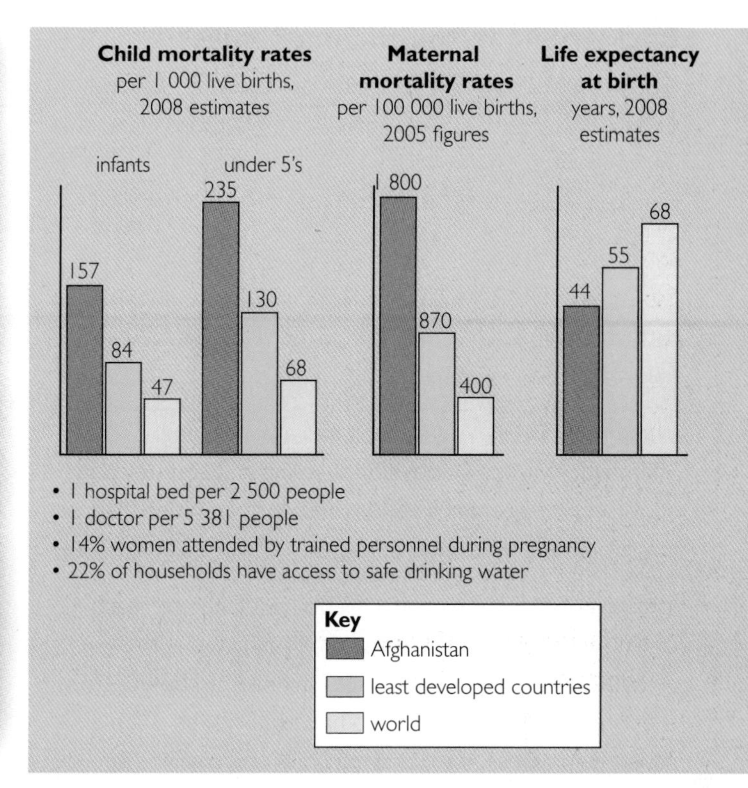

Figure 6.40 *Health indicators in Afghanistan* Source: Unicef/WHO ▲

Economic factfile

Afghanistan is one of the world's very poorest countries. In 1972 it was listed by the UN as being among the world's 25 poorest countries, and it has remained there ever since. Its economy is now growing, but only because of huge international aid and investment since the fall of the Taliban regime in 2001 (US$57 billion 2002-10). Its per capita income is among the worlds lowest. Decades of conflict put businesses off investing in economic growth and jobs.

- GDP (PPP$) in 2009 grew by 22.5% but from a very low base of $1000, and was 212th in the world
- 35% of people are unemployed; the same percentage live below the poverty line
- Growth is fragile; although Helmand province is the most productive area, political security there is worst (Figure 6.42)
- Formal economic data means little. Although agriculture accounts for 31% of formal GDP, all data exclude illicit opium production (Figure 6.42), which is the country's most significant crop
- Because the country is mountainous and arid, only 12% of land is arable. Farming is poor and employs 78.6% of people; industry employs just 5.7%.
- Poverty has led to deforestation for wood fuel and building materials, which in turn has caused soil erosion
- Water resources are small; only 4% of the country is irrigated. Most water reserves are used for irrigation.

Figure 6.41 *Economic statistics for Afghanistan* ▲

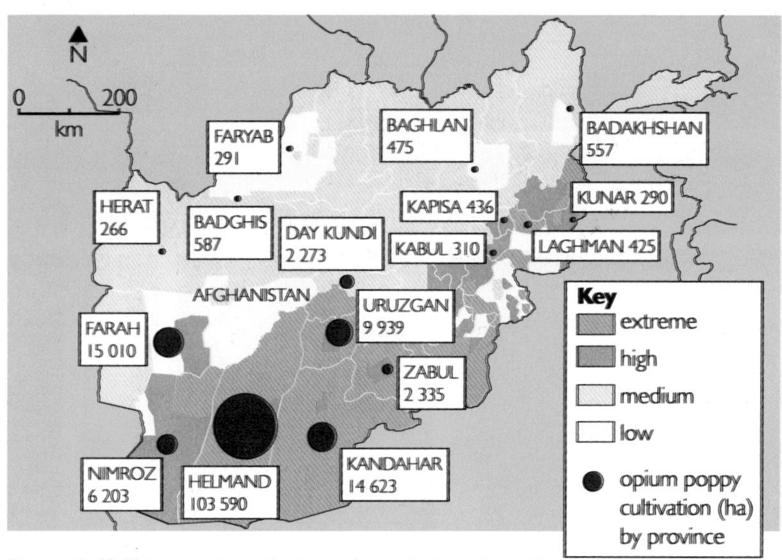

Figure 6.42 *This map shows the internal security in each province of Afghanistan compared to its estimated opium production* ▲

Background to the current conflict

Conflict is not new to Afghanistan. In the late 19th and early 20th centuries, Great Britain fought three wars with Afghanistan, each of which had two causes:

- the defence of its Indian empire
- preventing the Russian Empire from invasion.

Disputes between what is now Pakistan (then known as 'British India') and Afghanistan lasted until the 1970s when Pakistan became independent.

In the early 1970s, Afghanistan had a young king, Zahir Shah, who wanted to develop the Afghan economy and allow women greater freedom and opportunities than in traditional Islamic society.

This took hold in the capital, Kabul, but failed to reach more remote mountainous areas. Frustrated with slow growth, the Afghan Communist Party seized power from the King in 1978. Communists had always seen Islam as resisting progress, both economically and socially; for example, women in Afghanistan wore the burqa and were denied equal rights (Figure 6.43). Religious leaders bitterly opposed communists, and demonstrations and riots took place between rival groups.

In late 1979, the communist, atheist USSR sent forces and tanks into Afghanistan to occupy and prevent the collapse of communism. Their troops met resistance from Afghan guerrillas, known as the **mujahideen** (*'Islamic warriors'*). During Soviet occupation, hundreds of thousands of Afghans died fighting Soviet troops and millions of refugees fled to Pakistan and Iran.

The Soviet occupation provoked conflict between the USSR and USA. Unlike now, the USA – under a newly-elected President Reagan, a Republican – supported the anti-communist mujahideen with weapons and with recruiting Muslim fighters. With American funding and support they resisted Soviet troops, particularly in remote mountain areas, and the Soviets withdrew by 1989.

The Soviet departure left a power vacuum– civil war developed during the 1990s, bringing economic damage and decline. By the mid-1990s, the Taliban, a militant group among the mujahideen, gained control of much of the country and restored order. They enforced a programme of radical Islamic law, or **sharia**, which included restrictions on women such as forcing them to wear an all-covering burqa and denying educational opportunities to girls.

Figure 6.43 *Women in Afghanistan wearing the burqa* ▼

During this period, the Taliban gave protection to Saudi terrorist Osama bin Laden. Bin Laden established terrorist training camps in Afghanistan and organised attacks against US interests, e.g. the Hilton hotel in Nairobi, and on passenger aircraft. The most sensational occurred on 11 September 2001 when two passenger planes were hijacked and flown into New York's World Trade Towers, destroying them completely and killing nearly 3000 people.

In response to 9/11, the USA led an international military coalition against the Taliban, with forces by 2010 extending across the country (Figure 6.44). The assault included bombing raids to destroy terrorist training camps and kill terrorist leaders, with the aim of killing bin Laden. Coalition troops captured Afghan cities and supported the Northern Alliance, a group of Afghan resistance fighters. A new Afghan president, Hamid Karzai, was installed, rules on women's dress and schooling were eased, an Afghan army was trained and democratic elections were introduced for the first time.

On 1 May US time (2 May Pakistan time) US President Barack Obama announced that bin Laden had been killed. After years of intelligence gathering, he had been tracked to a secure compound in the Pakistani town of Abottabad and was shot in a raid by US Navy Seals.

Nonetheless, resistance has continued.

- The continued presence of foreign troops antagonises many supporters of reform.
- Taliban militants continue to attack forces and government targets, funded by opium trading (from poppies – used in manufacturing heroin).

Economic and social costs of war

Afghanistan

During the Soviet occupation of 1979-89, 870 000 Afghans were killed, three million maimed or wounded, a million were internally displaced and five million forced to flee the country. During the conflict beginning in 2001, deaths have been increasing (Figure 6.45).

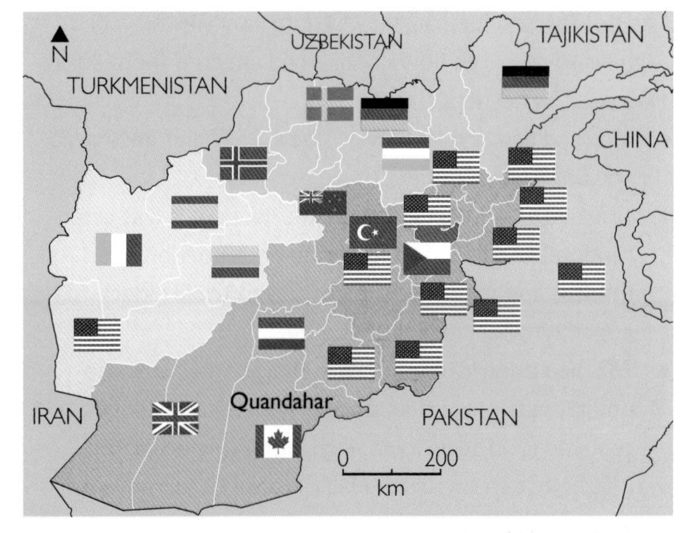

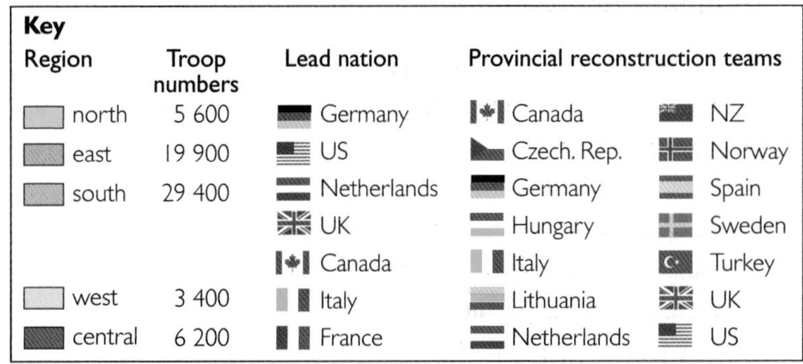

Figure 6.44 *The extent of international forces in Afghanistan in 2010* ▲

Figure 6.45 *Deaths and casualties among the Afghan population during the current conflict, 2007-10* ▼

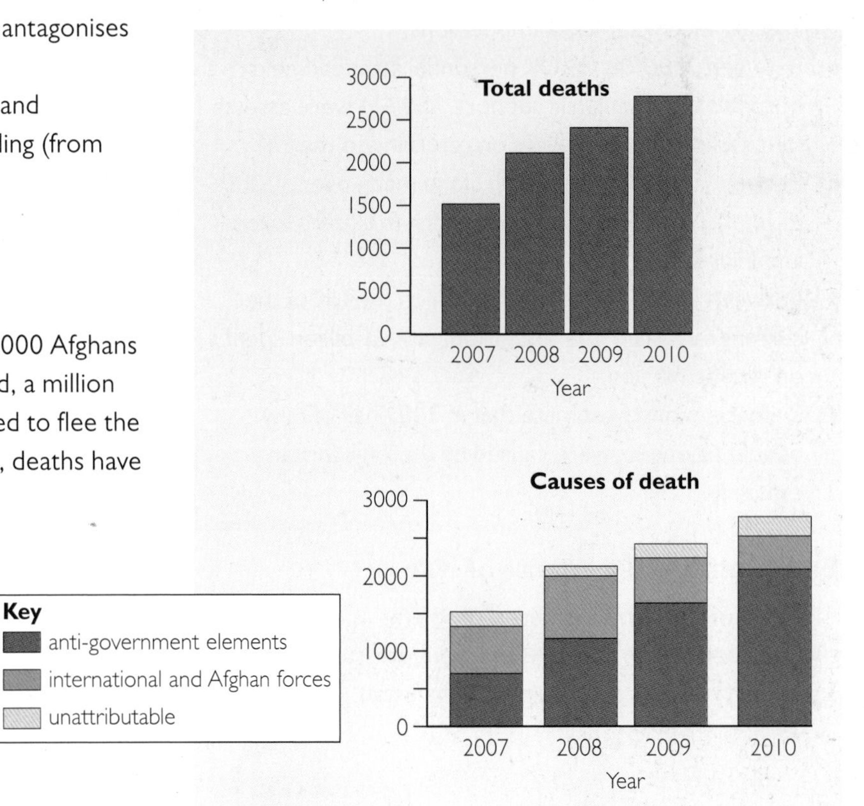

The not-for-profit aid agency Oxfam interviewed 700 Afghans to research how conflict had affected them during 1979-2009. They found that:

- 43% had had property destroyed, 25% land destroyed and 34% were robbed
- 13% had at some stage been imprisoned. It was often random, and linked to harassment, extortion and threats, and many had paid bribes to tribal elders negotiating for their safe release.
- 76% had been forced to leave their homes at some point; many more than once. This has significantly reduced food security in Afghanistan (Figure 6.46).
- 21% had been tortured. Half of these occurred in prison and were linked to ethnicity, political affiliations or, with many women, the actions of male family members.
- Only 1% received any compensation or apology (none came from those responsible)
- 70% believed that unemployment and poverty were major causes of the conflict
- Those blamed for the conflict were: the Taliban (36%); other countries (25%); Al Q'aeda (18%); international forces (18%); warlords (15%); and criminals (14%).

The UK

- Over 300 soldiers were killed in Afghanistan between 2001-10
- Between 2006-09, 218 soldiers suffered life-changing injuries and 50 suffered amputations
- Between 2006-08, 34 UK personnel attended field hospitals for psychiatric support, and 414 were assessed for psychological disorders on returning to the UK
- Between 2001-10, the British alone made over 30 000 payments for compensation for Afghan citizens killed and injured
- Between 2001-2009, the total cost to the UK of the current campaign was £12 billion, plus £1 billion spent on reconstruction
- Defence experts estimate that in 2009 half of new benefit payments were caused by the Afghanistan campaign.

Environmental impacts of war

Afghanistan faces environmental problems as a result of the current conflict from damage and from pressures caused by 3.5 million refugees returning to Afghanistan.

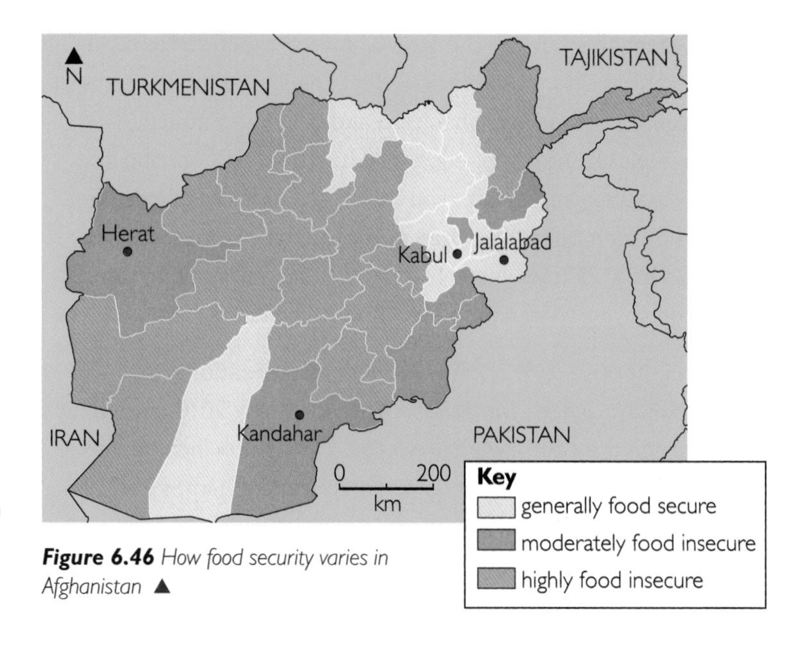

Figure 6.46 *How food security varies in Afghanistan* ▲

Water Problems

- **Disruption to water supply** – Only 25% of households have access to safe drinking water (Figure 6.47). Kabul at one stage was losing 60% of its water supply from leaks caused by war damage. In Herat, only 10% of 150 public taps were working.
- **Contamination** – Drinking water contains high concentrations of E.coli bacteria from sewage contamination. In Kabul, and Kandahar and Herat provinces, waste dumps were located close to drinking wells where heavy rains would wash waste into the city's rivers.

Waste Problems

- **Medical waste** – In Kandahar and Kabul medical waste and syringes are dumped in streets and wells
- **Health risks from pollution** are caused by damaged oil refineries and lead battery factories.

Figure 6.47 *Children in Afghanistan collecting clean drinking water* ▼

Loss of forest and biodiversity

- **Deforestation** – Coniferous forests in the north-west of Kabul halved in three decades. During Taliban rule, timber was logged for export. Pistachio woodlands, the nuts of which provided significant income for farmers, were cut for fuel and by military forces reducing ambush cover.
- **Impacts on ecosystems** – In the Amu Darya River, many hundreds escaping the conflict settled on unoccupied tugai forest islands, a unique ecosystem and refuge for rare species such as Eurasian otters, wild boar and waterbirds.
- **Hunting rare animals** – In the Wakhan Corridor, a remote area of the north, food shortages drove yurt-dwelling herders to hunt rare snow leopard, wolf, brown bear and Asian ibex for meat and fur, though they have since agreed to stop.

Where next?

International forces are likely to remain, attempting to reduce the drug trade and assist the Afghan national army in defeating the Taliban. The battle is uphill as many live in utter poverty. In spite of overseas investment, most aid to Afghanistan is military. New roads and services (e.g. electricity, water) remain targets for Taliban forces. Poverty compounds everything (Figure 6.48) as those living in poverty often support those offering better prospects – a gap filled by **fundamentalists**. While this persists, investment – the kind that generates growth – is unlikely. Many believe that *'there can be no peace without development, and no development without peace'*.

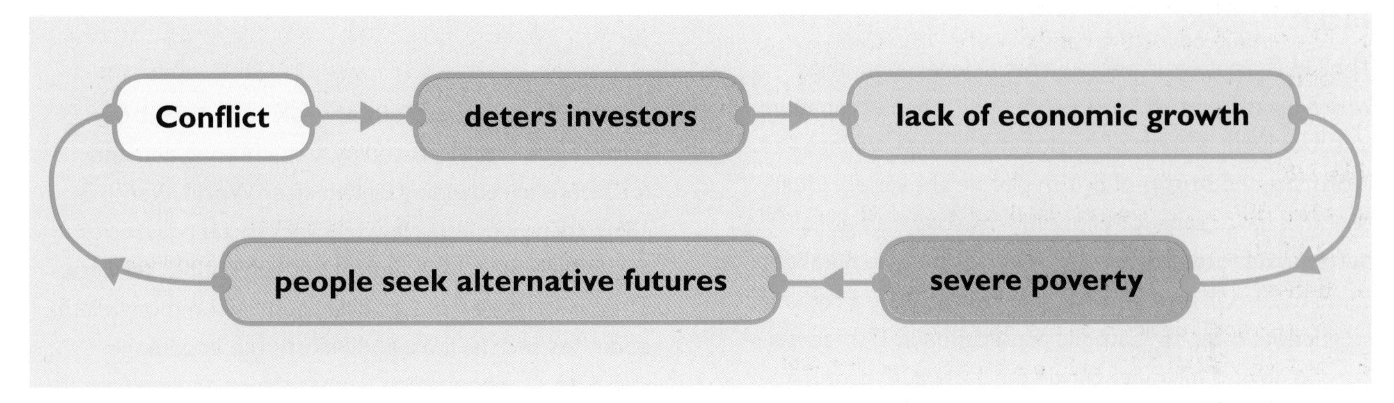

Figure 6.48 *The links between conflict and poverty* ▲

ACTIVITIES

1. In pairs, list possible reasons why people in Royal Wootton Bassett line the streets (Figure 6.34) when the bodies of the soldiers pass through their town.
2. Read the Afghanistan Factfile. In pairs:
   - **a** prepare a risk assessment for the British Army posed by physical and human factors
   - **b** identify which you believe are the three greatest difficulties for forces there.
3. Draw three Venn diagrams labelled social, economic and environmental impacts of conflict. Then, using information from pages 270–277 complete the diagram.
4. In pairs, discuss:
   - **a** why there are such differences in belief over women's rights in Afghanistan
   - **b** whether countries outside Afghanistan should fight for womens' rights there.
5. In 1000 words, discuss this statement: *'there can be no peace without development, and no development without peace'*.

Internet research

Visit the website 'onlinenewspapers.com' and select 2 or 3 Afghan newspapers. What is being reported about life there? What is the coverage of the conflict like in the newspapers? Is their reporting different from that of UK newspapers?

The challenge of multicultural societies

In this section you will learn about:

- the reasons for the development of multicultural societies
- the geographical distribution of cultural groupings
- issues related to multicultural societies

During May 2001, rioting took place in Oldham, Greater Manchester (Figure 6.49). These were the first disturbances in any British city for a decade and the most serious since 1981 when widespread rioting occurred in cities across the UK.

The media claimed that the cause of the Oldham riots was linked to the city's multicultural communities. During 2001, the British National Party (BNP) and other right-wing groups had stirred up Oldham's Asian neighbourhoods where social tensions had been simmering for some time.

Figure 6.49 *The riots in Oldham, 2001* ▲

How has the UK become a multicultural society?

The UK is a nation of people from different countries who have **migrated** here. From the Roman occupation to subsequent arrivals of Anglos and Saxons, Vikings and Normans, the origins of British people are varied. Most migrants have come for economic reasons, namely, a better and more prosperous life. Considerable numbers came to the UK to escape persecution, such as Protestant Huguenots escaping Catholic persecution in 17th-century Europe, Jewish migrants during the early 20th century, and Ugandan Asians escaping Idi Amin in the 1970s. As a result, the UK is now among the world's most multicultural countries.

Today, 'multicultural' has come to mean 'multi-ethnic', where people of different nations, religions and skin colour settle in a community (Figure 6.50). In the UK this has largely resulted from immigration policies which since 1945 have been closely tied to Britain's labour market.

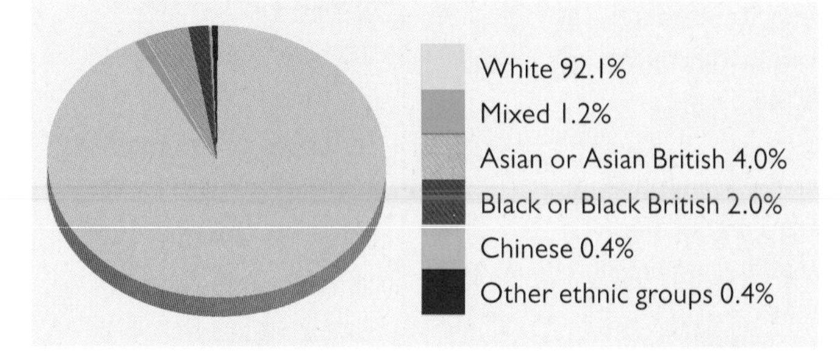

Figure 6.50 *This graph shows the proportions of the UK population from ethnic groups in 2001* ▲

1 From 1945 until the 1960s, the UK suffered **labour shortages** mainly caused by an increasing demand for labour in rebuilding Britain after World War II. By 1950, there were too few workers to keep factories in production, drive transport or run council services. The British government advertised in Commonwealth countries such as India and Pakistan to encourage migrants. Caribbean and Pakistani men arrived first, followed by Bangladeshis and Indians. Once they were in work, their families arrived, most settling in cheap terraced housing in manufacturing cities such as Manchester and East London.
2 UK membership of the European Union. The EU's 1992 Maastricht Treaty created free movement between people of member states. The political argument for this was that free labour movement was necessary to maintain economic growth and provide skilled workers.
3 The development of the '**knowledge economy**', particularly in London, has led to an expansion in the financial, legal, IT and media sectors of the economy. This expansion has required a pool of well-qualified skilled labour, which has drawn large numbers of overseas migrants attracted by high salaries. It has also created a demand for skilled construction workers who have arrived from eastern countries in the EU, such as Poland.

What does multicultural mean?

Most UK cities are multicultural, technically meaning that people from different countries or religions live there. The term 'multicultural' means that significant numbers of people differ from the majority, in that they:

- do not speak the majority language as their first language
- celebrate customs and festivals outside the calendar of the majority
- eat and, sometimes, dress differently from the majority, and
- observe different religious beliefs.

London is by far the UK's most multicultural area, with over 300 languages spoken there (Figure 6.51).

However, 'multicultural' can mean much more. For instance it can refer to:

- the extent to which different cultures co-exist – are they separate (or **segregated**) or have they adopted a 'majority culture' that blends with the rest? Have first- or second-generation children of migrants retained their cultural traits? Increasingly, UK politicians blur 'multiculturalism' with 'nationalism' – i.e. the extent to which new migrants accept that they are 'British'.
- the variety or types of communities – in many cities a degree of segregation of different ethnicities has occurred. Some groups live in '**enclaves**' and often have little understanding of the lifestyles of others.
- the extent to which people are 'different'. In 2008, the head of the UK's Commission for Racial Equality attacked 'multiculturalism', which defined people as 'different' – i.e., black, white, Muslim, Irish etc. – and *treated* them differently according to their ethnicity rather than their rights or responsibilities as citizens. He claimed that multiculturalism discouraged common goals, for example, defining someone as 'Muslim' identified and stereotyped them as 'different'.

Figure 6.51 *Multicultural London* ▶

Multicultural policies

Over time, three policies have emerged regarding the integration of ethnic groups into existing societies. These policies all centre on the question of whether different ethnicities should adopt the way of life of the majority.

Separation

Separation policy suggests that because people of different ethnicities have little in common with the majority population, they should be kept separate. This approach has influenced policies in several countries, notably Australia, which pursued a 'white Australia' migration policy in the 1960s, and South Africa, which until 1994 practised a policy of complete separation known as 'apartheid'.

Assimilation

Assimilation expects new migrants to lose their distinctiveness, such as their style of dress or beliefs, and adopt the culture of the host country.

Pluralism

Pluralism expects ethnic groups to participate and contribute to their host country, yet maintain their identity. It encourages communities which are different but not isolated. A pluralist society applies common values to all. Everyone in the society has the same rights and access to services; in turn, they are expected to both accept the society and actively participate.

The development of ethnic 'enclaves'

Migrant populations are highly urbanised (Figure 6.52), usually because of preference and access to jobs. For example, London contains the following concentrations of the UK's ethnic groups:

- 78% of Black Africans
- 61% of Black Caribbeans
- 54% of Bangladeshis
- 32% of White Irish people (3% of the population).

The Pakistani community is less concentrated in London – 19% of all Pakistani migrants live there compared to 21% in the West Midlands, 20% in Yorkshire and The Humber, and 16% in the North West.

Ethnic communities tend to concentrate, resulting in geographically separate areas, or **enclaves**. In most urban areas, migrant communities choose to live in such enclaves. Each consists of residents, businesses (e.g. specialist food shops), religious buildings, and community institutions. Two American sociologists, Portes and Wilson (1980) referred to ethnic enclaves as economic units, describing them as *'a tight community of buyers and sellers; the ethnic enclave commonly includes residences and meeting places but is foremost a community of businesses'.* Figure 6.53 shows how different ethnic groups have segregated into different geographical areas of London.

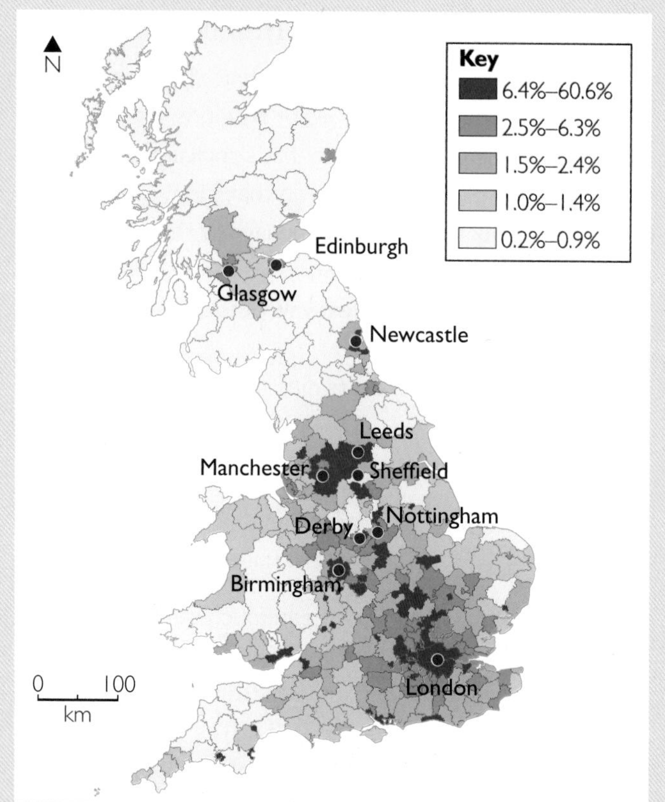

Urban region	% of all non-white residents
London region	45%
West Midlands	13%
South East	8%
North West	8%
Yorkshire and The Humber	7%
North East	<4%

Figure 6.52 *Urban concentrations of non-white populations in the UK* ▶

Figure 6.53 *A comparison of the concentrations of Indian and Black African populations in London* ▼

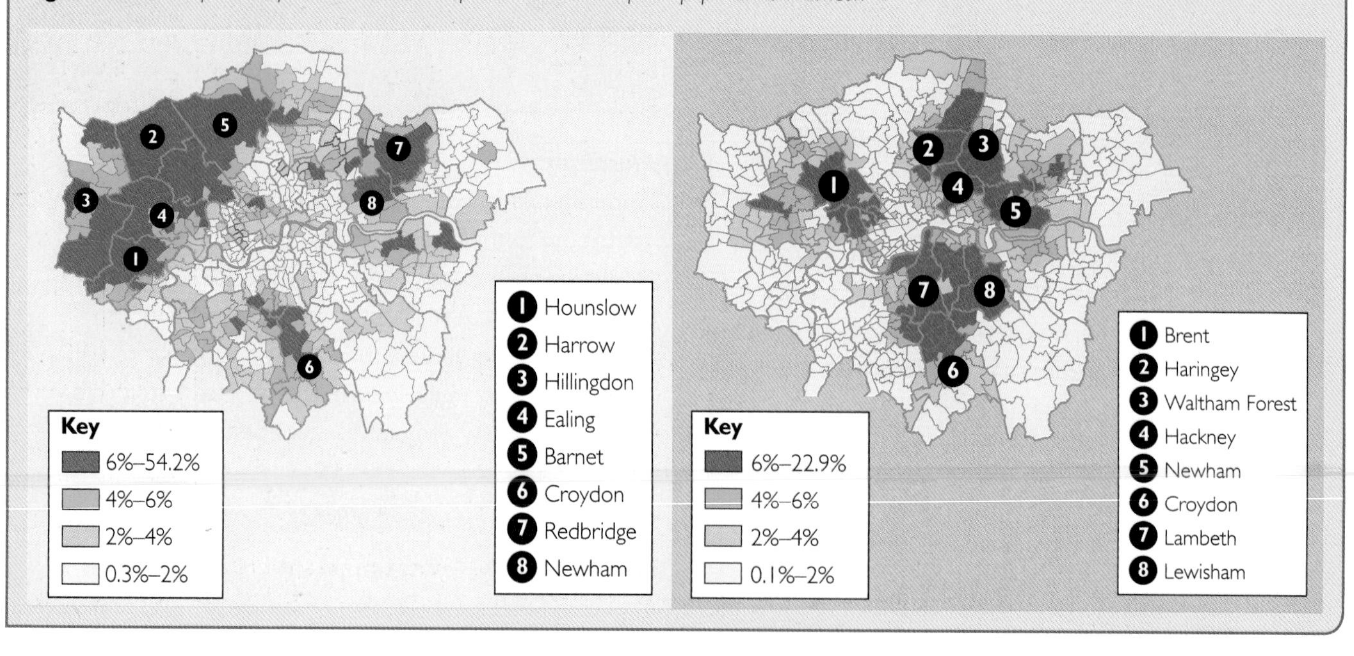

Enclaves were first recognised in an early sociological model of the city of Chicago by Ernest Burgess in the 1920s (Figure 6.54), a model which geographers have used extensively. Burgess recognised that in early-20th-century Chicago people of similar national (e.g. Italian) or ethnic-religious (e.g. Jewish) origin tended to live in close communities. Early communities in the USA were described by the nationality of early settlers, such as Little Italy in New York. Some still are – most major cities now have a Chinatown.

Historically, the clustering of different ethnic groups arose from either discrimination or poverty which prevented ethnic or religious minorities from settling in more prosperous parts of town. However, many enclaves also arose out of preference and economic motive, and at their best, represent the diversity of urban life – Bangladeshi restaurants in east London's Brick Lane, Bradford's 'curry trail', and Notting Hill's annual carnival are good examples. Ethnic enclaves in the UK are usually more integrated and harmonious than those in cities of the USA.

Some ethnic groups disperse to other urban areas as their economic situation improves and subsequent generations are born. This is true of London's Jewish population (Figure 6.55) which has moved from the poorest areas of east London into which they arrived in the early 20th century from Russia and Eastern Europe, outwards to Golders Green and suburbs of north London. Now, professional and business migrants from Europe, the USA and Australia settle in middle-class suburbs of London (e.g. Richmond-on-Thames or Hampstead).

But some ethnic enclaves persist, mainly because of the increased deprivation which has occurred among some ethnic communities. During the 1980s, many UK manufacturing companies closed their factories and moved overseas where labour was cheaper, causing mass unemployment. This particularly affected manufacturing areas in inner cities with large ethnic concentrations – the closures led to concentrated poverty. For example, 96% of Rochdale's Pakistani and 89% of its Bangladeshi communities live in former industrial areas which are among the most deprived areas in Greater Manchester. Unemployment there remains persistently higher than average. Black African migrants in east London similarly suffer high unemployment or low incomes, even when they are well-qualified.

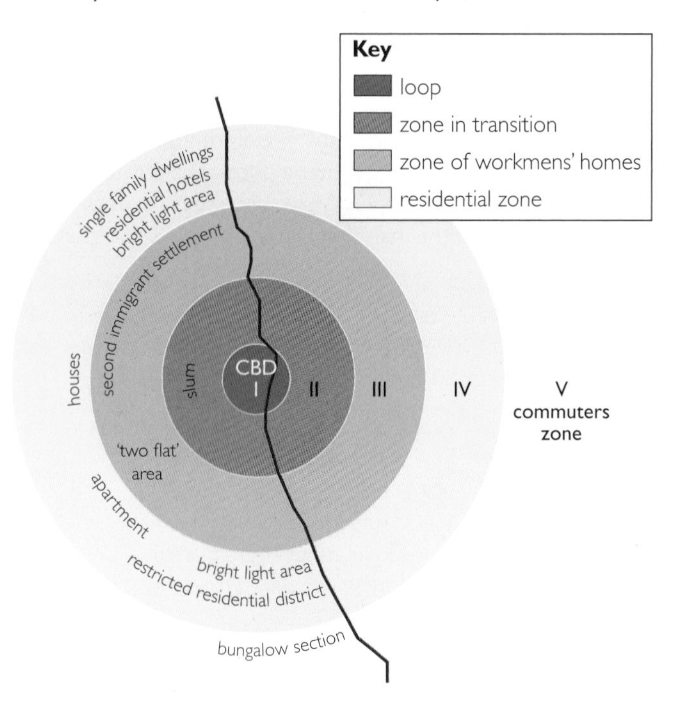

Figure 6.54 *The Burgess model from Chicago in 1924 reflecting 'ethnic enclaves' as they were perceived then* ▲

Figure 6.55 *Buildings that remain from early-20th-century Jewish communities in east London, near Brick Lane. This area is now occupied by the Bangladeshi community.* ▲

What challenges does multiculturalism pose?

Many people see nothing but positive benefits from living in a multicultural society, while others see problems. These opposing views are summarised in Figure 6.56.

Did you know?

There were substantial Indian and Black African communities in London in the 18th century as a result of immigration to Britain via trading companies such as the East India Company.

Did you know?

Research shows that bi- or multi-lingual children generally learn more quickly than those who speak just a single language.

Positive aspects of multiculturalism	Challenges posed by multiculturalism
• a better understanding of different cultures • increasing tolerance • anti-immigration political parties (e.g. BNP) have little impact	• protest groups form among those opposed to immigration, such as the English Defence League and the BNP, who often demonstrate intolerance towards others
• awareness of other religions and their values, for example, many schools have holidays for festivals, such as the Muslim holiday of Eid, as well as Christmas • greater diversity in cultural life, e.g. in street festivals (Notting Hill Carnival) and different cuisines (e.g. Indian and Thai)	• increased ethnic tensions, though rarely in areas with migrant communities. In London's Notting Hill area in 1956, riots between Caribbean immigrants and younger whites caused damage and 108 arrests. The annual August Carnival was created to reduce that tension.
• a well-integrated population provides a source of national pride and a positive self-image • London is often portrayed overseas as a city where multiculturalism works	• ethnic tension and violence, such as the riots in Brixton and Toxeth in 1981, and 2001 in Oldham and Bradford. However, many past riots have been caused by outsiders — Oldham's riots in 2001 were led by BNP supporters from Stoke-on-Trent.
• migrants bring business links and investment – in 2002 migrants brought £2.5 billion into the UK	• the economic impact of migrants is not evenly spread. Multiculturalism has the biggest economic impact in London but less in regions such as the South West.
• Home Office research shows that a 1% increase in population from migration creates a 1.5% increase in GDP • 12% of the UK's working age population is migrant, boosting the UK economy by £6 billion per year	• greater demand for housing, both privately rented and for sale, results in higher rents and house prices. The greatest house price increases since 2000 have occurred in expensive areas of London where many highly paid EU or US immigrants live.
• migrant workers fill job vacancies that are often low-paid (e.g. in catering and cleaning) or where there are shortages (e.g. doctors filling jobs in the NHS)	• impact on schools. A survey from Reading schools showed that over 150 languages were spoken by children at primary schools, causing a shift in staff resources.
• immigrants to the UK: o increase the percentage of the working population, an important factor as the host population ages o are 5% less likely than UK-born people to claim benefits o take less in public services than the UK-born population, but contribute to tax revenues at about the same rate o are 58% less likely to live in social housing	• some enclaves are socially more isolated, e.g. Bangladeshis in East London, where: o women are less likely to speak English o people have fewer educational qualifications o most people earn less than other migrants (typically earning under £150 a week compared to those from Australia averaging over £750)

Figure 6.56 *The positive and the challenging aspects of multicultural societies* ▲

Meeting the challenges of multiculturalism

How well does the UK meet challenges posed by multiculturalism? Immigrants not only contribute to the UK economy, but many multi-ethnic communities function well and are engaged in the wider community. For example, the council of the London borough of Newham has councillors and employees from a variety of ethnic groups, all of whom are involved in local decision-making. The council recognises that lower language skills, poor educational achievement and poverty are linked and must be tackled. Educational achievement in Newham has risen sharply in response to school improvement programmes there.

Why, therefore, do some communities – white or other – suffer tensions? The key cause is usually poverty resulting from the closure of manufacturing companies. It results in low achievers in the most deprived areas staying in low-paid work or in unemployment for lengthy periods. Pockets of low educational achievement and deprivation exist in all of Britain's cities. An under-class has emerged in some deprived areas that sees multiculturalism as the cause of their poverty. Political parties such as the BNP are most active in recruitment among the most deprived white communities. What they fail to see is that ethnic communities often share that same poverty, and that poverty is a complex problem.

Figure 6.57 *A BNP campaign poster distributed in London. By 'local', many people felt that they meant 'white' whereas many ethnic groups are second- or third-generation British and just as local as the white population.* ▲

ACTIVITIES

1. Using examples, outline the economic and social reasons why different peoples have migrated to the British Isles since Roman times.
2. Draw a table to show the strengths and disadvantages of **a**) segregation, **b**) assimilation and **c**) pluralism as policies relating to ethnic groupings.
3. In pairs, **a**) explain the reasons ethnic groups tend to live in ethnic enclaves. **b**) What benefits and problems arise from the development of such enclaves?
4. Use Figure 6.56 to draw a table showing the economic, social and cultural advantages and disadvantages that a multicultural society creates for the UK.
5. Write a 1000-word essay entitled 'Multicultural society – a good or bad thing?'

Internet research

Use news websites (e.g. news.bbc.co.uk or skynews.com) to research details of one urban riot since 2000 (e.g. Oldham 2001) and one from 1981 (e.g. Toxteth in Liverpool, Tottenham in London). Research a) causes, b) events leading to the riots, c) impacts of the riots. How far are the two similar or different?

Internet research

Identify one local council (e.g., an inner London borough such as Newham) in which there are large ethnic communities. Research the efforts made to integrate ethnic communities, using the following headings:

- Is there support for education, e.g. improving literacy?
- Are employment opportunities and training being provided?
- Are the needs of minority groups being met (e.g. multi-lingual council websites)?
- Is there support for festivals, holidays and celebrations?
- Is there acceptance of religious/traditional dress?

The challenge of global poverty

In this section you will learn about global initiatives to tackle poverty, including the Millennium Development Goals

The north-south divide

In 1981, a landmark report known as the Brandt Report was published about **global development**. The report identified a disparity between rich and poor countries which it called the **development gap**.

Although today issues surrounding global wealth are more complex, for some countries things have barely changed and the gap between rich and poor has widened. In 1980, 26 of the 30 poorest countries were in sub-Saharan Africa – this is still true today.

Figure 6.58 shows that countries with the lowest Gross Domestic Product (GDP) per capita are also poorest in terms of the Human Development Index (HDI – see pages 217 and 221). The reason that this pattern remains is largely due to debt.

The top ten	HDI Rank	GDP rank
Norway	1	3
Australia	2	10
Iceland	3	13
Canada	4	12
Ireland	5	9
Netherlands	6	8
Sweden	7	16
France	8	22
Switzerland	9	7
Japan	10	23
The bottom ten	**HDI Rank**	**GDP rank**
Guinea-Bissau	173	167
Burundi	174	179
Chad	175	152
Congo (Democratic Republic)	176	182
Burkina Faso	177	158
Mali	178	165
Central African Republic	179	176
Sierra Leone	180	175
Afghanistan	181	171
Niger	182	177

Figure 6.58 *The links between HDI and GDP. All ranks are out of 182 countries.* ▲

The origins of global debt

Most of the poorest countries are in sub-Saharan Africa and, until recently, almost all have had huge debts. Few have any hope of repaying the debt as interest alone exceeds the value of their exports. The causes of the debt are complex, but in most cases can be traced back to the 1970s:

1. Between the late 19th century and the 1960s, the majority of countries in Africa south of the Sahara had been **colonies** ruled by European powers. In the 1960s, most obtained their independence and were keen to invest in future economic growth.
2. In the 1970s, **OPEC** (Organisation of the Petroleum Exporting Countries) raised the price of oil twice – massively increasing its members' earnings. Members invested these earnings in Western banks.
3. The banks lent this money to developing countries – often for development projects such as dams, but also at times to finance conflict and oppressive regimes.
4. In the 1980s, global interest rates more than doubled, massively increasing repayments needed to manage these loans.
5. Many countries found themselves unable to meet debt repayments so unpaid interest was added to the original loans. Each year, debts grew as unpaid interest mounted. Banks grew nervous – would they get their money back?
6. To prevent a collapse of the world's banking systems, in the late 1980s and 1990s, the International Monetary Fund (IMF) developed a solution called **Structural Adjustment Packages (SAPs)**. This involved re-scheduling loans to make them more affordable. However, re-scheduling was only allowed if governments agreed to spending cuts set by the IMF. Without IMF approval, countries could not get any further credit – SAPs therefore became compulsory.
7. The biggest government spending items were usually health and education. Cutbacks imposed by the IMF affected both, with the greatest impact being on the poor.

Ways forward 1: The HIPC initiative

The Highly Indebted Poor Countries (HIPC) are the 38 least developed countries in the world with the greatest debt. 29 are in sub-Saharan Africa. In 1996, their poverty made them eligible for help from an IMF and World Bank programme to reduce debt. Further reductions came in 2005, when the UK held the presidency of the G8 (an association of the world's richest nations, consisting of the USA, Canada, Britain, France, Germany, Italy, Japan and Russia). Under an agreement made at the G8 in Gleneagles, the World Bank, the IMF and the African Development Bank cancelled loans (worth US$40 billion) owed to them by 18 HIPC countries (Figures 6.59 and 6.60). This single action saved those countries $1.5 billion annually in debt repayments. However, the countries were required to meet two conditions:

- each government had to demonstrate good financial management and a lack of corruption
- the money saved had to be spent on poverty reduction, education and healthcare.

By 2008, 27 of the 38 HIPCs had met the conditions for debt relief and had received US$85 billion in aid. But many African countries still owed US$300 billion and at present there is little chance of them being able to repay those debts. For countries affected by war (e.g. Sudan) or natural disasters, new loans are needed just as old debt is being cut. Debt campaigners believe that 24 more countries should be added to the list, including India and Indonesia.

"It is a splendid start, and one hopes that they will go on to cancel all debt for most of the countries – I gather it is about 62 countries – which are heavily indebted. But remember the West had a hand in promoting some of those leaders, because it suited them at the time."

– Reaction from Bishop Desmond Tutu to the Gleneagles G8 debt agreement in 2005

	1990	2005
Congo	19.0	2.3
Gambia	11.9	6.3
Côte d'Ivoire	11.7	2.8
Madagascar	7.2	1.5
Malawi	7.1	4.6
Ghana	6.2	2.7
Zambia	6.1	3.3

Figure 6.59 *Levels of debt in some sub-Saharan African countries. The figures show the percentage of GDP taken up by debt repayments in 1990 and in 2005, following large write-offs of debt by western countries.* ▲

Figure 6.60 *The countries eligible for debt relief in 2005 under the HIPC initiative* ▼

Key

HIPC countries that qualified for debt relief in 2005

1 Benin
2 Bolivia
3 Burkina Faso
4 Ethiopia
5 Ghana
6 Guyana
7 Honduras
8 Madagascar
9 Mali
10 Mauritania
11 Mozambique
12 Nicaragua
13 Niger
14 Rwanda
15 Senegal
16 Tanzania
17 Uganda
18 Zambia

Key

HIPC countries that in 2005 did not yet qualify for debt relief by meeting the required conditions

19 Afghanistan
20 Burundi
21 Cameroon
22 Central African Republic
23 Chad
24 Democratic Republic of the Congo
25 Republic of Congo
26 Côte d'Ivoire
27 The Gambia
28 Guinea
29 Guinea-Bissau
30 Haiti
31 Kenya
32 Kyrgyzstan
33 Liberia
34 Malawi
35 São Tomé and Príncipe
36 Sierra Leone
37 St. Lucia
38 Yemen

CASE STUDY

Debt cancellation in Uganda

The cancellation of $2 billion debt under the HIPC initiative has had major impacts in Uganda (Figure 6.61):

- Spending on public services has risen by 20% – with 40% extra being spent on education and 70% on healthcare, including the removal of fees for basic healthcare.
- Primary schooling is now free. Five million extra children now attend school. Enrolment rates for primary schooling increased from 62.3% in 2000 to 92% of girls and 94% of boys in 2006. Before debt relief, there were 20% fewer girls than boys in primary school, but now the numbers are almost even.
- 2.2 million people (nearly 10% of the population) have gained access to clean water. Girls benefit most as fetching water is usually the responsibility of women and girls, and is often a reason for girls not going to school.
- Its HDI improved from 0.48 in 2000 to 0.514 in 2009.

Figure 6.61 *Progress in Uganda following the beginnings of debt cancellation in the late 1990s. The years for each set of data are shown in brackets.* ▼

	Before debt cancellation	During and after debt cancellation
Population using an improved water source (%)	44 (in 1990)	64 (in 2006)
Population undernourished (% of total population)	24 (1990-92)	15 (2005)
Public expenditure on education (% of GDP)	1.5 (1991)	5.2 (2002-05)
Public expenditure on education (% of total government expenditure)	11.5 (1991)	18.3 (2002-07)
Adult literacy rate (% aged 15 and older)	56.1 (1985-95 average)	73.6 (2007)
Young adult literacy rate (% aged 15-24)	69.8 (1985-95 average)	88 (2003-07 average)
Total debt service expenditure (% of income from exports, plus net income from abroad)	81.4 (1990)	2 (2007)

Figure 6.62 *But there's still some way to go … all data are for 2006-07* ▼

Population without electricity (millions)	24.6
Population without safe water (%)	36
Estimated earned income, female (PPP$)	861
Estimated earned income, male (PPP$)	1256
Adult literacy rate, female (% aged 15 and older)	57.7
Adult literacy rate, male (% aged 15 and older)	76.8

Figure 6.63 *Primary school students in Uganda* ▼

Ways forward 2: the Millennium Development Goals

At a United Nations summit in 2000, eight Millennium Development Goals (MDGs) were agreed to provide a set of development targets for the world to reach by 2015. Every UN member state signed up to these targets, making it the largest-ever multinational attempt to rid the world of extreme poverty. The United Nations Development Programme (UNDP) was tasked with reaching the targets listed below:

The Millennium Development Goals

1. **Eradicate extreme poverty and hunger**
   - **a** Halve the proportion of people living on less than US$1 a day
   - **b** Full employment and decent work for all
   - **c** Halve the proportion of people suffering from hunger
2. **Achieve universal primary education** and ensure that all complete it
3. **Promote gender equality and empower women** – eliminate gender disparity in primary and secondary education by 2015
4. **Reduce child mortality** for those under five by two-thirds
5. **Improve maternal health**
   - **a** Reduce by three-quarters the maternal mortality ratio
   - **b** Achieve universal access to reproductive health by 2015
6. **Combat HIV and AIDS, malaria and other diseases**
   - **a** Halt and begin to reverse the spread of HIV/AIDS, malaria and other major diseases
   - **b** Provide universal access to treatment for HIV/AIDS for all who need it by 2010
7. **Ensure environmental sustainability**
   - **a** Integrate principles of sustainable development into country policies and programmes; reverse loss of environmental resources
   - **b** Achieve significant reductions in the rate of biodiversity loss by 2010
   - **c** Halve the proportion of people without sustainable access to safe drinking water and basic sanitation
   - **d** Improve significantly the lives of at least 100 million slum dwellers by 2020
8. **Develop a global partnership for development**
   - **a** Develop further an open, rule-based, predictable trading and financial system
   - **b** Address the needs of the least developed countries, of landlocked developing countries and of small island developing states
   - **c** Deal with debt problems of developing countries to make debt sustainable
   - **e** Provide access to affordable essential drugs in developing countries
   - **f** Make available benefits of new technologies, e.g. ICT

Progress in achieving the MDGs

There have been some success stories. 41 million more children have enrolled in primary school, 2 million more receive HIV/AIDS treatment, and economic growth averaged 6% in sub-Saharan African countries by 2008. However, these are not sufficient to meet the MDG targets by 2015, especially in sub-Saharan Africa. For example, in 2007:

- over 500 000 women died from treatable, preventable complications of pregnancy and childbirth. The odds of a woman dying from these in sub-Saharan Africa are 1 in 16 compared to 1 in 3800 in the developed world.
- 980 million people still live on less than US1$ per day.

To speed progress, a programme entitled 'Call to Action' was launched in 2005, led by UN Secretary-General Ban Ki-moon and UK Prime Minister Gordon Brown. Its focus was to create a coalition of governments working in developing countries with:

- the **private sector** to mobilise technology and job creation
- **professionals** in medicine to find cures for diseases that most affect developing countries (e.g. TB, Malaria), and in education
- **NGOs** to encourage people in developed countries to continue contributing to good causes and governments to help financially in meeting the MDGs
- **faith groups** who have the networks to mobilise millions in eliminating poverty
- **consumers** in the developed world, for example, to promote development by buying Fair Trade products.

CASE STUDY

Progress in Bangladesh?

After China and India, Bangladesh has the world's third-largest number of poor. Of its 150 million people (2007), half live in poverty, with 50 million in extreme poverty. With its large, rapidly growing population, Bangladesh faces by far the biggest financial challenge in meeting the MDGs – to do so would cost an average of US$14.5 billion **each year** between 2005-15.

However, a healthy 5% average growth in GDP allows the government to raise more money than many other countries – it needs US$45 per capita each year, or a total of US$7.5 billion. International aid is also helping.

Bangladesh faces huge challenges:

- in education where the number of teachers needs to rise from 350000 in 2005 to 815000 by 2015
- in health where the number of doctors will have to double to 58000, and nurses and midwives increase by a factor of four to 145000
- its surface water supplies are badly infected – 20000 people die each year from diarrhoea.

UK and UNDP Starts Project to Uplift 3 Million People from Extreme Poverty

To improve the livelihoods and living conditions of 3 million urban poor, especially women and girls, the UNDP and the UK Department for International Development (DFID) has initiated a US$120 million development project with the government of Bangladesh and city corporations. The project, which will continue until March 2015, covers 30 towns. It will:

- support the development of poverty reduction strategies at town level
- link people with local micro-finance bodies
- give access to a range of financial services to community groups for housing and business development
- help urban poor communities to create healthy living environments
- support poor families to acquire the skills they need to increase their income.

Figure 6.64 *Professor Muhammad Yunus, founder of the Grameen Bank in Bangladesh, visits a Grameen Bank Centre. This bank has pioneered the concept of small, or micro, loans to women, allowing them to start businesses and gain economic power. The bank also finances the Grameen Phone Project, which aims to equip every Bangladeshi village with a telephone.* ▶

Millennium Development Goal	Progress	Comment
1 Eradicate extreme poverty and hunger	Poverty is reducing and numbers living on $1 per day or less will halve by 2015.	In 2009, 36.3% of the population still lived below the poverty line.
2 Achieve universal primary education 3 Promote gender equality and empower women	Significant progress has occurred. There is now equality in school attendance between girls and boys. Average length of schooling is now 8 years.	Literacy rates are still low (48%) and show gender disparity (boys 54%, girls 41%).
4 Reduce child mortality	Mortality among under-1s and under-5s reduced by 2.8% and 2.3% respectively. If maintained, the under-5s rate will decrease by two-thirds by 2015.	Infant mortality was still 58 per 1000 in 2008, one of the world's highest rates. Less than one-third of births are attended by a skilled professional.
5 Improve maternal health	Maternal mortality rate fell from 574 to 360 deaths per 100000 live births between 1990-2001, but is still high and remains a challenge.	Target of 143 per 100000 won't be met – disappointing because such deaths are largely preventable. Causes include severe bleeding, eclampsia, unsafe abortions, and obstructed labour.
6 Combat HIV/AIDS, malaria and other diseases	Improvements in containing the spread and fatality of malaria and TB. Spread of HIV-AIDS remains low.	Measles cases have fallen dramatically. The lack of insecticide-treated nets is one of the key health issues in rural areas.
7 Ensure environmental sustainability	Considerable progress in ensuring safe drinking water and sanitation in urban areas.	Maintaining wetlands and biodiversity is a challenge due to population pressure and loss of land from flooding caused by cyclones. Some rare species (e.g. Bengal tigers) remain under threat.
8 Develop a global partnership for development	Significant reduction in debt has been achieved.	Under- and un-employment among youth remains a challenge.

Figure 6.65 *Progress in Bangladesh towards achieving the MDGs by 2008* ▲

ACTIVITIES

1 Draw a flow chart to show **a**) how Uganda and other LICs got into debt, and, **b**) how the debt climbed to such high levels.
2 In what ways are the eight MDGs linked? Draw a spider diagram showing the MDGs. Add arrows and label any links, e.g. between poverty and education.
3 What are the benefits for both LICs and HICs in cancelling all debt?
4 **a** Using a scale of 0 -10, score the progress Bangladesh has made in meeting each of the Millennium Development goals. How do you rate overall progress?
  b How likely does it seem that the MDGs will be met by 2015? Use examples to explain your answer.
5 Assume that Bangladesh achieves all of the MDGs by 2015. In pairs, discuss and answer these questions:
  - What should the next set of goals be for 2015-2030?
  - Will the development gap still be a matter of concern then?

Internet research

Research the overall progress by nations in meeting the MDGs. The following websites will be helpful:

- UK DfID (dfid.gov.uk/mdg)
- UNDP (undp.org/mdg)
- UN Millennium Goals (http://unmillenniumproject.org).

How does general progress compare with that of Bangladesh?

arbitration Where a dispute requires the intervention of a third party

assimilation The expectation that new migrants will lose their distinctiveness in favour of the culture of the host country

autonomy The right of self-government and self-determination

colony A country ruled by a more powerful country

conflict Disagreement caused by the actual or perceived opposition of needs, values, and interests between people

conflict continuum The span between non-violent conflict (e.g. political activity and debate) and direct action and violence (e.g. terrorism, insurrection, and war)

development gap The differences between poorer countries of the developing world (or LICs), and wealthier developed countries (or HICs)

diaspora The movement or migration of people with a common national and/or ethnic identity away from their perceived homeland

ethnic enclaves Communities of people from similar cultural or ethnic backgrounds, e.g. nationality, language, or religion

ethnicity A belief in a common cultural descent, similarities of physical type, or of customs, or because of memories of colonisation and migration; a group with similar beliefs and experiences

fundamentalism An uncritical belief in the rightness of one's own opinion, with a refusal to accept any other viewpoint

geopolitics Political conflict and how it varies over space

insurrection Violent uprising against civil authority, established government, or the rules of that government

LEQ (Loudness EQuivalent) A measure of noise around airports or other locations in decibels at different distances

lobbying The process by which individuals and organisations (including commercial companies) seek to win over politicians by promoting a viewpoint

localism Where issues or values bring local people together, e.g. against large organisations or governments

manifesto A programme which sets out the beliefs and programme of action for a political party and what it would do in office

Millennium Development Goals A set of eight development targets agreed by UN members in 2000 to be reached by 2015

multicultural Where significant numbers differ from the majority population, e.g. in terms of language, customs and festivals, food, dress, or religious beliefs

nation A cultural community, with five main dimensions: psychological, cultural, historical, territorial, and political

nationalism A belief in the value of a nation and its sense of identity for a population

pluralism The expectation that ethnic groups will participate and contribute to their host country, yet maintain their identity

public enquiry An official review ordered by the Planning Inspectorate of the UK government about a major planning issue

regionalism Where one region tries to take control of functions from a national government

separation Treating different ethnic groups separately assuming that they have little in common

separatism To advocate political separation from a larger majority on the basis of cultural, ethnic, religious, or political differences

Structural Adjustment Programmes (SAP) Plans imposed on governments seeking debt relief from the IMF

territory Where a geographic area is defined for a particular purpose, e.g. in exploration, or identifying resources such as oil

EXAMINER'S TIPS

Section B

1 (a) Study Figure 6.7 (page 252). Comment on the relationship between conflict and economic development. *(7 marks)*

(b) Explain why separatist groups emerge. *(8 marks)*

(c) Explain the part played by one separatist group you have studied in bringing about conflict. *(10 marks)*

(a) State the relationships you notice, illustrating each with an example.

(b) Focus on one separatist group and think about the reasons for its development and growth.

(c) Consider the part played, e.g. by the Kurds, in bringing about conflict. What pressures has your chosen group put upon a central government? How has the government reacted? Use examples.

2 (a) Study Figure 6.53 (page 280). Compare the concentrations of Indian and Black African populations in London. *(7 marks)*

(b) Using examples, explain why ethnic groups choose to live in communities, or 'enclaves'. *(8 marks)*

(c) Explain how far multicultural societies provide challenges for places you have studied. *(10 marks)*

(a) Stick to description – first of general patterns, and then examples of specific places.

(b) Think about the economic and social benefits to ethnic groups of living close together.

(c) The key part of this question is that you should show 'how far' the things you write about are challenges, great or small. Think about social challenges, then economic; to hit the top level you will need to assess how great the challenges are.

Section C

1 Referring to examples, explain the factors that cause internal differences to escalate into more serious conflicts. *(40 marks)*

2 Assess the economic, social, and environmental impacts of an international conflict that you have studied. *(40 marks)*

(1) The key point is not to show narrative about how a conflict develops, but to explain the factors that lead from an internal separatist conflict and disagreement (e.g. over space) into something more serious – for example, the factors (political, religious) that led an internal conflict in Afghanistan to became a major international dispute.

(2) Spending a minute or two on a good plan here will help you to develop an answer well, so that you cover economic, social, and environmental impacts. Remember that to hit the top levels you must assess – i.e. weigh up which impacts are greatest and least, and why.

Unit 4A: Geography fieldwork investigation

The exam paper

The assessment for Unit 4A involves a 1½ hour written exam worth 60 marks. The exam is worth 20% of your overall grade. There are two sections to the exam paper:

Section A: In this section you must answer three questions within 60 minutes. The questions are worth a total of 40 marks.

Your answers should relate to the geographical fieldwork investigation you undertook in preparation for the exam. The questions are both short- and extended-answer and cover different aspects of your fieldwork investigation, such as the reasons you chose particular forms of data presentation.

Section B: This shorter section (20 marks and a time limit of 30 minutes) covers the more generic aspects of the AQA geography specification with an emphasis on fieldwork skills. This section of the exam contains stimulus material, such as tables of data, and may include questions relating to statistical procedures.

You will not be able to take any notes or other materials into the exam with you.

In this chapter we will concentrate on the steps you need to take to ensure a successful fieldwork investigation, from choosing a topic through to analysing and presenting your findings. We will also go over the specific A2 level geography skills outlined in the AQA geography specification. All other skills are considered to be AS level and were covered in the Oxford University Press AQA AS textbook. A checklist of both the AS and A2 skills can be found on page 303.

Stages in a geographical investigation

Like all other scientific investigations, a good fieldwork investigation follows a series of logical stages. This helps ensure that the investigation is accurate, thorough and stands as a valid piece of research.

AQA has developed a list of fieldwork skills you must demonstrate as you carry out your investigation (Figure 7.1). These skills will be assessed in your exam – you may find it helpful to use Figure 7.1 as a checklist when you are revising. They will also help you conduct a good and valid fieldwork investigation.

By the end of your investigative fieldwork, you will be expected to:

- ☐ display an understanding of the purpose of the investigation and relevant spatial and conceptual background
- ☐ demonstrate knowledge and understanding of the geographical content, concepts and processes
- ☐ plan, construct and carry out sequences of enquiry
- ☐ show an awareness of the suitability of the data collected and the methods used
- ☐ be aware of the alternatives and evaluate methodology
- ☐ use this information in a straightforward way, presenting it in a different or more easily understood form, e.g. graphs, maps
- ☐ be familiar with alternative methods of data presentation/processing
- ☐ analyse, interpret and evaluate geographical information, issues and viewpoints and apply understanding in unfamiliar contexts
- ☐ draw conclusion(s) relating to the specific enquiry, understand their validity, limitations and implications for the study
- ☐ demonstrate an awareness of safety issues and risk assessment in geographical fieldwork
- ☐ select and use a variety of methods, skills and techniques to investigate questions and issues, reach conclusions and communicate findings
- ☐ use and understand your own experience of fieldwork and enquiry.

Figure 7.1 *The AQA fieldwork investigation skills* ▲

Stage 1: Identifying an appropriate question or hypothesis

The specification requires that you conduct:

> *'an individual investigation of a geographical argument, assertion, hypothesis, issue or problem'*

In choosing your investigation title you should consider the following points:

- your title must be geographical and linked to the specification
- there must be one or more clear connections to sound geographical theory, concepts or processes (such as the Bradshaw model of changes down the course of a river)
- primary data (data that is collected first-hand through observations or measurements) must be included and, where appropriate, also secondary data (data that is published or processed)
- tthe investigation must be based on a small manageable area of study such as a stretch of a small river or a small town. Consider that 1-2 days maximum should be needed for collecting the data.
- the investigation title must lend itself to the full development of a geographical investigation. For example, it should provide plenty of data for interpretation and statistical analysis.
- you must be able to conduct the investigation safely – no abseiling, wading up to your waist in fast flowing rivers, crossing motorways, etc!

Some examples of appropriate titles include:

How and why do river channel characteristics change with distance downstream in the River Horner?

Local climate characteristics differ between a north- and a south-facing slope on Exmoor

Beach characteristics vary between the east and west side of Porlock Bay

Do sand dune characteristics change with distance from the sea at Braunton Burrows?

Urban land use changes with distance from the peak land value intersection (PLVI) in Southampton

Counter-urbanisation has changed the characteristics of the village of Trull

Stage 2: Planning

One of the more common problems students find when conducting their geographical investigation is the lack of proper planning. Thorough, careful planning and forward thinking will prevent problems and help you achieve a successful outcome.

When planning your investigation you need to consider the following:

Data collection

- Where are you going to collect your data and when?
  - The choice of site(s) is very important and must be justified
  - You should also consider when to carry out your data collection. Early morning or late afternoon? On a weekday or weekend? Figure 7.2 below lists some factors you should consider when conducting a river or an urban study.
- What data are you going to collect and why?
  - You need to carefully consider what data you need to fully test your hypothesis or answer your question
  - Be careful not to collect so much data that you simply become overwhelmed
- What sampling strategies are you going to employ and why?
  - Make sure you choose an appropriate strategy
  - Consider whether you need to employ point, line (transect) or area (e.g. quadrat) sampling
  - Consider too whether you wish to choose **random** (chance), **systematic** (regular interval, e.g. every 10th person or every 10 metres) or **stratified** (biased, e.g. 25% sample points in each of four areas) sample locations
- What equipment is needed and how does it work?
  - You need to select appropriate equipment for an A2 level geography investigation. For example, the use of a rudimentary float to measure river velocity might be fine in Year 7 but at A2 level you should ideally use a more scientific instrument such as a flow meter.
  - The equipment should be reasonably easy to use, not too expensive and should be of good enough quality to provide you with accurate and reliable results
  - Take time to practise using the equipment before you start your investigation
  - Consider having 'Plan B' equipment (e.g. a float in case your flow meter does not work)

Considerations	River	Urban
Where?	• Is the river an appropriate (safe) size to enable me to collect the data I need? • How can I get there? • Does it have the appropriate characteristics for my study, such as exhibiting marked changes with distance downstream? • Can I gain safe (and legal!) access to an appropriate number of sites (i.e. about 8-10 if conducting a downstream study)?	• Where do I go to collect my data and why? • Are some locations more accessible than others? • Is the settlement big enough? Is it too big? • How safe is the urban area? Are some areas too risky for me to go to?
When?	• Will the flow rate be suitable and safe (not too low at mid-summer or too high in winter)? • Will access be available (e.g. use of fields for crops or livestock)? • Will the time of year affect biological indicators, such as vegetation, etc? • Will the weather be ok?	• Will the time of year affect my results (tourism)? • Will the time of day (rush hour) affect my results? • Is the day of the week important (e.g. Sunday compared with Monday)? • Will the weather be ok?

Figure 7.2 *Some factors to consider when collecting data for rivers and urban studies* ▲

Risk assessment

It is a legal requirement to carry out a risk assessment. A risk assessment aims to identify the potential risks of your fieldwork investigation and then minimise their likelihood of occurring.

To start, think about the risks that could occur on the day you conduct your fieldwork investigation (ideally, you should pay a preliminary visit to your study site to see for yourself what the likely risks are). Then, identify ways of minimising the risks – Figure 7.3 below shows part of a typical risk assessment of a fieldwork site.

It is also important that you consider what to do if the something does happen, for example, if you get a tick bite or the weather turns wet and foggy. There is a huge amount of helpful advice on the Internet and your school or college will be able to help you too.

Figure 7.3 *The potential risks and some possible solutions of a fieldwork investigation* ▲

Data presentation and analysis

It might seem strange to consider how you are going to present and analyse your data at the start of your investigation, but it is very important that you do so. In order for you to produce academic diagrams and interpret your data using statistical techniques, you need to have the appropriate type and amount of data. For example, if you plan to use Spearman's rank correlation test, you will need to be sure that you collect two sets of data that can be ranked, such as velocity and hydraulic radius, ideally with between 10–20 readings of each.

Stage 3: Data collection

When collecting data in the field you need to try to ensure that the results you obtain are accurate and reliable. For example, make sure that you take several readings at a particular site and keep double-checking that your results are accurate and reliable. Check the equipment regularly. Exam questions often focus on 'accuracy and reliability' so keep a record of what you do to address these issues.

Don't forget to take photos and draw field sketches. These provide a good record of what you did and also help to put your fieldwork sites in geographical context. For example, the photograph in Figure 7.3 could be used to help explain certain trends or anomalies in the results of a river study.

Try to include secondary data if possible. This might include maps, extracts in books or articles in magazines, statistics or other students' work. Ideally, having another study to compare with your own is an excellent strategy.

Stage 4: Data presentation

Raw data is of very little value when it comes to data interpretation. It needs to be processed and displayed so that it makes sense and reveals any trends and patterns. Look online or in any fieldwork textbook and you will find a vast range of presentation methods available to you, such as histograms, kite diagrams and dispersion graphs. At A2 level you need to select your methods with care and make sure they are drawn accurately and are appropriate to your study.

For example:

- if you carry out a transect in a town, data should be displayed in a linear manner along a scaled line as opposed to a single pie chart
- points on a scatter graph should not be joined up to form a line graph but should have a best fit line to show the trend
- spatial data (e.g. pH readings at a number of sites on sand dunes) is most effectively shown on a base map.

Make sure that all diagrams are complete. Don't forget to include:

- a clear and focused title
- north arrow pointer and scale bar on a map
- labelled axes on graphs
- annotations on field sketches and photos
- a key, if appropriate.

Stage 5: Interpretation and analysis

In this section of your investigation you need to make some written comments about your results. To do this effectively you should:

- Make use of appropriate statistical techniques to analyse your data. Consider using measures of central tendency (e.g. mean) or dispersion. If you are examining a relationship between two variables then you should consider using Spearman's rank correlation test. If you are examining associations or looking to identify significant differences between data sets, you should consider using the Chi-squared or Mann Whitney U tests (see below).
- Describe your results. Consider patterns and trends rather than simply quoting lots of numbers. But be sure to use numbers to support any points you make.
- Look for any exceptions (anomalies) to the general trends and explain why they have occurred
- Attempt to explain the patterns and trends you have observed using geographical terminology and referring to concepts and processes as appropriate
- Try to make statements that link your data sets together
- Make connections back to your title.

Significance levels – don't forget!

With all statistical tests, the calculated result is only as good as the interpretation. Significance tables should be used to assess the probability that the result was due to chance. For most geographical investigations a significance level of 0.5% is reasonable, i.e. there is a 95% likelihood that the result was **not** caused by chance. Any test of association or correlation (e.g. Spearman's rank correlation test) must be based on a sound geographical concept or theory.

Chi-squared test

The Chi-squared test is used to test the difference between grouped data sets. The two sets of data can be either two sets of raw data or a comparison between actual observed values and theoretical 'expected' values representing equal chance.

There are a number of important aspects of this test:

- Data must be in the form of frequencies, grouped or categorised
- The total number of observations must exceed 20
- The expected frequency for any one group must exceed 5
- The categories should not have a directly causal link.

Mann Whitney U test

The Mann Whitney U test is used to test the difference between two sets of data. It is often carried out after drawing a dispersion graph, which might suggest a possible difference between two sets of data, for example, the sizes of pebbles collected on two beaches.

There are a number of important aspects of this test:

- It measures the difference around median values
- Raw individual data must be used rather than grouped data
- There should be between 5-20 values, although it is possible to use 20+
- The two data sets do not need to have an equal number of values.

Stage 6: Conclusion and evaluation

Your conclusion should pull together all of your results so you can address the question or hypothesis in your title. To do this, briefly run through the main thrust of your results, but make sure that you focus on the title.

To evaluate your investigation you should:

- consider whether there were any inaccuracies in the data collection
- suggest how data collection could have been improved (e.g. more sites, more data, different types of equipment, etc.)
- discuss the accuracy of your results, bearing in mind any problems with the data collection
- assess the validity of the conclusion, i.e. would a different person doing the same investigation get exactly the same results as you did?

8 Unit 4B: Geographical issue evaluation

In this chapter you will learn about:

- achieving success in the Unit 4B examination
- making the most of the Advance Information Booklet (AIB)

Unit 4B contrasts with the other units in the A2 examination because it is an issue evaluation exercise. The exam is based on an issue introduced in an Advance Information Booklet (AIB) released well in advance. It tests a range of geographical skills, knowledge, and understanding. Issue evaluation is the complete test of the complete geographer. Thorough preparation is the key to success – see Figure 8.1. The Unit 4B exam demands:

- application of knowledge and understanding gained throughout the AS and A2 courses
- further research and wider 'reading round' the issue
- the ability to present and analyse data from the AIB
- data interpretation skills
- consideration of evidence from alternative points of view
- an awareness of limitations in the AIB data and resources
- the identification of potential conflicts – and how to overcome or at least reduce them
- the capacity to come to and justify a decision
- appreciation of the impacts of that decision.

▼ **Figure 8.1** *Prepare to succeed*

Before the examination		
8 weeks to go	**4 weeks to go**	**2 weeks to go**
Become familiar with the pre-release material (AIB) through reading, re-reading, and then reading again! Treat everything as relevant.	Extend your knowledge through targeted wider reading. In part, this will be sign-posted in the AIB as 'Ideas for further study'. It may include reference to fieldwork or extended internet research.	Be absolutely clear on how additional information could be collected using fieldwork or internet research if directed. Don't rely on vague references – refer to fieldwork you carried out in your course or the websites you visited.
Identify the overlapping themes (for example, Human and Physical Geography) and match this to content in your file notes and textbook(s).	Identify different interest groups, and outline arguments for and against. Rank answers in order of importance. A conflict matrix may help (see Figure 8.2).	Develop a sense of empathy – evaluate different options or decisions from these different points of view. Prepare balanced answers.
Create a glossary of any words or phrases that are unfamiliar – the examiner will assume that you understand all the concepts and any underpinning theory.	Invest time interpreting statistics – identify trends and anomalies. Avoid simple data repetition. Instead make the figures work for you (for example, calculate differences or percentage change).	Check that your notes convey a real 'sense of place' – in other words, your notes are not just generic, about any place, but might include place references or named local groups. Make full use of all map evidence in support. Make it 'real'.

▼ **Figure 8.2** *An example of a conflict matrix*

Conflict matrix for a proposed new bypass				
	Local government	Sustrans (an NGO)	Local environmental pressure group	Local business
Local government		?	?	✓
Sustrans (an NGO)	?		✓	✗
Local environmental pressure group	✓	✓		✗
Local business	✓	✗	✗	

Note: Conflicts may then be ranked in order of importance of social, economic, or environmental concerns.

Key ✓ = no conflict ✗ = conflict ? = possible conflict

In the examination

DOs

- **Do** read the whole question paper before answering.
- **Do** follow the guidance in the AIB.
- **Do** take note of references to Item or Figure numbers in **bold**. It is likely that you will need to make explicit reference to them throughout an answer.
- **Do** treat everything as relevant (in particular, every artwork will have been carefully prepared and presented).
- **Do** think as a geographer (consider differences and contrasts over time and space).
- **Do** support opinions with clear evidence. (This can be qualitative, such as a specific quote, or quantitative, such as a re-worked figure or statistic.)
- **Do** work methodically and use the lines in the answer booklet as a guide on how much to write.
- **Do** treat the question paper as a single exercise – ideas will be developed through the entire question paper and there will be opportunities to make links and connections to issues and/or answers considered earlier, i.e. think **synoptically** (Figure 8.3).
- **Do** remember that the mark range will indicate whether the question is point or level marked (Figure 8.4). Most questions will be level marked.

▼ ***Figure 8.3*** *Thinking and working synoptically*

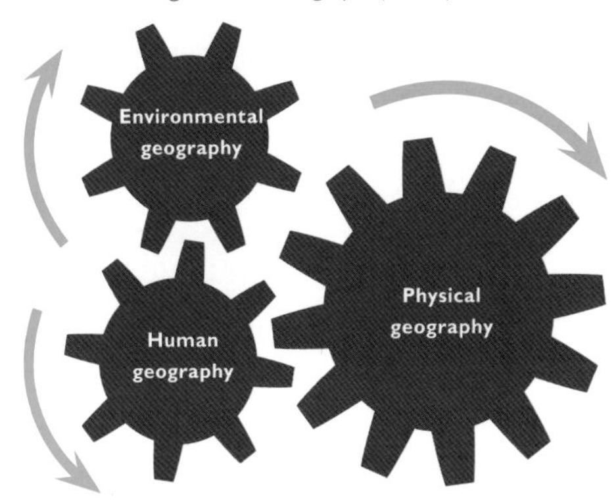

synoptic To present a general view or summary; this will only be achieved by considering links, inter-relationships and applying geographical understanding to both familiar and less familiar contexts.

DON'Ts

- **Don't** leave preparation to the last minute. (Why else would the AIB be released over two months in advance of the examination?)
- **Don't** panic when the case study is unfamiliar – it is supposed to be!
- **Don't** be careless in your use of statistics – remember to be precise and accurate.
- **Don't** over-quote from the AIB. (This shows little or no understanding and may trap your answer in Level 1.)
- **Don't** second-guess the questions and pre-prepare a narrow set of answers.
- **Don't** spend too long answering early questions – remember good time management.
- **Don't** rush the reading of the question and misinterpret or confuse command words.

▼ ***Figure 8.4*** *Point and level marking*

Point marking: 1–4 marks.

A mark is awarded for each clearly developed and illustrated point.

Level marking: 5 or more marks.

Level 1 (Basic): attempts the question to some extent, but is over-simplified, and lacks clarity and real-world exemplification.

Level 2 (Clear): answers the question well, is accurate, and demonstrates a clear understanding of the topic. Appropriate exemplification is used.

Level 3 (Detailed): answers the question very well, is detailed, and demonstrates a thorough understanding of the topic. Appropriate exemplification is used to support all points made.

Issue evaluation exercise: Switzerland's integrated public transport

The following information about public transport in Switzerland is arranged as a series of 'Items', just as it would be in the Advance Information Booklet.

The Activities on page 302 are in the style of the Unit 4B examination, with appropriate weighting of marks allocated.

Item 1: Background

Switzerland is renowned for its stunning scenery, one of the highest living standards in the world, low unemployment, and efficient public services.

But Switzerland's environment faces severe pressures from agriculture, industry, tourism, and transport. These pressures, which stem from high population densities, a high level of economic activity, and the country's location in the heart of Europe, have deeply concerned the Swiss people.

In consequence, for over 30 years, ambitious environmental policies, not least to combat pollution, have been promoted by the central government. These have been implemented wholeheartedly by the regional cantons (federal states) and communes. Switzerland's environmentally-friendly transport policies, focusing on efficient integrated public transport, help promote more sustainable national economic progress.

But a highly-integrated public transport system is expensive to construct and maintain. Switzerland is helped by two things: it is one of the richest countries in the world, and is relatively small. Even so, some people in Switzerland would prefer to pay lower taxes, or would prefer the government to spend money in other areas.

Item 2: Switzerland's integrated public transport

Public transport in Switzerland operates as a single integrated network. The system is regarded as one of the finest in the world. Measured in numbers of journeys per person, Switzerland is second only to Japan in public transport usage.

Three international airports (Zurich, Geneva, and Basel) and another 61 airports and airfields (including Sion, St. Moritz, and Gstaad, used by smaller aircraft) are linked by enviable transport alternatives. All railways, bus services, and tramways operate within a timetable developed to ensure that almost any point in the country can be accessed once per hour.

A national tariff system called Direct Travel ensures that, in most cases, one ticket is enough for any single journey – even if numerous railway, bus, and, on occasions, cable and ship operators are involved.

The significance of all this is demonstrated in over 2 million citizens carrying half-fare travel cards, with another 1 million on regional travel passes. A further 350 000 carry travel cards entitling them to unlimited travel on all services anywhere in the country. Furthermore, the punctuality of the system is legendary – 95% of arrivals are achieved with less than 5 minutes' delay and 75% with less than 1 minute delay!

Most notable is that the Swiss have arguably 'lost their love for cars.' In national referendums they have even voted for higher petrol prices and lower speed limits on motorways! This may seem incomprehensible to other Europeans, but to the Swiss they represent sensible, environmentally-friendly initiatives to promote sustainable 'ecological forms of mobility' and reduce air pollution.

Did you know?

The total for Unit 4B is typically 60 marks, in a 90-minute examination. This equates to one mark for every one and a half minutes.

Did you know?

In Unit 4B, there are 2 lines per mark in the question/answer booklet on which to write your answer. Use this as a good indication on how much to write.

Item 3: The railway system

Switzerland has been described as 'a nation of train passengers'. Public transport, not the car, is the preferred choice for millions of business people, commuters, families, students, and politicians. Nowhere in the world is the rail network used as intensively (Figure 1). For example, over 2000 trains pass per day through Zurich – more than one per minute! The key is that all service providers unite in a so-called Verkehrsverbund (transport association), with the same tickets valid on trains, tramways, buses, and even ships.

Switzerland's main railways are totally electrified and operated by the nationally-owned Swiss Federal Railways.

There are also a number of privately-owned railway companies – but the Swiss confederation, cantons, and communes hold a vast majority of their capital. They also subsidise their infrastructure and operation.

The network is further extended by narrower-gauge lines, including cogwheel railways in narrow alpine valleys such as in Valais and the Bernese Oberland. These are particularly scenic and so most of their passengers are tourists – both foreigners and Swiss. Among them are the famous Glacier Express linking Zermatt, Andermatt, and St. Moritz (Figure 2).

▲ ***Figure 1*** *Switzerland has the highest train density in Europe with excellent Intercity connections, including extended S-Bahn networks of fast metropolitan area trains*

▲ ***Figure 2*** *The Glacier Express – 'the world's slowest express train'*

Item 4: The bus system

For any long-distance route within Switzerland there is at least one train per hour from early morning to midnight. There are no public long-distance buses. This is because trains have greater capacity, are comfortable, faster, cheaper, and more reliable in that they are not affected by traffic congestion.

Switzerland's bus service is therefore complementary rather than competing. Cross-country buses are operated on routes with very little traffic and almost every village can be reached by a regional bus service several times a day – most of them at least once per hour.

Swiss Post operates many of these services with their distinctive yellow Post buses (Figure 3). Again, schedules and tickets are combined into the integrated public transport system. Locals and tourists alike can simply plan journeys online, getting all necessary transfer information, and even print-outs of through-fare tickets.

▲ ***Figure 3*** *Post bus outside Sion railway station, Valais*

Item 5: The road network

▲ *Figure 4 Car-free Zermatt, Valais*

More than 1600 km of motorways (interstate expressways) and 70 000 km of other highways and second-class roads form a dense road network – a greater investment than in public transport infrastructure!

But planning policies continue to 'gently persuade' drivers to consider more sustainable transport alternatives. Only one in two Swiss inhabitants now owns a car. Indeed, some popular resorts, such as Zermatt, have even gone so far as to go car-free.

Elsewhere, however, the main east-west (A1) and north-south (A2) motorways and major Swiss cities remain congested. Finding a parking space can be difficult, not least where bus-lanes and reduced car parking places (such as in Geneva) limit capacity.

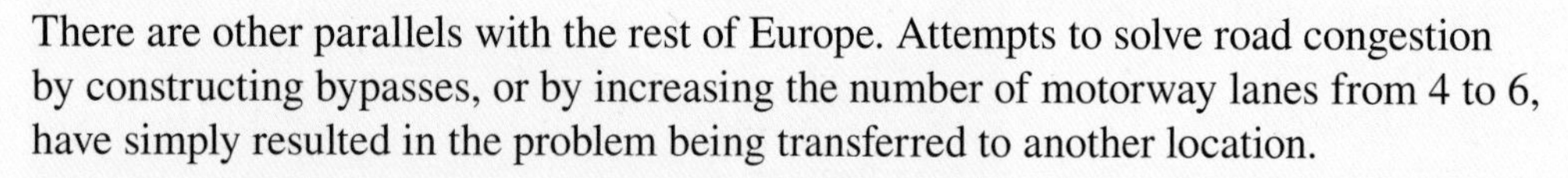

There are other parallels with the rest of Europe. Attempts to solve road congestion by constructing bypasses, or by increasing the number of motorway lanes from 4 to 6, have simply resulted in the problem being transferred to another location.

One particular traffic bottleneck is the Alps. Since the opening of the Saint Gotthard road tunnel (16.3 km long), road traffic crossing the Alps has increased considerably – causing heavy air pollution. Consequently, an ambitious target to reduce road freight through the Alps is now enshrined in the Swiss constitution. The New Alpine Rail Link, or Neue Eisenbahn-Alpentransversale (NEAT), aims to move 90% of all goods in transit through Switzerland by 2018.

Item 6: Ideas for further study

You can find out more about integrated public transport in Switzerland, including NEAT, at the following websites:

http://www.urbanrail.net/eu/zur/zurich.htm

http://www.high-end-travel-switzerland.com/Train-Travel-in-Switzerland.html

http://www.alptransit.ch/pages/img/projekt/The_new_Gotthard_rail_link.pdf

You can find out more about public transport in the UK at: http://www.dft.gov.uk/

ACTIVITIES

1. Outline a graphical technique you might use to present comparative data on passenger numbers for different modes of transport and explain why you chose the technique. *(7 marks)*
2. Using three headings – **Social**, **Economic**, and **Environmental** – describe the advantages of integrated public transport. *(15 marks)*
3. You should have carried out research into integrated public transport in Switzerland using one or more websites. Critically evaluate **one** or **more** of the websites that you visited. *(8 marks)*
4. 'Voting for higher petrol prices and lower speed limits on motorways is an illustration of how the Swiss promoted sustainable ecological forms of mobility.' Using all the resources provided, evaluate the evidence that Swiss transport policies are some of the most environmentally friendly in the world. *(15 marks)*
5. 'An integrated transport solution for the UK is as unrealistic as it is unattainable.' Discuss this opinion with reference to transport provision in both the UK and Switzerland. *(15 marks)*

AQA Geography skills checklist

In your fieldwork investigation exam you will be required to show your understanding of a range of basic, investigative, cartographic, graphical, applied ICT and statistical skills. You will be expected to demonstrate knowledge of the suitability of the different skills, as well as their limitations.

Basic Skills

To include:

- annotation of illustrative material, base maps, sketch maps, OS maps, diagrams, graphs, sketches photographs etc.
- use of overlays
- literacy skills.

Investigative Skills

To include:

- identification of aims, geographical questions and issues, and effective approaches to enquiry
- identification, selection and collection of quantitative and qualitative evidence, including the use of appropriate sampling techniques, from primary sources (including fieldwork) and secondary sources
- processing, presentation, analysis and interpretation of evidence
- drawing conclusions and showing an awareness of the validity of conclusions
- evaluation, including further research opportunities
- risk assessment and identification of strategies for minimising health and safety risks in undertaking fieldwork.

Cartographic Skills

To include:

- atlas maps
- base maps
- sketch maps
- Ordnance Survey maps at a variety of scales
- maps with located proportional symbols – squares, circles, semi-circles, bars
- maps showing movement – flow lines, desire lines and trip lines
- choropleth, isoline and dot maps
- weather maps – including synoptic charts
- detailed town centre plans.

Graphical Skills

To include:

- line graphs – simple, comparative, compound and divergent
- bar graphs – simple, comparative, compound and divergent
- scatter graphs – and use of best fit line
- pie charts and proportional divided circles
- triangular graphs
- radial diagrams
- logarithmic scales
- dispersion diagrams
- kite diagrams.

ICT Skills

To include:

- use of remotely sensed data – photographs, digital images including those captured by satellite
- use of databases, eg census data, Environment Agency data; meteorological office data
- use of geographical information systems (GIS)
- presentation of text and graphical and cartographic images using ICT.

Statistical Skills

To include:

- measures of central tendency – mean, mode, median
- measures of dispersion – interquartile range and standard deviation
- Spearman's rank correlation test
- application of significance level in inferential statistical results
- comparative tests – Chi-squared, Mann Whitney U Test.

Developing extended writing skills

What is extended writing?

In your AQA Geography A2 exams, 'extended writing' takes the form of essays. The length of these essays and the marks available varies between the different papers:

- Unit 3 – essays are marked out of 40
- Unit 4A – the longer part questions are marked out of 12
- Unit 4B – the longer part question is marked out of 15

How is extended writing marked?

Extended writing is marked using Levels Marking. This means that you need to reach particular thresholds in your answer, e.g. by the detailed use of case studies or by making sure that you are really focusing on the question in hand. When writing extended answers, it is very much about the quality of what you write, not the quantity!

Levels Marking criteria

Level 1: Basic attempts to answer the question. Simple description/explanation. Lack of clarity.

An answer at this level is likely to:

- display a basic understanding of the topic
- make one or two points without the support of appropriate exemplification and/or application of principle
- demonstrate a simplistic style of writing, perhaps lacking close relation to the terms of the question and/or failing to communicate the complexity of subject matter
- lack organisation, relevance, and specialist vocabulary
- demonstrate deficiencies in legibility, spelling, grammar, and punctuation which detract from the clarity of meaning

Level 2: Well-answered question and accurate use of language

An answer at this level is likely to:

- display a clear understanding of the topic
- make one or two points with the support of appropriate exemplification and/or application of principle
- give a number of characteristics, reasons, and/ or attitudes ('more than one') where the question requires it
- provide detailed use of case studies
- give responses to more than one command, e.g. 'describe and explain ...'
- demonstrate a style of writing which matches the requirements of the question and acknowledges the potential complexity of the subject matter
- demonstrate relevance and coherence, with appropriate use of specialist vocabulary
- demonstrate legibility of text, and qualities of spelling, grammar, and punctuation which do not detract from the clarity of meaning

Level 3: Very effectively and clearly answered question

An answer at this level is likely to:

- display a detailed understanding of the topic
- make several points with the support of appropriate exemplification and/or application of principle
- give a wide range of characteristics, reasons, and/or attitudes, etc.
- provide highly detailed accounts of a range of case studies
- respond well to more than one command
- demonstrate evaluation, assessment, and synthesis throughout
- demonstrate a sophisticated style of writing incorporating measured and qualified explanation and comment as required by the question and reflecting awareness of the complexity of the subject matter and any incompleteness or tentativeness of explanation
- demonstrate a clear sense of purpose so that the responses are seen to closely relate to the requirements of the question with confident use of specialist vocabulary
- demonstrate legibility of text, and qualities of spelling, grammar, and punctuation which contribute to complete clarity of meaning

Deconstructing the question

This is the first step in writing a successful essay. At A2, questions will be quite demanding and they may at first reading appear quite daunting. However, you should take a deep breath, count to ten, and read them through carefully – and read them more than once!

Here is a typical question that you might be asked in Unit 3:

> Discuss the importance of the theory of plate tectonics in reducing the earthquake hazard.

Notice that the question has a command word, 'discuss'. It has a focus that is drawn from the specification ('earthquake'). And it asks you to focus on a particular aspect or slant ('importance of the theory of plate tectonics in reducing the hazard'). To answer this question successfully, you must adapt your knowledge about earthquakes and plate tectonics to focus on the question and make sure that you enter into some discussion where you present different (often opposing) points of view.

Common command words at A2

Account for
Analyse
Assess
Comment on
Compare
Contrast
Define
Describe
Discuss
Evaluate
Examine
Explain
How far?
Illustrate
Justify
List
Outline
To what extent?

For an explanation of these command words, see page 311.

What are the characteristics of a successful essay?

A successful essay should have the following features:

- Clear focus on the question in hand, both in terms of the geographical content and in addressing the command word(s) correctly.
- Clear and logical structure, showing evidence of planning.
- Use of detailed and appropriate case studies.
- Use of clear and informative diagrams where appropriate (generally, students are very reluctant to use diagrams, yet examiners are keen to reward their use).
- Use of correct and appropriate geographical terminology.
- Correct use of spelling, grammar, and punctuation.
- For longer essays (Unit 3), the use of a clear introduction and conclusion.

The value of making a plan!

You should never start a piece of extended writing without first making a plan. This might take 3-5 minutes.

- Start by deconstructing the question so that you are clear about what you need to include.
- Make a list or draw a spider diagram (it doesn't matter which) of what you think you need to include in your essay. Include key geographical terms and case studies. Write down any facts and figures that you remember.
- Now re-read the question to make sure you have interpreted it correctly.
- Rearrange your points so that it your plan is logical and fits the requirements of the question.
- Look at the question once more to check that you have clearly interpreted the thrust of the question and the meaning of the command word(s).
- Begin your answer.
- Read through your answer at the end to correct any errors!

Extended writing in Unit 3

In Sections A and B, the questions are marked out of 7, 8, and 10. The 10-mark question requires a piece of extended writing – think of it as a short essay!

- You will come across command words such as 'discuss' and 'evaluate' – so make sure you know what they mean.
- Time is limited – so get on with answering the question as soon as you have made a quick plan.
- Don't write more than a sentence or so as an introduction and conclusion.

In Section C you will be expected to write a fully developed essay. This is worth 40 marks and will be a very demanding exercise. It is essential that you prepare yourself thoroughly by writing a detailed plan and that you make a wise choice of essay title to answer. Remember that you must not answer from the same options chosen in Sections A and B.

The key aspects that examiners will be looking for are as follows:

- Strong evidence of thorough, detailed, and accurate knowledge and critical understanding of concepts and principles and of specialist vocabulary.
- Explanations, arguments, and assessments or evaluations that are direct, logical, perceptive, and purposeful and that show both balance and flair.
- A high level of insight, and an ability to identify, interpret, and synthesise a wide range of material with creativity.
- Evidence of maturity in understanding the role of values, attitudes, and decision-making processes.
- Examples that are well developed and sketch maps and/or diagrams that are fully integrated in the answer.
- An answer that is fully synoptic. This means that it draws upon a range of geographical themes and aspects in answering the question.

Extended writing in Unit 4A

In Unit 4A, there are a number of questions involving extended writing, each worth 12 marks. These questions require short, well-focused, punchy essays. There is no need for lengthy introductions and conclusions.

- Be sure to respond correctly to the command word (e.g. 'Describe the location of your fieldwork investigation …').
- Make sure you focus precisely on the requirements of the question.
- You should refer to your fieldwork investigation throughout your essay if required to do so by the title.

Extended writing in Unit 4B

In Unit 4B, a number of short essay answers are required for questions worth 8 and 10 marks, and there is one question worth 15 marks. The same principles apply as for other pieces of extended work – you must do what the command word says and focus rigorously on the requirements of the question.

As Unit 4B is based on pre-released material (text, data, maps, diagrams) you will often be required to refer to evidence from the resources in your answers. Make sure that you do this. Take time to plan your answer carefully, to make sure that you answer all parts of the question.

11 Exams: how to be successful

No matter how much you enjoy geography, preparation for the exam is essential. To be successful, you not only need to know and understand the geography that you have been studying, but you also need to know how you will be examined, what kinds of questions you will come up against, how to use what you know, and what you will get marks for. That is where this chapter can help.

What is the AQA A2 Geography specification all about?

The AQA specification aims to give you a broad understanding of the physical and human themes in Geography, in a contemporary setting. It includes new ideas, e.g. 'Contemporary conflicts and challenges', and offers a new twist to old geographical favourites, e.g. 'Development and globalisation'. Like the AS, it provides a balance between your own physical, human, and/or environmental interests and key geographical options that provide you with the knowledge, understanding, and skills to ready you for further study in higher education or employment – and, especially, it gives you understanding of contemporary issues affecting the world.

What are the A2 Units?

The A2 specification consists of two units – Units 3 and 4 – each of which has different options within it.

Unit 3 Contemporary Geographical Issues has two themes, physical and human. Within each theme are three options. You must study **three** of these in total, including at least **one** physical and **one** human.

The three **physical** themes, from which you or your teachers select at least one, are:

- Option 1: Plate tectonics and associated hazards
- Option 2: Weather and climate and associated hazards
- Option 3: Ecosystems: change and challenge

The three **human** themes, from which you or your teachers select at least one, are:

- Option 4: World cities
- Option 5: Development and globalisation
- Option 6: Contemporary conflicts and challenges

Unit 4 has two options, from which you – or, more likely, your teachers – need to select one:

- **Unit 4A Geography Fieldwork Investigation**
  This unit gives you the chance to develop an area of the subject into a detailed fieldwork study. It might be something you have studied already, e.g. one of the themes at AS, or something completely new. In the exam, you are not allowed to take your fieldwork investigation into the exam room, but you will be asked to analyse and evaluate it in the questions set, as well as to demonstrate fieldwork skills.

- **Unit 4B Geographical Issue Evaluation** is about an issue, which you study using a pre-release resource booklet which is released 2 months prior to the exam to help you research into an area which is based on – or extends from – the specification content.

The more you study the themes in the two A2 units, the more you should spot links between them – and also between them and the two AS units. For example:

- Unit 3 (above) and AS Unit 1 (Physical and Human Geography)
- Unit 4A or 4B (above) and AS Unit 2 (Geographical Skills).

▶ *The Daintree Rainforest in northern Queensland, Australia, is a unique ecosystem (explored in Unit 3 'Ecosystems: change and challenge'), but faces threats which require management as more and more tourists visit the area. Studying such issues will prepare you well for Unit 4 if you study the Geographical Issue Evaluation*

▶ *GM crop protests in France. GM crops are claimed to fix food supply problems (that you may have studied in AS 'Food supply issues'), but also cause conflict (explored in Unit 3 'Contemporary conflicts and challenges'); it could even provide a focus for fieldwork in rural areas, preparing you for Unit 4 if you do the Geography Fieldwork Investigation*

How will you be assessed?

There are two exams for A2 Geography; there is no coursework. This table shows what the exams are like:

Unit	Assessment information	Marks and method of marking
Unit 3 Contemporary Geographical Issues	This exam is 2 hours and 30 minutes long, and consists of three sections, A, B, and C. **Sections A (Physical Options) and B (Human Options)** each consist of three questions about the Options in Unit 3. You have to select and answer **one** from Section A and **one** from Section B. **Each question** in Sections A and B has three parts (a, b, and c), worth 25 marks in total – giving a combined mark for Sections A and B of 50. Part (a) will normally be about a resource, such as a map, diagram, or data and is normally worth 7 marks. Part (b) will normally require you to answer based on your own knowledge and understanding and is normally worth 8 marks. Part (c) also assesses your knowledge and understanding of case studies you have investigated, and is normally worth 10 marks. **Section C** is an essay question. It consists of six questions, each worth 40 marks, from which you have to select **one**.	There are 90 marks in total for this exam, which are then converted to a UMS mark out of 120. All questions are level marked.
Unit 4A Geography Fieldwork Investigation OR Unit 4B Geographical Issue Evaluation	You do either 4A or 4B – not both! The exam for each of these is 1 hour and 30 minutes long. **Unit 4A** This exam has two sections, A and B. It consists of structured short and extended questions in each section. • Questions in **Section A** – on which you are advised to spend one hour – are based on fieldwork you have carried out, and its findings. • Questions in **Section B** – on which you are advised to spend 30 minutes – are about data handling skills. **Unit 4B** This exam is based on a pre-release resource booklet (normally about 12-16 pages long, which may include an Ordnance Survey map), based on a geographical theme. The booklet will be pre-released as advance information about two months before the exam. There is no restriction on the preparation you carry out using this pre-release booklet, but you must not take it into the exam with you. • The exam itself consists of compulsory questions, ranging normally from short structured questions worth about 4 marks, up to extended writing for about 15 marks. These are all based on the pre-release resource booklet, and range from data handling skills to discussion analysis of the issue in the booklet.	There are 60 marks for each of these options, which are then converted to a UMS mark out of 80. Shorter questions are point marked, whilst longer questions are level marked.

What will the exam questions be like?

The table on the previous page explained what type of questions you will get in the exams.
Here are examples from each unit:

Sample structured question for Unit 3 Section A (Weather and climate and associated hazards)

2 Study Figure 2, a map showing the impacts of Hurricane Katrina in 2005. (The map is not shown here for the purposes of this sample question.)

(a) Using Figure 2, comment on the impacts of Hurricane Katrina in September 2005. *(7 marks)*

(b) Explain the varied causes of tropical revolving storms. *(8 marks)*

(c) With reference to one tropical climate that you have studied, explain its characteristics. *(10 marks)*

(Total 25 marks)

Sample structured question for Unit 3 Section B (World cities)

4 Study Figure 4. (The map is not shown here for the purposes of this sample question.) Figure 4 is a world map showing the percentage of population living in urban areas by country.

(a) Comment on the variation in urban populations shown in Figure 4. *(7 marks)*

(b) Explain the causes of rapid urbanisation being experienced in many developing countries. *(8 marks)*

(c) Assess the effects of counter-urbanisation on large cities. *(10 marks)*

(Total 25 marks)

Sample essay question for Unit 3 Section C (Contemporary conflicts and challenges)

6 With reference to one recent separatist conflict, assess the geographical causes of the demands for separatism on the area(s) involved.

(Total 40 marks)

Sample question for Unit 4A

1 Describe the aims of your fieldwork investigation and explain how the locations in which you collected data were relevant to your aim(s). *(12 marks)*

2 (a) Justify one of the methods of data collection that you used for this investigation. *(6 marks)*

(b) Evaluate the effectiveness of this method, and suggest one or more improvements that you would make if you were to repeat the investigation. *(12 marks)*

(Total 30 marks of the 60 for the exam)

Sample question for Unit 4B

The advance resource booklet consisted of about 12 pages of materials about sustainability and transport issues in Guildford, including the use of cycle paths. (The resource booklet is not shown here.)

1 Explain why the government is trying to encourage the planners of urban areas to make transport systems more suitable for walking and cycling. *(10 marks)*

2 Refer to one or more areas that you have studied.

(a) Evaluate the suitability of the transport network for people walking or cycling to and from school or college. *(10 marks)*

(b) With reference to the area(s) that you referred to in 2(a), suggest how provision for walking and cycling in the area around the school or college could be improved. *(8 marks)*

(Total 28 marks of the 60 for the exam)

Making the step to A2

The biggest step up between AS and A2 is the requirement for extended writing. Whereas some questions at AS carry a few marks and require short answers, at A2 the questions range between 7 marks – in part (a) of Section A in the Unit 3 exam – and a full-blown 40-mark essay in Section C. Therefore, it is important that you develop your writing skills so that you can earn the full marks.

Developing extended writing

Most candidates can write at length. The difference is whether they can:

- understand the question
- plan their answer, so that it stays focused
- keep track of time
- learn and use case-study examples to support their argument(s)
- organise their essay

The following steps will help you to prepare for this.

Understanding the question

When reading an exam question, check the command words. They tell you what the examiner wants you to do. The table gives you some of the most commonly used A2 command words. The A2 command words are different from those at AS. For instance, you are unlikely to find 'describe' at A2, so refresh your knowledge of likely command words using this list.

The key rule is to read the exam question carefully and answer the question that has been set – not the one that you hoped would be set! Therefore, read and interpret the question to work out exactly what is being asked. Review the question through its key words (see the yellow box on the right).

Command words – these have distinct meanings, which are listed opposite.
Theme or option – this is what the question is about. The examiner who wrote the question will have tried to narrow the theme down, so that you do not write everything you know about the theme.
Focus – this shows how the theme has been narrowed down, e.g. the environmental impacts of development in China.
Case studies – look to see if you are asked for specific examples.

Here is an extract from a question that has been interpreted using the key words above. Part (a), about the resource stimulus, has been left out.

Command words

- *'Suggest the possible impacts.' You must give more than one possible impact.*
- *'Evaluate.' Having identified the various threats to the ecosystem, you need to weigh them up and show their relative importance.*

Theme or option

- *This question is from 'Ecosystems: change and challenge', and looks specifically at the factors and processes that threaten ecosystems.*

Question:

(b) Suggest the possible impacts of alien species on ecosystems. *(8 marks)*

(c) Evaluate the relative importance of global and local threats to one named ecosystem. *(10 marks)*

Focus

- *Part (a) asks you to 'Suggest the possible impacts' of alien species invasions on ecosystems.*
- *Part (b) asks you to give a named example of an ecosystem and 'Evaluate the relative importance' of threats to it.*

Case studies

- *Choose one named ecosystem only.*

▼ *Command words and how these might be used at A2*

Command word	What it means	Example question
Account for	Explain the reasons for. Marks are given for explanation, rather than description.	Account for the high rate of urbanisation in developing countries.
Analyse	Identify the main characteristics and rate the factors with respect to importance.	Analyse the social, environmental, and economic impacts of one international conflict that you have studied.
Assess	Examine closely and 'weigh up' a particular situation, e.g. strengths and weaknesses, for and against.	Using examples, assess the view that environmental impacts of conflict are usually the most serious.
Comment on	This is asking you to assess a statement. You need to put both sides of the argument.	Comment on the view that out-of-town shopping brings more problems than solutions to retailing in large cities.
Compare	Identify similarities and differences between two or more things.	Compare the impacts of tectonic hazards that you have studied in different locations.
Contrast	Identify the differences between two or more things.	Contrast the impacts of two or more tectonic events that you have studied.
Discuss	Similar to assess.	'The causes of the development gap largely lie in the debt burdens of the world's low-income countries.' Discuss.
Evaluate	The same as assess.	Evaluate this statement. 'The debate about whether aid or investment is better for developing countries misses the point – it is the relief of debt that matters most.'
Examine	You need to describe and explain.	Examine the attempts to manage the threats to an ecosystem you have studied.
Explain	Give reasons why something happens.	Explain how the desire to develop ecosystems often threatens them.
How far?	You need to put both sides of an argument.	How far are the conflicts in the economic development of an ecosystem you have studied a case of economic versus environmental interests?
Illustrate	Use specific examples to support a statement.	Illustrate the ways in which the most economically developed countries use resources as economic weapons.
Justify	Give evidence to support your statements.	Rank the threats affecting an ecosystem that you have studied and justify these rankings.
Outline	You need to describe and explain, but more description than explanation.	Outline the ways in which environmental issues can affect the economic development of a country that you have studied.
To what extent?	The same as 'How far?'	To what extent do you agree with the view of the World Bank that 'many countries are poor because they do not use their resource potential'?

Planning an answer

All the research shows that students who plan get more marks than those who do not. This is because:

- they stay focused – having a plan stops them from wandering off-track
- they do not suffer from 'memory blanks', in which they forget what to say

So, even if you never have before, learn to plan now! Planning need not be complex, nor take long. Take roughly 5-10% of the exam time to plan your answers. A 35-minute answer will require 2-3 minutes of planning. People plan differently – some make a list, some use spider diagrams, some make notes in the margin. Do what works for you. The plan gets marked – so don't cross it out!

Keeping to time

Like planning, keeping to time not only makes sense, but will help you earn marks. Bear in mind that you earn most marks in the first half of an answer. Prolonging the answer beyond its time slot will progressively earn you fewer and fewer marks. Therefore, it is better to draw a slightly incomplete question to a close and start a new one – you'll earn more marks that way.

Generally, use these rules:

- Work out how long a complete question should take to answer, and then how long each sub-question should be given.
- Work on a basis of 5-10% planning time, 80-85% writing time, and 10% checking time.

Learning and using case studies

Case studies are in-depth examples, of which there are a number in this book. Some are brief (2 pages), while others are much longer (4-6 pages). Apply them to help you answer the exam questions.

However, in comparison with AS, two main differences apply when using case studies at A2:

- One example is occasionally enough at AS (e.g. the study of one coastal stretch under threat from erosion). But, at A2, you need to use a range of examples. For instance, one example of a development project in an answer about 'Development and globalisation' will not normally be enough.
- Make sure that you are not just describing or reeling off points you have learned. At A2, you will need to **apply** the material to the question. Here is an example where you would need to use – and apply – a case study to answer a question on 'Ecosystems: change and challenge':

> **(b)** Assess the importance of global and local threats to one named global ecosystem.

For this question, knowing examples is not enough. You have to:

- understand and be able to write about the range of threats, which might mean one set of threats in one location, to which are added threats from another
- be able to say – with evidence – which threats are greater, and which are lesser. Ideally, you would not just say this at the end, but would have built up an answer, progressing from those which are lesser to those which are greater, or vice-versa.

So, knowledge by itself is not enough. Here is an example of a 40-mark essay question on 'World cities':

> Referring to examples, assess the role of different schemes in attempting to regenerate urban areas.

This question involves:

- using more than one example – you will limit your marks if you restrict yourself to just one
- saying what the different schemes are, and what makes each different
- assessing the role that each plays (i.e. how important each scheme is) in regenerating urban areas that you have studied.

This does not mean that you cannot limit yourself to one place. The guidance below shows how you could use examples of three types of regeneration used in London and covered in the 'World cities' chapter in this book.

There are some things you can do to help yourself learn and use case studies.

A portfolio of case studies

Build up a 'portfolio' of case studies, so that you know which part of the Specification and content the case studies fit. A grid like this might help:

Option name	Subject content	Case studies
World cities	Urban regeneration	• Market-led – Docklands and Canary Wharf • Sports-led – London's 2012 Olympic and Paralympic Games • Retail-led – Westfield Centre, Stratford

The grid works like an index – or pigeon hole. If you fill it in, you will know exactly which case study and theory is appropriate for each part of an option. It also helps to make links between options.

Remembering case studies

Look again at the question above from 'World cities'. There are three case studies of regeneration in this book that you could use to answer this question:

- Market-led – Docklands and Canary Wharf
- Sports-led – London's 2012 Olympic and Paralympic Games
- Retail-led – Westfield Centre, Stratford.

One way of remembering these is, in each case, to:

- draw a spider diagram with each type or method of regeneration and details about it
- extend the spider with how regeneration took place (with examples), and identify the role of each in achieving regeneration
- use a large or small '+' around the outside to show how big its role has been – that would help you to assess each type.

Learn the case study!

Finally, make sure you learn the detail.

- Get basic details right – learn key facts and figures. Accuracy and detail will gain marks.
- Be careful with dates and places; there is little worse than starting out with 'London's Docklands have helped to regenerate west London'.

Common inaccuracies include: incorrect location, poor knowledge of physical processes, lack of terminology (e.g. 'soaks into the ground', instead of 'infiltration'), lack of factual detail (e.g. 'lots of' instead of actual data), poor spelling, and poor grammar and punctuation.

Organising an essay

Beware! Very good students can underachieve on essays. Most commonly, they show excellent knowledge and understanding – in amazing depth – but run out of time, or fail to answer the question. A weaker student who actually answers the question can do much better.

Essay practice – including timed essays – is essential. Essays are assessed in Section C of Unit 3, where you need to be able to write a formal essay for 40 marks. These guidance points apply less to Unit 4, where writing needs to be more focused and to the point about your fieldwork (Unit 4A) or the issue (Unit 4B).

Here are some handy tips to add to those about extended writing. The guidance is based on the following essay question on 'World cities'.

> **6** With reference to examples, assess the geographical impacts of urban regeneration. *(40 marks)*

1 Introduce the essay

Define the key terms, set out the context to which the question is referring (e.g. that regeneration can have a range of effects – from economic, environmental, and social), and outline your argument.

2 Arrange the main body of the essay

Argue or show knowledge of both sides. Do not just write 'whatever comes next'. Organise your examples so that you:

- for instance, classify those on one side of an argument from those on the other – perhaps into social, economic, and environmental
- consider which arguments are strong on each side and which are weaker
- progress towards an answer, e.g. from factors that strongly support the argument, to those that support it less so.

In each case, keep coming back to the title, e.g. 'This example shows how effective this kind of regeneration can be ...'

3 Summarise your answer to the question in a conclusion

Here you should answer the question fully – showing where the balance of the argument lies, or evaluating which are the greatest impacts of regeneration or which schemes are most effective, for example. Wait until then – do not give it all away in the introduction!

Preparing for the synoptic resources in Unit 4B

Four weeks before the Unit 4B exam, you will receive the pre-release resource booklet. This is considered in more detail in chapter 8, so is not discussed here. But do note the following:

- The resource booklet will form the basis of lesson work in the weeks leading up to the exam. Lessons will be like normal, except that you will be going through the pre-release resource materials instead of learning new taught material. Four weeks is plenty of time to grasp the issue and absorb the detail.
- The booklet itself will give you enough detail to answer the exam questions, but will always give you opportunities to research extra websites or to carry out your own research. Clearly, you will benefit if you follow these up. For example, the question on Guildford's transport on page 309 shows that it will help if you've carried out your own research into the issues in the resource booklet.

To succeed with the questions in Unit 4B:

- Read the pre-release materials thoroughly. You will not be allowed to take any notes into the exam, so be sure to keep notes and remember the points made in class to help you prepare.
- Use evidence in the resource materials (photos, diagrams, tables of statistics, etc.) to support your answers.
- Plan what you want to say at the start of every question. You do not need to do much, just make a few notes.
- Keep coming back to the question and make sure that you answer it.
- Keep to a rigid time schedule – if you have 90 minutes for 60 marks, then take about 10 minutes for planning, 70 minutes for writing and 10 minutes for reading, checking, and adding any other points that come to mind.

How to be synoptic!

The questions in both Units 3 and 4 are synoptic, i.e. they are intended to draw out what you have learnt across the whole of the AS and A2 course. You will be expected to make links between topics. For instance, the example of 'World cities' links in with many options, e.g.

- in AS, 'Population change', inner city areas – their population, and issues facing them
- in A2, 'Development and globalisation' – how urban areas in the developing world may face particular problems.

You could therefore refer to examples where you may have come across instances affecting 'World cities'. These examples need only be brief – a couple of sentences, with names or data – but they will help you to earn maximum marks. You will not achieve maximum marks unless you do this.

How are exam papers marked?

Examiners mark your exam papers, and are given clear guidance about how to do it. You are rewarded for what you know and can do, and are not penalised for anything you have left out. Because most questions at A2 are extended writing, they are level marked.

- Questions with between 1 and 4 marks (in Unit 4) are point marked – that is, each correct point earns a mark.
- Levels are used for questions worth 5 marks or more; two levels are used for those questions worth between 5 and 8 marks, and three levels for those worth between 9 and 15 marks. If your answer matches the best qualities in the top level of the mark scheme, you will get full marks.

The following sections tell you about level marking, and show you how mark schemes are constructed.

Level marking Unit 3, Sections A and B

Questions in Sections A and B are set in three parts – (a), (b), and (c) – and are worth 7, 8, and 10 marks respectively. This means that all questions are level marked.

Look at this typical 7-mark question and the mark scheme set for it:

> **4** Study Figure 4. (For the purposes of this sample question, the map is not shown here.) Figure 4 is a world map showing the percentage of population living in urban areas by country.
>
> **(a)** Comment on the variation in urban populations shown in Figure 4. (7 marks)

Level	Mark	Descriptor
Level 1	1-4	Simple statements about geographical distribution of percentages, e.g. 'high in Europe'. Some basic attempts at commentary, but with little geographical place terminology or reference to data.
Level 2	5-7	More detailed statements which refer to specific locations on the map and data, using evidence from the map. Top level answers identify anomalies (e.g. high rate of urban population in India compared to south-east Asia as a whole).

Now consider this 10-mark question where three levels are used:

> **(c)** Assess the effects of counter-urbanisation on large cities. *(10 marks)*

Level	Mark	Descriptor
Level 1	1-4	The structure is poor or absent. There are one or two basic ideas explaining the results of counter-urbanisation. Lacks understanding of the impacts, e.g. may only describe population trends without identifying services and employment involved. Explanations are over-simplified and lack clarity. Geographical terminology is rarely used. Frequent grammar, punctuation, and spelling errors.
Level 2	5-7	Some structure is present. Explains the results of counter-urbanisation with some clarity. There is some understanding about the impacts of counter-urbanisation. It uses some geographical terminology. Some explanations are clear, but this varies. May be limited in range of examples used. There are some grammar, punctuation, and spelling errors.
Level 3	8-10	Well structured. There is a sound explanation of the impacts of counter-urbanisation, with an understanding of several impacts on people, the economy, and the urban environment. Good use of geographical terminology. Explanations are clear. Grammar, punctuation, and spelling errors are rare.

Level marking Unit 3, Section C

These questions are much longer; you answer one question and have to write a long essay for which 40 marks are available. Marks are awarded for the following:

- the basic grasp of the subject matter
- evidence of understanding and explanation
- the organisation of the answer
- quality of geographical language.
- evidence of using synoptic material from other parts of the course.

Each of the four levels includes descriptors of these characteristics. To reach the highest marks at Level 4, you need to demonstrate a high standard in each of these qualities.

Level	Mark	Descriptor
1	1-10	The answer shows a basic grasp of concepts and ideas, but points lack development or depth. Explanations are incomplete and arguments partial and lack coherent organisation or reasoned conclusions. Examples are superficial. There is no evidence of synopticity.
2	11-20	The answer is relevant and accurate, and shows reasonable knowledge and critical understanding of concepts and principles, with some use of specialist vocabulary. Arguments are not fully developed and the organisation of ideas and the use of examples and general theories show imbalances. Some ability to identify, interpret, and synthesise some of the material. Limited ability to understand the roles of values, attitudes, and decision-making processes. Sketch maps/diagrams are not used effectively. Evidence of synopticity is limited.
3	21-30	Sound and frequent evidence of thorough, detailed, and accurate knowledge and critical understanding of concepts and principles, and of specialist vocabulary. Explanations, arguments, and assessments or evaluations are direct, logical, purposeful, and generally balanced. Some ability to identify, interpret, and synthesise a range of material. Some ability to understand the roles of values, attitudes, and decision-making processes. Examples are developed and sketch maps/diagrams are used effectively. There is strong evidence of synopticity.
4	31-40	Strong evidence of thorough, detailed, and accurate knowledge and critical understanding of concepts and principles, and of specialist vocabulary. Explanations, arguments, and assessments or evaluations are direct, logical, perceptive, purposeful, and show both balance and flair. There is a high level of insight, and an ability to identify, interpret, and synthesise a wide range of material with creativity. Evidence of maturity in understanding the role of values, attitudes, and decision-making processes. Examples are well-developed and sketch maps/diagrams are fully integrated. The answer is fully synoptic.

How to gain marks and not lose them

Gaining marks instead of losing them can be just a matter of technique. You might get a lower mark than you really ought to, not because you don't understand the material, but because you don't use what you know to answer the question properly. In your exams, choose the right question, use your case studies, apply your knowledge, and write clear, well-ordered answers.

List of LICs, MICs, and HICs

Based on the World Bank classification of economies by income, 2010

LIC – low income country   MIC – middle income country   HIC – high income country

East Asia and the Pacific

Cambodia	LIC
China	MIC
Fiji	MIC
Indonesia	MIC
Kiribati	MIC
North Korea	LIC
Laos	LIC
Malaysia	MIC
Marshall Islands	MIC
Micronesia	MIC
Mongolia	MIC
Burma (Myanmar)	LIC
Palau	MIC
Papua New Guinea	MIC
Philippines	MIC
Samoa	MIC
Solomon Islands	MIC
Thailand	MIC
Timor-Leste	MIC
Tonga	MIC
Vanuatu	MIC
Vietnam	LIC

Europe and Central Asia

Albania	MIC
Armenia	MIC
Azerbaijan	MIC
Belarus	MIC
Bosnia & Herzegovina	MIC
Bulgaria	MIC
Georgia	MIC
Kazakhstan	MIC
Kosovo	MIC
Kyrgyzstan	LIC
Latvia	MIC
Lithuania	MIC
Macedonia	MIC
Moldova	MIC
Montenegro	MIC
Poland	MIC
Romania	MIC
Russia	MIC
Serbia	MIC
Tajikistan	LIC
Turkey	MIC
Turkmenistan	MIC
Ukraine	MIC
Uzbekistan	LIC

Latin America and the Caribbean

Argentina	MIC
Belize	MIC
Bolivia	MIC
Brazil	MIC
Chile	MIC
Colombia	MIC
Costa Rica	MIC
Cuba	MIC
Dominica	MIC
Dominican Republic	MIC
Ecuador	MIC
El Salvador	MIC
Grenada	MIC
Guatemala	MIC
Guyana	MIC
Haiti	LIC
Honduras	MIC
Jamaica	MIC
Mexico	MIC
Nicaragua	MIC
Panama	MIC
Paraguay	MIC
Peru	MIC
St Kitts and Nevis	MIC
St Lucia	MIC
St Vincent and the Grenadines	MIC
Surinam	MIC
Uruguay	MIC
Venezuela	MIC

Middle East and North Africa

Algeria	MIC
Djibouti	MIC
Egypt	MIC
Iran	MIC
Iraq	MIC
Jordan	MIC
Lebanon	MIC
Libya	MIC
Morocco	MIC
Syria	MIC
Tunisia	MIC
Yemen	LIC

South Asia

Afghanistan	LIC
Bangladesh	LIC
Bhutan	MIC
India	MIC
Maldives	MIC
Nepal	LIC
Pakistan	MIC
Sri Lanka	MIC

Sub-Saharan Africa

Angola	MIC
Benin	LIC
Botswana	MIC
Burkina Faso	LIC
Burundi	LIC
Cameroon	MIC
Cape Verde	MIC
Central African Republic	LIC
Chad	LIC
Comoros	LIC
Congo, Dem. Rep. of	LIC
Congo, Rep. of	MIC
Cote d'Ivoire	MIC
Eritrea	LIC
Ethiopia	LIC
Gabon	MIC
Gambia, The	LIC
Ghana	LIC
Guinea	LIC
Guinea-Bissau	LIC
Kenya	LIC
Lesotho	MIC
Liberia	LIC
Madagascar	LIC
Malawi	LIC
Mali	LIC
Mauritania	LIC
Mauritius	MIC
Mozambique	LIC
Namibia	MIC
Niger	LIC
Nigeria	MIC
Rwanda	LIC
Sao Tome and Principe	MIC
Senegal	LIC
Seychelles	MIC
Sierra Leone	LIC
Somalia	LIC
South Africa	MIC
Sudan	MIC
Swaziland	MIC
Tanzania	LIC
Togo	LIC
Uganda	LIC
Zambia	LIC
Zimbabwe	LIC

HICs – OECD* countries

Australia
Austria
Belgium
Canada
Czech Republic
Denmark
Finland
France
Germany
Greece
Hungary
Iceland
Irish Republic
Italy
Japan
South Korea
Luxembourg
Netherlands
New Zealand
Norway
Portugal
Slovakia
Spain
Sweden
Switzerland
UK
USA

HICs – others

Andorra
Antigua and Barbuda
Aruba
Bahamas, The
Bahrain
Barbados
Bermuda
Brunei
Cayman Islands
Croatia
Cyprus
Equatorial Guinea
Estonia
Israel
Kuwait
Liechtenstein
Malta
Monaco
Oman
Puerto Rico
Qatar
San Marino
Saudi Arabia
Singapore
Slovenia
Taiwan
Trinidad and Tobago
United Arab Emirates

This 2010 listing is according to 2008 GNI per capita:
LIC – US$975 or less
MIC – US$976 - US$11 905
HIC – US$11 906 and above

The World Bank actually uses four groups: 'low income', 'lower middle income', 'upper middle income', and 'high income' – this book combines the two middle income groups into a single 'MIC' group to allow a more convenient three-tier division of world economies.

* OECD – Organisation for Economic Co-operation and Development: an international organisation of 34 countries with the aim of stimulating economic progress and world trade; most members are HICs.